Foundation Mathematics for Engineers and Scientists with Worked Examples

Foundation Mathematics for Engineers and Scientists with Worked Examples covers fundamental topics in mathematics required for science and engineering disciplines. It is primarily designed to provide a comprehensive, straightforward, and step-by-step presentation of mathematical concepts to engineers, scientists, and general readers. It moves from simple to challenging areas, with carefully tailored worked examples of different degrees of difficulty. Mathematical concepts are deliberately linked with appropriate engineering applications to reinforce their value and are aligned with topics taught in major overseas curriculums.

This book is written primarily for students at Levels 3 and 4 (typically in the early stages of a degree in engineering or a related discipline) or for those undertaking foundation degree, Higher National Diploma (HND), International Foundation Year (IFY), and International Year One (IYO) courses with maths modules. It consists of seven parts:

- Basic Concepts in Mathematics
- Coordinate Geometry
- Algebraic Expression and Equations
- Surds
- Indices and Logarithms
- Polynomials
- Trigonometry

Each chapter is devoted to a topic and can be used as a stand-alone guide with no prior knowledge assumed. Additional exercises and resources for each chapter can be found online. To access this supplementary content, please go to www.dszak.com.

Shefiu Zakariyah is a Chartered Engineer (CEng) and a senior member of IEEE (SMIEEE), with over a decade of experience in teaching, content development and assessment, and technical training and consulting.

Foundation Mathematics for Engineers and Scientists with Worked Examples

Shefiu Zakariyah

Routledge
Taylor & Francis Group

LONDON AND NEW YORK

Cover image: © Shutterstock

First published 2025
by Routledge
4 Park Square, Milton Park, Abingdon, Oxon OX14 4RN

and by Routledge
605 Third Avenue, New York, NY 10158

Routledge is an imprint of the Taylor & Francis Group, an informa business

British Library Cataloguing-in-Publication Data
A catalogue record for this book is available from the British Library

Library of Congress Cataloging-in-Publication Data
Names: Zakariyah, Shefiu, author.
Title: Foundation mathematics for engineers and scientists with worked
examples / Shefiu Zakariyah.
Description: Abingdon, Oxon ; New York, NY : Routledge, 2025.
Identifiers: LCCN 2023044209 | ISBN 9780367462901 (hardback) | ISBN
9780367462895 (paperback) | ISBN 9781003027928 (ebook)
Subjects: LCSH: Engineering mathematics–Problems, exercises, etc. |
Engineering mathematics–Study and teaching (Higher)–Great Britain.
Classification: LCC TA333 .Z35 2024 | DDC 620.001/51–dc23/eng/20240320
LC record available at https://lccn.loc.gov/2023044209

ISBN: 978-0-367-46290-1 (hbk)
ISBN: 978-0-367-46289-5 (pbk)
ISBN: 978-1-003-02792-8 (ebk)

DOI: 10.1201/9781003027928

Typeset in Times

by Deanta Global Publishing Services, Chennai, India

Access the Support Material: www.routledge.com/9780367462901

To the stimuli

Parents (late), sister (late), brother (late), and grandparents (late)

and

Wife, children, siblings, and friends

Contents

About the Author

Shefiu Zakariyah is a Chartered Engineer (Engineering Council UK) with more than 15 years of experience in research, and teaching engineering and mathematics courses within the higher education sectors. In this period, he has worked for Loughborough University, the University of Greenwich, the University of Cambridge, the University of Malaya, and the University of Namibia. Currently, Shefiu is an Associate Lecturer with the University of Derby, an Ofqual subject matter specialist, a Materials Developer for NCUK, and a technical consultant. He also serves as a Professional Registration Advisor (PRA) and Professional Registration Interviewer (PRI) for the Institution of Engineering and Technology (IET) supporting engineers and technicians to gain professional recognition. Shefiu served as an executive member of the IET Manufacturing Network and played a significant role in the Laser Institute of America, serving on the technical subcommittee for Control Measures and Testing (TSC-4). Furthermore, Shefiu was a member of the standard subcommittee responsible for the Safe Use of Lasers in Research, Development, and Testing (ANSI SSC-8-Z136-8). Shefiu has worked with learners from pre-university to postgraduate levels and is committed to making complex concepts accessible.

How to Use This Book

Users

- A detailed explanation is provided for each concept using simple and digestible language so that you can relax and enjoy studying this book. This is essential in order to enrich a self-study style of learning and to meet any need for remote study. If you decide to skip any part, you can head to the summary section where you will find a succinct yet rich overview of the chapter.

- You are advised to attempt the worked examples before checking the solution. This is to measure your understanding and check if there is any gap. Additional practice for each chapter is available at www.dszak.com.

- Where applicable, alternative method(s) to solve a problem are provided in a box following the first method introduced.

- Key formulas are numbered as $(\mathbf{X.Y})$, where $\mathbf{X}$ is the chapter number and $\mathbf{Y}$ represents the ordinal position of the equation in the chapter. In other words, an equation that is numbered as $(\mathbf{2.10})$ is the tenth equation in **Chapter 2**. A similar style is used for figures and tables.

- Important facts, terms, and formulas are clearly given in **bold type** (and placed in a box as $\boxed{\textbf{Formula OR Equation}}$) for emphasis and ease of reference.

- If you encounter a term or symbol, which might have been introduced but not explained in the section you are reading, the author has provided a range of useful resources at the beginning and end of this book, including a glossary and mathematical symbols, that you can easily refer to. Also, the index is useful to cross-reference other applications and usage of the same term or symbol.

- Whilst chapters are designed to be stand-alone, it will however be helpful in a few cases that you read related chapter(s) or section(s). Also to mention that this book does not follow the order of any syllabus, as it is intended for a wider audience.

- The first three chapters are designed to lay a foundation in arithmetic and algebra, equipping the reader with the essential principles and building a robust base from which learners can expand their mathematical understanding and skills. For those already familiar with the subject matter, a quick review of these sections may suffice.

- Having read this book and attempted the worked examples, it is advised that you try more examples from other similar textbooks or sources. This is to ensure that you have now fully grasped and can apply the concepts explained.

Instructors

- A set of PowerPoint slides for each chapter is provided online at www.dszak.com. Instructors may modify these materials for educational purposes, provided that proper credit is given.

- Also, a bank of questions is available for adaptation with due acknowledgement.

Acknowledgements

This book has come to fruition due to unparalleled support and contributions from many people that it becomes impossible to mention them all, I'm sincerely grateful for their help and hope that they will accept this general expression of gratitude. I want to start by thanking the staff at Routledge, Taylor & Francis, for their confidence and support throughout this process, especially: Tony Moore (Senior Editor), Lillian Woodall (Project Manager at Deanta), Gabriella Williams (then Editorial Assistant), Frazer Merritt (then Editorial Assistant), and Aimee Wragg (Editorial Assistant). I can't repay your hard work, dedication, and patience.

My friends and colleagues in the teaching profession and industries have been very helpful and generous with their time in carefully reviewing this book and giving me constructive feedback. In this regard, I am very grateful to (listed in alphabetical order): Dr Abdelhalim Azbaid El Ouahabi (*PhD*), Dr Aliyu Ahmad (*PhD*), Burket Ali (Head of Engineering and Computing, DMUIC), Dr Paul Richard Hammond (*MSc, PhD*), Dr Richard Welford (*DPhil*), Dr Sarbari Mukherjee (*MSc, PhD*), Mouna Mohamed Lemine, Salma Okhai (*BSc* Combined Science, PGCE (Secondary) (DMUIC), Dr Shamsudeen Hassan (*PhD, CEng, MEI*), and Tim Wilmshurst (*MSc, FIET, CEng*).

Finally but most importantly, I owe a lot to my family who have been my source of inspiration and motivation. I would like to thank my wife: Khadijah Olaniyan (*B.A.*, Loughborough University), and my daughters and sons (fondly called 'Prince(ss)' and 'Doctor') for everything, including proofing and helpful tips.

Abbreviations

Term	Explanation
BEDMAS	Brackets-Exponent-Division-Multiplication-Addition-Subtraction
BIDMAS	Brackets-Index-Division-Multiplication-Addition-Subtraction
BODMAS	Brackets-Order-Division-Multiplication-Addition-Subtraction
d.p.	Decimal Places
HCF	Highest Common Factor
LCM	Lowest Common Multiple
LHS	Left-Hand Side
RHS	Right-Hand Side
s.f.	Significant Figures
SI	Système International d'unités (International System of Units)
STEM	Science, Technology, Engineering, and Mathematics

Greek Alphabet

Letter		Name
Upper Case	**Lower Case**	
A	α	Alpha
B	β	Beta
Γ	γ	Gamma
Δ	δ	Delta
E	ϵ or ε	Epsilon
Z	ζ	Zeta
H	η	Eta
Θ	θ or ϑ	Theta
I	ι	Iota
K	κ	Kappa
Λ	λ	Lambda
M	μ	Mu
N	ν	Nu
Ξ	ξ	Xi
O	o	Omicron
Π	π	Pi
P	ρ or ϱ	Rho
Σ	σ or ς	Sigma
T	τ	Tau
Y	υ	Upsilon
Φ	ϕ or φ	Phi
X	χ	Chi
Ψ	ψ	Psi
Ω	ω	Omega

Mathematical Operators and Symbols

Sign	Name	Sign	Name
=	equal to	$\mathbb{N}$	natural number
≠	not equal to	$\mathbb{Z}$	set of integers
+	addition (or plus)	$\mathbb{R}$	set of real numbers
−	subtraction	$\mathbb{Q}$	set of rational numbers
±	plus or minus	$\mathcal{R}$	real part
∓	minus or plus	$\mathfrak{I}$	imaginary part
×	multiplication (or times)	%	percentage
÷	division	"	arcsecond
/	division slash	°F	degrees Fahrenheit
∞	infinity	°C	degrees Celsius
≡	identical to	$\lvert k \rvert$	modulus of k, absolute value of k or determinant of k
≢	not identical to	∗	asterisk operator (or multiply by)
$\sqrt{}$	Radical sign (or square root)	•	bullet operator or dot operator (or multiply by)
$\sqrt[3]{}$	cube root	⋯	horizontal ellipsis
$\sqrt[4]{}$	fourth root	⋮	vertical ellipsis
$\sqrt[n]{x}$	nth root of x	∴	therefore
!	factorial	∵	because
∝	proportional to	$\sum$	summation (or sigma)
~	similar to/approximately	∷	proportion
≈	almost equal to	:	ratio (or such that)
≃	asymptotically equal to	log	logarithm to base 10
≄	not asymptotically equal to	ln	natural logarithm (or logarithm to base e)
≉	not almost equal to	→	tend to
<	less than	∅ or {}	empty set
≮	not less than	∈	element of (or belongs to)
>	greater than	∉	not an element of
≯	not greater than	∠	angle
≤	less than or equal to	∡	measured angle
≰	neither less than nor equal to	∟	right angle
≥	greater than or equal to	⦟	right angle with arc
≱	neither greater than nor equal to	⊿	right triangle
⪇	less than but not equal to	⊥	perpendicular to (orthogonal to)
⪈	greater than but not equal to	∥	parallel to
≪	much less than/far less than	∦	not parallel to
⋘	very much less than/far far less than	⩨	equal and parallel to
≫	much greater than/far greater than	δ	delta
⋙	very much greater than/far far greater than		
Δ	increment/change		

Physical Quantities

TABLE 1
Fundamental (or Base) Quantities

Quantity Name	Unit Name	Unit Symbol
Length (l)	metre	m
Mass (m)	kilogram	kg
Time (t)	second	s
Electric current (I)	ampere	A
Temperature (θ)	kelvin	K
Amount (n)	mole	mol
Luminosity (L)	candela	cd

TABLE 2
Some Common Derived Quantities

Quantity Name	Unit Name	Unit Symbol
Area (A)	metre squared	m^2
Volume (V)	metre cubed	m^3
Density (ρ)	kilogram per metre cubed	kgm^{-3}
Linear velocity (u or v)	metre per second	ms^{-1}
Linear acceleration (a)	metre per second squared	ms^{-2}
Force (F)	newton	N
Pressure (P)	newton per metre squared or Pascal	Nm^{-2} or Pa
Frequency (F)	hertz	Hz
Time period (T)	second	s
Energy (E)	joule	J
Work (W)	joule	J
Power (P)	watt	W
Voltage (V)	Volt	V
Electric charge (Q or q)	coulomb	C
Resistance (R)	ohm	Ω
Conductance (G)	Siemens or per ohm or mho	S or Ω^{-1}
Reactance (X)	ohm	Ω
Impedance (Z)	ohm	Ω
Admittance (Y)	Siemens or per ohm or mho	S or Ω^{-1}
Capacitance (C)	farad	F
Inductance (L)	henry	H

Prefixes Denoting Powers of Ten

Prefix	Symbol	Exponential Form	Equivalent	Value
peta-	P	10^{15}	1 000 000 000 000 000	one quadrillion
tera-	T	10^{12}	1 000 000 000 000	one trillion
giga-	G	10^{9}	1 000 000 000	one billion
mega-	M	10^{6}	1 000 000	one million
kilo-	k	10^{3}	1 000	one thousand
hecto-	h	10^{2}	100	one hundred
deca-	da	10^{1}	10	one ten
deci-	d	10^{-1}	0.1	one-tenth
centi-	c	10^{-2}	0.01	one-hundredth
milli-	m	10^{-3}	0.001	one-thousandth
micro-	μ	10^{-6}	0.000 001	one-millionth
nano-	n	10^{-9}	0.000 000 001	one-billionth
pico-	p	10^{-12}	0.000 000 000 001	one-trillionth
femto-	f	10^{-15}	0.000 000 000 000 001	one-quadrillionth

NOTE

- From power -3 down and $+3$ up, the powers are all in multiple of three.
- Positive powers have capital letter symbols except for deca, hecto, and kilo, while negative powers have small letter symbols.
- 'kilo' is the only small letter symbol in the positive multiples of three.

1 Fundamental Arithmetic

Learning Outcomes

Once you have studied the content of this chapter, you should be able to:

- Identify different types of numbers
- Determine prime factors of numbers
- Compute the Highest Common Factor (HCF) of numbers
- Compute the Lowest Common Multiple (LCM) of numbers
- Use precedence rules in mathematical operations
- Use negative numbers
- Express numbers in specified decimal places
- Express numbers in required significant figures
- Express numbers in scientific and engineering notations

1.1 INTRODUCTION

Central to mathematics is the practice of counting and processing numbers, which is indispensable in the current digital age. Arithmetic, derived from the Greek word **'arithmos'**, meaning 'number', is a branch of mathematics that deals with numbers and the relationships that exist between them. This chapter focuses on this important branch and covers a range of topics relating to this. Some of the concepts introduced here will be explained in greater detail at a later stage in this book and others will simply be encountered and applied.

1.2 NUMBERS

Numbers are vital to numeracy, and there are various classifications and types of numbers, which we are set to discuss.

1.2.1 COMMON NUMBERS

Under this category, we have the following:

1) Natural numbers   These are numbers that we arbitrarily use, also known as **whole numbers**. The symbol $\mathbb{N}$ is used to denote natural numbers. Examples are:

$$1, 2, 3, 4, 5, ...$$

DOI:10.1201/9781003027928-1

The operator within the square brackets [$\cdots$], i.e., three dots, is used to show that the pattern continues. In this case, the numbers increase by one unit each time.

Zero [**0**] is included in this category if it is considered as part of natural counting and excluded when it is viewed as absence of a number. Sometimes, the term 'whole numbers' is reserved for when zero is included in natural numbers and thus they are not entirely synonymous.

Natural numbers are endless. The sum of two or more natural numbers is another natural number; the same can be said about the product. However, this is not always true for subtraction and division. To illustrate, we will use 3 and 4. The results of addition ($3 + 4 = 7$) and multiplication ($3 \times 4 = 12$) are also natural numbers, but subtraction ($3 - 4 = -1$) and division ($\frac{3}{4} = 0.75$) are not.

---

2) Integers

These are natural numbers and their mirrors. In other words, we have positive integers (or whole numbers) and negative integers (or whole numbers with a negative sign before them).

The symbol $\mathbb{Z}$ is used to denote integers, and $\mathbb{Z}^+$ and $\mathbb{Z}^-$ indicate positive and negative integers respectively. Examples are:

$$1, \quad -1, \quad 2, \quad -2, \quad 3, \quad -3, \quad 4, \quad -4, \quad 5, \quad -5,...$$

Alternatively, we can write this compactly as

$$\pm 1, \qquad \pm 2, \qquad \pm 3, \qquad \pm 4, \qquad \pm 5, \, ...$$

The operator $\pm$, pronounced as '**plus or minus**', is used to combine two similar numbers into one. For example, ± 1 is a shorthand for $+1$ and -1. You will also find the operator $\mp$, pronounced as '**minus or plus**'. This is just to say that the negative comes before the positive.

Again, zero [**0**] is included or excluded in this category for the same reason stated above. When included, it represents the line of symmetry or the mirror line for positive and negative integers. Whatever the case, zero is neither a positive nor a negative integer, i.e., it is a neutral integer.

---

3) Real numbers

These are all possible numbers between $-\infty$ and $+\infty$. The symbol ∞ is called **infinity** and is a mathematical way of representing 'endlessness' or 'impossibility'. Therefore, any number that can be written or said is either less than $+\infty$ or more than $-\infty$.

Real numbers are denoted using the symbol $\mathbb{R}$. We an write that $-\infty < \mathbb{R} < +\infty$, reads '**minus infinity is less than $\mathbb{R}$, $\mathbb{R}$ is less than plus infinity**'. We will come across more of this sign in Chapter 10 on inequalities. Examples of real numbers are:

$$-7.1 \qquad -8.45 \qquad 60 \qquad \sqrt{3}$$

4) Complex numbers

These are numbers that are not real. They generally have two parts: a) a real part, denoted as $\Re$, and b) an imaginary part, denoted by $\Im$.

The symbol z (different from the symbol of an integer) is generally used to denote a complex number and it takes the general form $z = x + jy$ or $z = x + iy$. Both j and i are used for the imaginary part. Examples of complex numbers:

$$3 + j4 \qquad 0.2 - j0.5 \qquad 6 + j10 \qquad -1 - j$$

Complex numbers are central to many operations in science and engineering. For example, we use them when working with alternating current, and this is covered in *Advanced Mathematics for Engineers and Scientists with Worked Examples* by the same author. Whilst the square of any real number is always a non-negative number, this is not true for a complex number. In other words, a complex number multiplied by itself can result in a negative real number.

5) Rational numbers

These are numbers that can be written as a ratio of two integers. In other words, they can be written as a fraction $\frac{x}{y}$, such that x and y are integers, excluding 0 (provided 0 is not included in the real number because $0 = \frac{0}{y}$ and $\infty = \frac{x}{0}$).

The symbol $\mathbb{Q}$ is used to denote a rational number. Examples are:

$$-1.1 = -\frac{11}{10} \qquad -0.4 = -\frac{4}{10} = -\frac{2}{5} \qquad 2 = \frac{2}{1}$$

6) Irrational numbers

Opposite to rational numbers are irrational numbers (also called non-rational numbers), these are numbers that cannot be written as an exact fraction $\frac{x}{y}$, where x and y are integers, excluding 0.

The symbol $\mathbb{P}$ is used to denote irrational numbers. This category of numbers can be surds, non-recurring decimals, or special numbers. Examples are:

$$\sqrt{2} = 1.414213\ldots \qquad e = 2.718281\ldots \qquad \pi = 3.141592\ldots$$

From the above, we note that real numbers are the direct opposite of imaginary numbers as rational numbers are to irrational numbers. Additionally, there is an overlap in the classification such that a number belongs to two or more classes. Figure 1.1 shows the relationship between the above types of numbers.

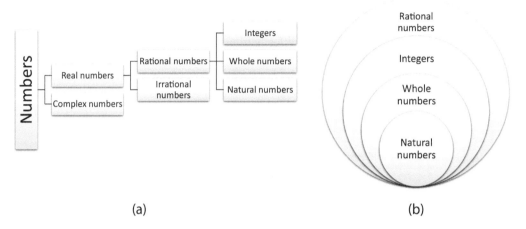

(a)   (b)

FIGURE 1.1 Numbers: (a) hierarchical relationship between the types of numbers, and (b) Venn diagram showing the relationship between the various types of rational numbers.

1.2.2 SPECIAL NUMBERS

There are other numbers that we should be aware of, which include:

1) Negative numbers These are numbers with a negative sign before them. They can be regarded as a 'deficiency'. So -2 is a deficiency of 2. Let's think about this in terms of currency, $-£2$ means a deficiency (or debt) of £2. Having this at the back of our mind may be helpful in dealing with algebra or operations involving negative values. Note that the negative integers mentioned above are special negative numbers. As a result, $-\frac{3}{4}$, $-\frac{11}{3}$, -0.25, and -5.76 are negative numbers but not negative integers.

---

2) Even numbers These are numbers (integers) that are divisible by 2 without leaving any remainder (or fractional component). So, 2, 4, and 6 are even numbers while 3, 7, and 9 are not. As a rule, any number that ends with 0, 2, 4, 6, or 8 is an even number.

---

3) Odd numbers These are the opposite of even numbers. In other words, any number that is not exactly divisible by 2 is an odd number. 1, 3, 5, 7, and 9 are odd numbers. Any two or more digit number is odd if its last right digit is 1, 3, 5, 7, or 9.

---

4) Prime numbers These are natural numbers that are divisible by one and themselves without a remainder and not divisible by any other number. On this basis, 2, 3, 5, and 7 are prime numbers. On the other hand, 4 is not, because it is divisible by 1, 2, and itself (4). Similarly, 8 is not a prime number, because we can divide it by 1, 2, 4, and itself (8) without a remainder.

The list of prime numbers is endless but below are the ones (25 in total) between 0 and 100.

$$2, 3, 5, 7, 11, 13, 17, 19, 23, 29, 31, 37, 41,$$
$$43, 47, 53, 59, 61, 67, 71, 73, 79, 83, 89, 97$$

Notice that only 2 is the even number in the list, the others are odd numbers.

1.3 FACTORS AND MULTIPLES

A factor (or factors), say f, of a number, say N, are numbers that can divide N without leaving any remainder. Informally, we say that the number 'goes into it' without a remainder. Mathematically, $\frac{N}{f} \in \mathbb{Z}$, where $\in$ stands for 'a member of'. In other words, when a number is divided by its factor, it always results in an integer. For example, 1, 2, 3, and 6 are the factors of 6, because $\frac{6}{1} = 6$, $\frac{6}{2} = 3$, $\frac{6}{3} = 2$, and $\frac{6}{6} = 1$.

On the other hand, multiples of a number M are numbers obtained when M is multiplied by natural numbers 1, 2, 3, etc. For example, the multiples of 3 are 3, 6, 9, 12, 15, etc. We can therefore say that the multiple of a number is either equal or greater than the said number, but the factor of a specific number is either equal or less than the number. Another inference is that the factors of a number are finite whilst the multiples of a number are infinite.

Let's take a quick break and try some examples.

Example 1

For each of the following numbers:

　i) Write down all the factors.
　ii) State the first five multiples.

a) 3　　　　　**b)** 8　　　　　**c)** 10　　　　　**d)** 15　　　　　**e)** 20

What did you get? Find the solution below to double-check your answer.

Solution to Example 1

a) 3
Solution

i)	Factors of 3 are:	**1, 3**
ii)	The first five multiples of 3 are:	**3, 6, 9, 12, 15**

b) 8
Solution

i)	Factors of 8 are:	**1, 2, 4, 8**
ii)	The first five multiples of 8 are:	**8, 16, 24, 32, 40**

c) 10
Solution

    i)   Factors of 10 are: **1, 2, 5, 10**

    ii)   The first five multiples of 10 are: **10, 20, 30, 40, 50**

---

d) 15
Solution

    i)   Factors of 15 are: **1, 3, 5, 15**

    ii)   The first five multiples of 15 are: **15, 30, 45, 60, 75**

---

e) 20
Solution

    i)   Factors of 20 are: **1, 2, 4, 5, 10, 20**

    ii)   The first five multiples of 20 are: **20, 40, 60, 80, 100**

1.3.1 PERFECT NUMBER

The term perfect number simply refers to a number that is exactly half of the sum of its factors. For example, the factors of 6 are 1, 2, 3, and 6. Let's add these factors together:

$$1 + 2 + 3 + 6 = 12$$

The summation of the factors obtained above (i.e., 12) is double of 6, the original number. Hence, 6 is a perfect number.

We can equally say that a perfect number is equal to the sum of all its factors excluding itself. Thus, 28 is a perfect number because the sum of its factors excluding itself is 28, i.e.:

$$1 + 2 + 4 + 7 + 14 = 28$$

To proceed, we will now discuss three essential terminologies that fall under factors and are central to most operations and simplifications that we encounter in mathematics.

1.3.2 PRIME FACTORS

This is a factor that is also a prime number. For example, 21 has 1, 3, 7, and 21 as its factors. Among these, 3 and 7 are prime numbers and they are called the **prime factors** of 21.

Any non-prime number can be expressed as the product of its prime factors. For example, 21 can be written as $21 = 3 \times 7$. Table 1.1 provides further examples of numbers and their respective factors and prime factors.

TABLE 1.1

Numbers, Their Factors, and Prime Factors

Number	Factors	Prime Factor(s)	Expression (as a Product)
9	1, 3, 9	3	$9 = 3 \times 3$
12	1, 2, 3, 4, 6, 12	2, 3	$12 = 2 \times 2 \times 3$
15	1, 3, 5, 15	3, 5	$15 = 3 \times 5$
21	1, 3, 7, 21	3, 7	$21 = 3 \times 7$
27	1, 3, 9, 27	3	$27 = 3 \times 3 \times 3$
33	1, 3, 11, 33	3, 11	$33 = 3 \times 11$

Follow these steps to express a number (even or odd) as a product of its prime factors:

Step 1: Start with 2 and divide the number repeatedly with 2 until it is no longer divisible by 2.

Step 2: Move to the next prime number (i.e., 3) and repeat step 1 above.

Step 3: Repeat step 2 above until the remaining number is a prime number.

Step 4: Divide the prime number by itself to get 1.

Step 5: Multiply all the prime numbers used together.

Let's try some examples.

Example 2

Write the following numbers as a product of their prime factors.

a) 36   **b)** 168   **c)** 231   **d)** 420

What did you get? Find the solution below to double-check your answer.

Solution to Example 2

HINT

In this example, the steps are given in (a), but not shown for the remaining questions.

a) 36
Solution

Step 1: We start with **2** and divide **36** by **2** to obtain **18**

Step 2: We repeat step **1** and obtain **9**.

Step 3: **9** cannot be divided by **2** so we choose **3**, and obtain **3**.

Step 4: We repeat step **3** and obtain **1**.

Step 5: The numbers on the LHS (Left Hand Side) are the product of **36**

2	36
2	18
3	9
3	3
	1

$$\therefore 36 = 2 \times 2 \times 3 \times 3$$

---

b) 168
Solution

2	168
2	84
2	42
3	21
7	7
	1

$$\therefore 168 = 2 \times 2 \times 2 \times 3 \times 7$$

---

c) 231
Solution

3	231
7	77
11	11
	1

$$\therefore 231 = 3 \times 7 \times 11$$

---

d) 420
Solution

2	420
2	210
3	105
5	35
7	7
	1

$$\therefore 420 = 2 \times 2 \times 3 \times 5 \times 7$$

1.3.3 HIGHEST COMMON FACTOR (HCF)

The HCF of two or more numbers is the largest factor that is common to both or all of them.

Follow these steps to find the HCF of two or more numbers:

Method 1

> **Step 1:** Determine the factors of the numbers.
> **Step 2:** Look at the largest factor that is common to them. This is the HCF of the numbers.

Method 2

> **Step 1:** Express each number as a product of its prime factors.
> **Step 2:** Take all the prime factors that are common to them.
> **Step 3:** Multiply them together. This product is the HCF of the given numbers.

Method 1 will be appropriate if the numbers involved are relatively small in value, otherwise use Method 2. We will find HCF helpful when factorising expressions.

Let's try some examples.

Example 3

Find the HCF of the following sets of numbers.

a) 30 and 42   **b)** 66 and 105   **c)** 60, 126, and 132   **d)** 84, 112, 120, and 300

What did you get? Find the solution below to double-check your answer.

Solution to Example 3

---

a) 30 and 42
Solution

Number	Expression (as a Product)	Common Factors
30	$\mathbf{30 = ②\times③\times5}$	2 and 3 are common to the two numbers.
42	$\mathbf{42 = ②\times③\times7}$	

There is no other factor that is common to the numbers 30 and 42. Therefore, their HCF is:

$$2\times3 = 6$$

ALTERNATIVE METHOD

Since we have two relatively small numbers, we can use another option as:

Number	Factors
30	①, ②, ③, 5, ⑥, 10, 15, 30
42	①, ②, ③, ⑥, 7, 14, 21, 42

1, 2, 3, and 6 are common factors of both 30 and 42. Obviously, 6 is the highest of them. Therefore, the HCF for these numbers is 6.

b) 66 and 105
Solution

Number	Expression (as a Product)	Common Factors
66	$66 = 2 \times ③ \times 11$	3 is the ONLY number that is common to
105	$105 = ③ \times 5 \times 7$	both numbers.

Therefore, the HCF for these numbers is 3.

ALTERNATIVE METHOD

Since we have two relatively small numbers, we can use another option as:

Number	Factors
66	1, 2, ③, 6, 11, 22, 33, 66
105	1, ③, 5, 7, 15, 21, 35, 105

Obviously, 3 is the only number that appears in both lists. Therefore, their HCF is 3 as before.

c) 60, 126, and 132
Solution

Number	Expression (as a Product)	Common Factors
60	$60 = ② \times 2 \times ③ \times 5$	2 and 3 are the only common factors to
126	$126 = ② \times 3 \times ③ \times 7$	the three numbers 60, 126, and 132.
132	$132 = ② \times 2 \times ③ \times 11$	

Therefore, their HCF is

$$2 \times 3 = 6$$

Alternatively, attempt this question using method 1 by listing the factors of each number and determine (by inspection) the highest factor that is common to them all.

d) 84, 112, 120, and 300
Solution

Number	Expression (as a Product)	Common Factors
84	**84** $= ② \times ② \times 3 \times 7$	
112	**112** $= ② \times ② \times 2 \times 2 \times 7$	2 is common to these numbers in two
120	**120** $= ② \times ② \times 2 \times 3 \times 5$	separate instances.
300	**300** $= ② \times ② \times 3 \times 5 \times 5$	

Therefore, the HCF for these numbers is

$$2 \times 2 = 4$$

NOTE

Different sets of numbers may share the same HCF. The set of numbers in Examples 3(a) and 3(c) have the same HCF although their elements are not the same. In fact, it is also possible that a set of numbers has no HCF. For example, 10 and 21; the same can be said about the three numbers 43, 97, and 125.

1.3.4   LOWEST COMMON MULTIPLE (LCM)

The LCM of two or more numbers is the smallest number both or all of them can be divided by without any remainder. For example, the LCM of 2 and 3 is 6. This is because the smallest number that 2 and 3 can go in (without a remainder) is 6. Others include 12 and 18, but they are not the LCMs since we have a smaller value. Notice that the LCM for 2 and 3 can be obtained by multiplying the two numbers together, i.e., $2 \times 3 = 6$. However, this is not always true. For example, the LCM of 3 and 6 is 6, but their product is 18, i.e., $3 \times 6 = 18$.

It should also be noted that it is impossible to find HCM (Highest Common Multiple). This is because multiples of a number or numbers are infinite and finding the highest multiple that is common to two or more numbers will not be feasible.

Follow these steps to find the LCM of two or more numbers:

Method 1

   Step 1: Find the multiples of the numbers.
   Step 2: Look at the smallest multiple that is common to them. This is the LCM of the two numbers.

Method 2

   Step 1: Express the numbers as products of their prime factors.
   Step 2: Take all the prime factors (both common and uncommon).
   Step 3: Multiply the prime factors together. This is the LCM of the numbers.

If a prime factor appears in one of the numbers, this is considered as a single count, and if it appears once in all the numbers it is also considered as a count.

LCM will be very useful when simplifying fractions as will be shown in Chapter 2. Let's try some examples.

Example 4

Find the LCM of the following sets of numbers.

a) 3 and 4 **b)** 6 and 10 **c)** 3, 8, and 12 **d)** 2, 3, 6, and 9

What did you get? Find the solution below to double-check your answer.

Solution to Example 4

a) 3 and 4
Solution

Number	Multiples
3	**3, 6, 9, ⑫, 15, 18, 21, ㉔, ...**
4	**4, 8, ⑫, 16, 20, ㉔, 28, ...**

Note
From the list given above, we can see that 12 and 24 are common multiples of both 3 and 4. Obviously, 12 is the lower of the two. Therefore, the LCM of these numbers is 12. The LCM can also be obtained if we simply multiply the two numbers ($3 \times 4 = 12$), but this is not always the case as previously mentioned above (see the next example).

Alternative Method

As these are small numbers, we can easily express them as a product of their prime factors as shown below.

Number	Expression (as a Product)	Common Multiples
3	**3** = 3	= **3**
4	**4** = 2×2	= $\mathbf{2 \times 2}$

Their LCM is

$$\mathbf{2 \times 2 \times 3 = 12}$$

---

b) 6 and 10
Solution

Number	Multiples
6	**6, 12, 18, 24, ㉚, 36, 42, 48, 54, (60), ...**
10	**10, 20, ㉚, 40, 50, (60), ...**

From the list given above, we can see that 30 and 60 are common multiples to both 6 and 10. Obviously, **30** is the lower of the two and therefore is the LCM of the two numbers.

Notice that multiplying both numbers results in 60, i.e., 6×10. Although this is a common multiple of the given numbers, it is however not the lowest.

ALTERNATIVE METHOD

Since these are small numbers, we can easily express them as a product of their prime factors as shown below.

Number	Expression (as a Product)	Common Multiples
6	**6** $= 2 \times 3$	$= 2 \times 3$
10	**10** $= 2 \times 5$	$= 2 \quad \times 5$

Their LCM is

$$2 \times 3 \times 5 = 30$$

---

c) 3, 8, and 12
Solution

Number	Expression (as a Product)	Common Multiples
3	**3** $= 3$	$= \qquad 3$
8	**8** $= 2 \times 2 \times 2$	$= 2 \times 2 \times 2$
12	**12** $= 2 \times 2 \times 3$	$= \quad 2 \times 2 \times 3$

Their LCM is

$$2 \times 2 \times 2 \times 3 = 24$$

ALTERNATIVE METHOD

Number	Multiples
3	$3, 6, 9, 12, \ 15, \ 18, 21, \ \textcircled{24}, 27, \ 30, ...$
8	$8, 16, \textcircled{24}, 32, 40, 48, 56, ...$
12	$12, \ \textcircled{24}, \ 36, 48, 60, ...$

From the list given above, we can see that 24 and 48 are common multiples to 3, 8, and 12. Obviously, 24 is the lower of the two. Therefore, the LCM of these numbers is 24.

Again, you will notice that if we multiply the three numbers together, we will get 288, i.e., $3 \times 8 \times 12$. This is a common multiple of the given numbers, however it is not the lowest but a very large number. This further confirms that multiplying the numbers together is not a good approach, though it is sometimes not avoidable.

---

d) 2, 3, 6, and 9
Solution

Number	Expression (as a Product)	Common Multiples
2	$2 = 2$	$= 2$
3	$3 = 3$	$=\ \ \ 3$
6	$6 = 2 \times 3$	$= 2 \times 3$
9	$9 = 3 \times 3$	$=\ \ \ 3 \times 3$

Their LCM is

$$2 \times 3 \times 3 = 18$$

ALTERNATIVE METHOD

Number	Multiples
2	2, 4, 6, 8, 10, 12, 14, 16, (18), 20, 22, 24, 26, 28, 30, 32, 34, (36), 38, ...
3	3, 6, 9, 12, 15, (18), 21, 24, 27, 30, 33, (36), 39, ...
6	6, 12, (18), 24, 30, (36), 42, ...
9	9, (18), 27, (36), 45, ...

From the list given above, we can see that 18 and 36 are common multiples to 2, 3, 6, and 9. Obviously, 18 is the lower of the two. Therefore, their LCM is 18.

1.4 MATHEMATICAL OPERATORS

A mathematical operator is generally a symbol that is used to perform a task on a number or numbers. For example, the operator denoted by $\sum$ is called 'summation', and it is used to carry out the addition of numbers. Some operators act on a number or numbers while others show a relationship between numbers or variables. Yet, there are operators without a defined symbol, though they will be known by the position or how they are written. For example, the '**power**' of a number is known because it is relatively smaller than the variable or number that is acted upon and appears as a superscript.

The main operators are provided in Table 1.2, more operators are given in the Table of Mathematical Operators at the beginning of this book.

When more than one of the operations in Table 1.2 occur in a single expression, there is a priority order which must be followed. This is captured in a single mnemonic **BIDMAS**, which is summarised in Table 1.3.

From Table 1.2, it is noted that multiplication and division share the third order, which means that either one can be operated first, though preference should be given to the operator on the left. The same can be said about addition and subtraction, which occupy the fourth order.

Other mnemonics include **BEDMAS** and **BODMAS**. **E** and **O** are **E**xponent and **O**rder (or **O**thers) respectively. The three letters (I, E, and O) in the listed mnemonics refer to the same operation.

Parentheses or round brackets '()', square or box brackets '[]', curly brackets or braces '{}', or any other similar operator is used to instruct that whatever is within the brackets must be computed first. It is irrelevant whether any of the lower-order operators is within the brackets.

TABLE 1.2

Illustrating Mathematical Operators

Operator	Symbol	Note
Addition	+	The word '**sum**' is also used to imply addition. Examples: • $3 + 7 = 10$ • $x + 5x = 6x$
Subtraction	–	The word '**minus**' is also used to mean subtraction. Examples: • $20 - 14 = 6$ • $13m - 5m = 8m$
Multiplication	×	The words '**times**', '**product**', and '**of**' are all used to denote multiplication. Other symbols include asterisk ($*$) and dot ($\cdot$). Examples: • $3 \times 7 = 21$. Alternatively, we can write the same operation as $3 * 7 = 21$ or $(3) \cdot (7) = 21$. To avoid confusion, a dot as a multiplication sign is not often used with numbers which is why we need the brackets, otherwise it could be mistaken for a decimal number 3.7 (three point seven). • $2 \times 5x = 10x$
Division	÷	The word '**over**' is used to mean division. Forward slash (/) is also used to represent division. Examples: • $35 \div 5 = 7$. Alternatively, we can write the same operation as $\frac{35}{5} = 7$ or $^{35}/_5 = 7$ or 35/5 = 7. • $4m \div m = 4$. We will cover this type further in Chapter 3 when dealing with algebra.
Roots	$\sqrt[n]{}$	This operator will be discussed shortly.
Powers		This does not have a given sign or symbol, but it is known by placing it at the top-right corner of a number or a variable. • $3^2 = 9$ • m^2

TABLE 1.3
Priority Order of Mathematical Operation Explained

Priority Order	Letter	Operator
First	B	Brackets
Second	I	Indices/powers or roots
Third	D	Division
	M	Multiplication
Fourth	A	Addition
	S	Subtraction

A number that is immediately outside the brackets implies that the content of the brackets must be multiplied by that number. For example, in 2(3 + 5), the result of the content within the brackets, i.e., 8, must be multiplied by 2. Ideally, 2(3 + 5) should be written as $2 \times (3 + 5)$, but $\times$ is often omitted for brevity.

When you open a bracket, it must be closed with its pair. Therefore '(' must be followed by ')' and not another bracket, e.g., ']'. When there are layers of 'containers', different brackets can be used to distinguish between them. For example:

$$2\,[3 - (2x + 5)]$$

This said, it is fine to use the same type of brackets when nesting, as used routinely in programming. Let's try some examples to illustrate what we've covered so far.

Example 5

Without using a calculator, evaluate each of the following:

a) $5 + 4 - 7 \times 4$
b) $24 \div 2 - 3 \times 2 + 7$
c) $3^2 - 13 + 2(2 + 6)$

d) $4 \times 5 - 2^2 + \sqrt{25}$
e) $(7 \times 4) \div (6 - 4)^2$
f) $\dfrac{(19 - 7) \times 5}{4 + (18 \div 3)}$

What did you get? Find the solution below to double-check your answer.

Solution to Example 5

a) $5 + 4 - 7 \times 4$
Solution

$$5 + 4 - 7 \times 4 = 5 + 4 - 28$$
$$= 9 - 28$$
$$= -\mathbf{19}$$

NOTE

Let's assume that we are not using **BIDMAS** and work from left to right as

$$5 + 4 - 7 \times 4 = 9 - 7 \times 4$$
$$= 2 \times 4$$
$$= \mathbf{8}$$

Surprisingly, and of course mistakenly, the answer obtained is 8! This is because we performed addition, then subtraction, and finally multiplication, which is not in line with the order of mathematical operation.

b) $24 \div 2 - 3 \times 2 + 7$
Solution

$$24 \div 2 - 3 \times 2 + 7 = 12 - 6 + 7$$
$$= 6 + 7$$
$$= \mathbf{13}$$

c) $3^2 - 13 + 2(2 + 6)$
Solution

$$3^2 - 13 + 2(2 + 6) = 9 - 13 + 2(8)$$
$$= 9 - 13 + 16$$
$$= \mathbf{12}$$

d) $4 \times 5 - 2^2 + \sqrt{25}$
Solution

$$4 \times 5 - 2^2 + \sqrt{25} = 4 \times 5 - 4 + 5$$
$$= 20 - 4 + 5$$
$$= \mathbf{21}$$

e) $(7 \times 4) \div (6 - 4)^2$
Solution

$$(7 \times 4) \div (6 - 4)^2 = (28) \div (2)^2$$
$$= 28 \div 4$$
$$= \mathbf{7}$$

f) $\dfrac{(19-7)\times 5}{4+(18\div 3)}$

Solution

$$\frac{(19-7)\times 5}{4+(18\div 3)}=\frac{(12)\times 5}{4+(6)}=\frac{60}{10}$$
$$=6$$

1.5   NEGATIVE NUMBERS

Earlier we introduced negative numbers and distinguished them from negative integers, which are whole numbers. When we carry out addition and subtraction on two or more numbers, the result can either be a negative or a positive number depending on the numbers involved. However, when multiplication or division is carried out, the (sign of the) result is primarily determined by the sign of the numbers involved. Table 1.4 summarises this.

From Table 1.4, we can conclude that when the sign is the same for the two numbers, the answer is positive and when the sign is different, the answer is negative. This only applies to both multiplication and division.

Let's try some examples for addition and subtraction.

Example 6

Without using a calculator, work out each of the following:

a) $4-23$ **b)** $-30+16$ **c)** $-35-65$ **d)** $8-10-13$ **e)** $-47+14+11$

TABLE 1.4
Rules of Multiplying and Dividing Negative Numbers

	Operation	Result	Example
Multiplication	(positive) × (positive)	positive	$(3)\times(4)=12$
	(positive) × (negative)	negative	$(3)\times(-4)=-12$
	(negative) × (positive)	negative	$(-3)\times(4)=-12$
	(negative) × (negative)	positive	$(-3)\times(-4)=12$
Division	(positive) ÷ (positive)	positive	$(30)\div(6)=5$
	(positive) ÷ (negative)	negative	$(30)\div(-6)=-5$
	(negative) ÷ (positive)	negative	$(-30)\div(6)=-5$
	(negative) ÷ (negative)	positive	$(-30)\div(-6)=5$

What did you get? Find the solution below to double-check your answer.

Solution to Example 6

a) $4 - 23$
Solution

$$4 - 23 = -\mathbf{19}$$

b) $-30 + 16$
Solution

$$-30 + 16 = -\mathbf{14}$$

c) $-35 - 65$
Solution

$$-35 - 65 = -\mathbf{100}$$

d) $8 - 10 - 13$
Solution

$$8 - 10 - 13 = 8 - 23 = -\mathbf{15}$$

e) $-47 + 14 + 11$
Solution

$$-47 + 14 + 11 = -47 + 25 = -\mathbf{22}$$

Let's try some examples for multiplication and division.

Example 7

Without using a calculator, work out each of the following:

a) $7 \times (-3)$ **b)** $(-5) \times 12$ **c)** $(-4) \times (-35)$ **d)** $15 + (-20)$
e) $(-32) - (-19)$ **f)** $(-70) \div 5$ **g)** $13 \div (-2)$ **h)** $(-48) \div (-4)$

What did you get? Find the solution below to double-check your answer.

Solution to Example 7

a) $7 \times (-3)$
Solution

$$7 \times (-3) = -\mathbf{21}$$

b) $(-5) \times 12$
Solution

$$(-5) \times 12 = -\mathbf{60}$$

c) $(-4) \times (-35)$

Solution

$$(-4) \times (-35) = \mathbf{140}$$

---

d) $15 + (-20)$

Solution

This appears as though there is no multiplication (or division) involved. However, the presence of the brackets should be viewed as a multiplication. Let's show the working:

$$15 + (-20) = 15 + 1(-20) = 15 + 1 \times -20 = 15 - 20$$

$$\therefore \mathbf{15 + (-20) = -5}$$

---

e) $(-32) - (-19)$

Solution

$$(-32) - (-19) = -32 + 19 = \mathbf{-13}$$

---

f) $-70 \div 5$

Solution

$$(-70) \div 5 = \mathbf{-14}$$

---

g) $13 \div (-2)$

Solution

$$13 \div (-2) = \mathbf{-6.5}$$

---

h) $(-48) \div (-4)$

Solution

$$(-48) \div (-4) = \mathbf{12}$$

1.6 PRESENTING NUMBERS

The format with which numbers are presented is essential in science and engineering. This is particularly important when working with irrational numbers and providing answers to a long calculation. Here we will look at the format in terms of precision and layout.

1.6.1 DECIMAL PLACES

Irrational numbers (and answers often obtained from complex calculations) are generally approximated. One of the ways to do this is by writing them correctly to a particular number of decimal place(s). 1 decimal place means one digit (or more precisely one non-zero digit) after a decimal point, 2 decimal places mean two digits after the decimal point, and so on. For example, 2.1, 46.01, and 101.100 are numbers given to 1, 2, and 3 decimal places respectively.

In general, the term **decimal place** is shortened to **d.p.** The decimal place(s) of a number can be changed from a higher decimal place to a lower one; say, from 3 decimal places to 2 d.p. or 1 d.p. In this case, all excess digits to the right are discarded by applying the rules below:

Rule 1 If the digit(s) that will be dropped/eliminated is less than or equal to 4, then no change applies to the last retained digit after the decimal. For example, 7.64 (2 d.p.) and 23.745 (3 d.p.) becomes 7.6 (1 d.p.) and 23.7 (1 d.p.) respectively. This process involves '**rounding down**' and the number is said to have been 'rounded down'.

Rule 2 If the digit(s) that will be dropped/eliminated is greater than or equal to 6, then the last retained digit is increased by 1. For example, 30.162 (3 d.p.) and 0.4701 (4 d.p.) become 30.2 (1 d.p.) and 0.5 (1 d.p.). This process involves '**rounding up**' and the number is said to have been 'rounded up'.

Rule 3 If the digit to be eliminated is 5, then one of the acceptable practices (known as '**round to even**') is to round up or down such that the last retained digit is an even digit. This is possibly because digit 5 (and its family, such as 0.50, 50, 500, etc.) is right in the middle. For example, 10.35 (3 d.p.) and 0.65 (2 d.p.) become 10.4 (1 d.p.) and 0.6 (1 d.p.) respectively.

Note that 0.51, 0.502, 0.5003, etc., are not included in this, as they are not exactly in the middle. As a result, 3.451 (3 d.p.) and 0.85001 (5 d.p.) become 3.5 (1 d.p.) and 0.9 (1 d.p.) respectively.

It is also possible to increase the number of decimal places, say from 1 decimal place to 2, 3, etc., decimal places. To do this, we simply add an equivalent number of zeros. For example, 5.4 (1 d.p.) can be written as 5.40 (2 d.p.), 5.400 (3 d.p.), 5.4000 (4 d.p.), etc.

When a decimal number does not end, this is called **recurring decimal** (as against a **terminating decimal**) and three dots after the last number to the right are used to indicate recurrence. For example, **1.33333**... is a decimal number which continues endlessly. In this case, it is sufficient to write **1.3̇** to say that **3** repeats infinitely.

Sometimes, two or more digits are repeated; in this case, two dots are used in two ways. The first style is when two dots are placed on two non-adjacent digits, showing that the digits between the numbers (including the numbers with the dots) are repeated. For example, **0.7̇126̇ = 0.71267126**.... In the second style, a dot is placed on two adjacent numbers to indicate their repetition. For example, **2.14̇5̇ = 2.1454545**.... The recurring digit(s) can come straight after the decimal point, as seen in **1.3̇** and **0.7̇126̇** or not as in **2.14̇5̇**.

Consequentially, the use of the decimal point makes it easy to multiply and divide by 10, 100, 1000, etc. When a number is multiplied by 10, 100, 1000, etc., we move the decimal point to the right in a number of places equal to the number of zeros in the number. In other words, when you multiply by 10, the decimal point is moved one place to the right; when a number is multiplied by 100, we move the decimal point two places to the right; and when we multiply by 1000, the decimal place is moved three places to the right. Division is carried out in the same way except that the decimal point is moved to the left. This is illustrated in Figure 1.2 for multiplication by 10 and division by 100.

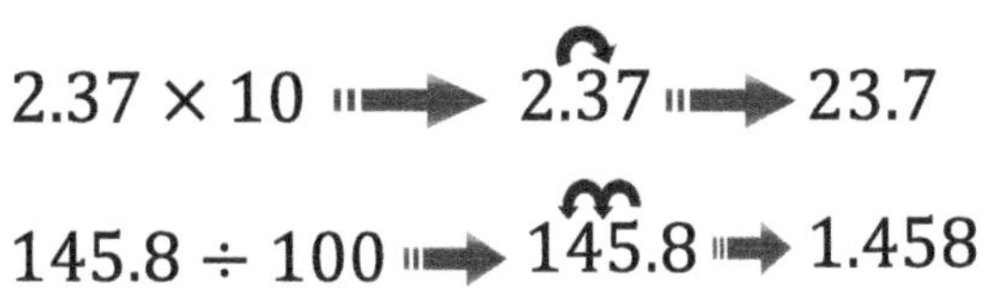

$$2.37 \times 10 \Longrightarrow 2.3\overset{\frown}{7} \Longrightarrow 23.7$$

$$145.8 \div 100 \Longrightarrow 1\overset{\frown\frown}{45}.8 \Longrightarrow 1.458$$

FIGURE 1.2 How multiplying by or dividing a number with powers of ten affects the position of a decimal point illustrated.

Let's try some examples.

Example 8

Write the following numbers in 1 decimal place.

a) 0.456   **b)** 0.204   **c)** 0.097   **d)** 1.450   **e)** 6.951   **f)** 20.4501   **g)** 19.95

What did you get? Find the solution below to double-check your answer.

Solution to Example 8

a) 0.5   **b)** 0.2   **c)** 0.1   **d)** 1.4   **e)** 7.0   **f)** 20.5   **g)** 20.0

NOTE
Generally, 1.450 will be given as 1.5 (in 1 d.p.) but, using round-to-even method, this is 1.4.

Another set of examples to try.

Example 9

Write the following amounts in 2 decimal places.

a) £5.1   **b)** £0.769   **c)** £2.750   **d)** £31.4651   **e)** £67   **f)** £45.565

What did you get? Find the solution below to double-check your answer.

Solution to Example 9

a) £5.10   **b)** £0.77   **c)** £2.75   **d)** £31.47   **e)** £67.00   **f)** £45.56

NOTE
Generally, £45.565 will be given as £45.57 (in 2 decimal places) but, using round-to-even method, this is £45.56.

Another set of examples to try.

Example 10

Write the following numbers in 3 decimal places.

a) 3.05   **b)** 1   **c)** 0.3259   **d)** 1.03   **e)** 0.0356

What did you get? Find the solution below to double-check your answer.

Solution to Example 10

a) 3.050   **b)** 1.000   **c)** 0.326   **d)** 1.030   **e)** 0.036

NOTE
Example 10(e) is in 3 s.f. but 4 d.p.

Let's try another set of examples to illustrate multiplying and dividing by 10s.

Example 11

Work out the following:

a) 1.39 × 10   **b)** 0.072 × 100   **c)** 24.85 ÷ 100   **d)** 56 ÷ 1000

What did you get? Find the solution below to double-check your answer.

Solution to Example 11

HINT

- All we need to do is to move the point to the right for multiplication and move it to the left if it is division.
- The number of times we move it to the right/left should be equal to the number of zeros such that 10 is moved once, 100 twice, and so on.

a) 1.39 × 10
Solution

$$1.39 \times 10 = \mathbf{13.9}$$

b) 0.072 × 100
Solution

$$0.072 \times 100 = \mathbf{7.2}$$

c) 24.85 ÷ 100
Solution

$$24.85 \div 100 = \mathbf{0.2485}$$

d) 56 ÷ 1000
Solution

$$56 \div 1000 = \mathbf{0.056}$$

1.6.2 SIGNIFICANT FIGURES

Numbers can be required to be presented in a particular significant figure, which implies the number of 'significant' digits to be written. What is considered significant may vary but the following notes would be helpful.

Rule 1 All non-zero digits are significant. In other words, all digits are significant except zero.

Rule 2 Zero before a non-zero digit is not significant, but zeros after non-zero digits are significant. For example, the zeros in the following numbers are not significant: 03, 005, 0.2, and 0.04. Conversely, the zero(s) in 30, 500, 2.0, and 4.00 are significant.

All the procedures mentioned for decimal places are equally applicable here.

Let's try some examples.

Example 12

Present the following numbers in 1 significant figure.

a) 0.456   **b)** 0.204   **c)** 0.097   **d)** 1.450   **e)** 6.951   **f)** 20.4501   **g)** 19.95

What did you get? Find the solution below to double-check your answer.

Solution to Example 12

a) 0.5   **b)** 0.2   **c)** 0.1   **d)** 1   **e)** 7   **f)** 20   **g)** 20

This is a repeat of Example 8. Notice that Example 12(a)–12(c) have the same answer for both 1 decimal place and 1 significant figure. 12(e) is the same in value for both cases but differs in format. This occurs because one is presented as **7.0**(1 d.p.) and the other is **7** (1 s.f.).

Note that 20 is the answer to Example 12(f) and not 2; this will be unambiguous if the number can be expressed in scientific notation. The same can be said of (g).

Another set of examples to try.

Example 13

Present the following numbers in 3 significant figures.

a) 2.1   **b)** 0.030569   **c)** 32750   **d)** 1.4650   **e)** 99.9504   **f)** 2

What did you get? Find the solution below to double-check your answer.

Solution to Example 13

a) 2.10 **b)** 0.0306 **c)** 32800 **d)** 1.46 **e)** 100 **f)** 2.00

NOTE
Answer to Example 13(d) can be given as 1.47 in another convention.

1.7 REPRESENTING SMALL AND LARGE NUMBERS

In science and engineering, we work with numbers that could be extremely small or forbiddingly large. The charge on an electron or the mass of the earth can be written but, if it is done, will be difficult to process or read.

One way to handle this is to use a prefix such as micro, milli, kilo, and mega among others. This approach will however change the unit. For example, a length of 0.000000023 m can be compactly written as 23 nm. Notice that the unit of the length changed from metre to nanometre. 'Nano' is a prefix, which is equivalent to 10^{-9} or one billionth.

The other option is to write the number and multiply the same by 10 to the power of x, where x is a positive or negative integer. This is a special notation and is of two styles: **scientific notation** and **engineering notation**.

1.7.1 SCIENTIFIC NOTATION

This is also known as **standard form** or **standard notation**. A number (N), small or large, written in scientific form takes the form:

$$\boxed{N = A \times 10^x}$$ (1.1)

where

- A, called the **mantissa**, is a number that is in the interval $1 \le A < 10$, meaning that 1 is included and 10 is excluded. Note that $\le$ (denotes 'less than or equal to') and $<$ (means 'less than') are some operators that form the basis of an inequality that will be covered in Chapter 10.
- x is any integer, positive or negative.

Let's run through some points about this standard notation.

Note 1 When x is negative, it indicates that the number is small and less than 1. The more the negative x the smaller the number.

Note 2 When x is positive, it indicates that the number is large and greater than 1. The more the positive x the bigger the number.

Note 3 When the decimal point in the number (N) is moved to the right, the power x reduced by 1 for every movement. It however increases by 1 for every unit step movement to the left.

Scientific notation is very helpful in removing the ambiguity that can be associated with the significant figure of a number. For example, 200 can be said to be in 1, 2, or 3 s.f., depending on its origin. It could be 240 (in 2 s.f.) and rounded to 200 (1 s.f.); it may be 203 (3 s.f.) and rounded to 200 (2 s.f.). It may also be 200.4 (4 s.f.) and rounded to 3 s.f. To remove this ambiguity, 200 will be written as:

- $200 = 2 \times 10^2$ Correct to 1 s.f.
- $200 = 2.0 \times 10^2$ Correct to 2 s.f.
- $200 = 2.00 \times 10^2$ Correct to 3 s.f.

1.7.2 ENGINEERING NOTATION

Engineering notation is a special form of scientific form such that a number (M) can also be written as:

$$\boxed{M = B \times 10^y} \tag{1.2}$$

where

- B, called the **mantissa**, is a number that is in the interval $1 \leq B < 1000$, meaning that 1 is included and 1000 is excluded.
- y is any integer that is a multiple of 3 (i.e., 3, 6, 9, 12, ...).

Engineering notation is the same, in principle, as scientific notation (Section 1.7.1) with the exception of the extra condition attached to the power and the fact that its mantissa can be up to 1000, i.e., up to 3 digits to the left of a decimal point. This notation makes it easy to apply prefixes, as they (with the exclusion of the prefixes between 10^{-2} and 10^2) are all generally multiples of 3.

Let's try some examples.

Example 14

Express the following in scientific and engineering notations.

a) 33 000   **b)** 625 000   **c)** 2023   **d)** 0.047   **e)** 0.00000189   **f)** 0.000000264

What did you get? Find the solution below to double-check your answer.

Solution to Example 14

a) 33 000
Solution

    i)   Scientific notation $33\ 000 = 3.3 \times 10\ 000 = \mathbf{3.3 \times 10^4}$

    ii)   Engineering notation $33\ 000 = 33 \times 1000 = \mathbf{33 \times 10^3}$

b) 625 000
Solution

    i)   Scientific notation $625\ 000 = 6.25 \times 100\ 000 = \mathbf{6.25 \times 10^5}$

    ii)   Engineering notation $625\ 000 = 625 \times 1000 = \mathbf{625 \times 10^3}$

c) 2023
Solution

    i)   Scientific notation $2023 = 2.023 \times 1000 = \mathbf{2.023 \times 10^3}$

    ii)   Engineering notation $2023 = 2.023 \times 1000 = \mathbf{2.023 \times 10^3}$

NOTE
Here, there is no difference in representing the number in either notation.

d) 0.047
Solution

    i)   Scientific notation $0.047 = 4.7 \div 100 = 4.7 \div 10^2 = \mathbf{4.7 \times 10^{-2}}$

    ii)   Engineering notation $0.047 = 47 \div 1000 = 47 \div 10^3 = \mathbf{47 \times 10^{-3}}$

NOTE
Notice in scientific notation how we changed 10^2 to 10^{-2} as we changed $\div$ to $\times$. Alternatively, we can use $\div 10^2$ instead of $\times 10^{-2}$. Let's show the working for this too.

$$0.047 = 4.7 \times \frac{1}{100} = 4.7 \times \frac{1}{10^2} = \mathbf{4.7 \times 10^{-2}}$$

We will soon learn, in Chapter 11, that $\frac{1}{10^2} = 10^{-2}$ is based on a law of indices. In the meantime, we can use either $\div 10^2$ or $\times 10^{-2}$ for the same purpose.

e) 0.00000189
Solution

    i)   Scientific notation $0.00000189 = 1.89 \div 1000000 = 1.89 \div 10^6$
                                                   $= \mathbf{1.89 \times 10^{-6}}$

    ii)   Engineering notation $0.00000189 = 1.89 \div 1000000 = 1.89 \div 10^6$
                                                   $= \mathbf{1.89 \times 10^{-6}}$

NOTE
Again, there is no difference in representing the number 0.00000189 in either notation.

---

f) 0.000000264
Solution

     i)   Scientific notation $0.000000264 = 2.64 \div 10\,000\,000 = 2.64 \div 10^7$
$$= 2.64 \times 10^{-7}$$

     ii)   Engineering notation $0.000000264 = 264 \div 1\,000\,000\,000 = 264 \div 10^9$
$$= 264 \times 10^{-9}$$

1.7.3   OPERATIONS OF NUMBERS IN SCIENTIFIC AND ENGINEERING NOTATIONS

When numbers are expressed in scientific or engineering notations, we can still carry out basic operations.

1) Multiplication

When numbers in either of these notations are multiplied, we need to multiply the mantissa and add the powers of 10. This is illustrated below.

General expression	$(a \times 10^x) \times (b \times 10^y) = (a \times b) \times 10^{x+y}$
Examples	• $(2 \times 10^4) \times (3 \times 10^5) = (2 \times 3) \times 10^{(4+5)} = 6 \times 10^9$ • $(1.2 \times 10^{-7}) \times (5 \times 10^{-2}) = (1.2 \times 5) \times 10^{(-7-2)} = 6 \times 10^{-9}$

2) Division

When a number in either of these notations is divided by another similar number, we divide the mantissa and subtract the power of the divisor from the power of the dividend. This is illustrated below.

General expression	$(a \times 10^x) \div (b \times 10^y) = (a \div b) \times 10^{x-y}$
Examples	• $(7 \times 10^6) \div (2 \times 10^4) = (7 \div 2) \times 10^{(6-4)} = 3.5 \times 10^2$ • $(8 \times 10^{-5}) \div (4 \times 10^7) = (8 \div 4) \times 10^{(-5-7)} = 2 \times 10^{-12}$

3) Addition and Subtraction

To carry out addition or subtraction of numbers expressed in either of the two notations, we need to:

Step 1: Express both numbers in the same power if they are not already the same.
Step 2: Add/subtract the mantissas.
Step 3: Multiply the answer by the new common power to the numbers, which is obtained in step 1 above.

This is illustrated below.

General expression $\left(a \times 10^{x}\right) \pm \left(b \times 10^{y}\right) = \left(a_1 \pm b_1\right) \times 10^{z}$

where a_1 and b_1 are newly derived mantissa when the numbers are expressed in a new power z. The new power z is such that $z = x$ or $z = y$. In other words, the power of one of the numbers is made as the reference and does not change. Consequently, either a_1 or b_1 will be the same as its respective old equivalent.

Examples   a) When the power is the same
- $\left(4.5 \times 10^3\right) + \left(2.3 \times 10^3\right) = (4.5 + 2.3) \times 10^3 = 6.8 \times 10^3$
- $\left(6.45 \times 10^{-5}\right) - \left(3.21 \times 10^{-5}\right) = (6.45 - 3.21) \times 10^{-5}$
$$= 3.24 \times 10^{-5}$$

b) When the power is NOT the same
- $\left(7.6 \times 10^3\right) + \left(2.9 \times 10^2\right) = \left(7.6 \times 10^3\right) + \left(0.29 \times 10^3\right)$
$$= (7.6 + 0.29) \times 10^3 = 7.89 \times 10^3$$
- $\left(1.2 \times 10^{-2}\right) - \left(8 \times 10^{-3}\right) = \left(12 \times 10^{-3}\right) - \left(8 \times 10^{-3}\right)$
$$= (12 - 8) \times 10^{-3} = 4 \times 10^{-3}$$

NOTE
In the second scenario when the power is not the same, we simply changed one of the powers to align with the other.

Let's try some examples.

Example 15

Without using a calculator, simplify the following and express the final answer in scientific and engineering notations.

a) $\left(8 \times 10^3\right) + \left(7 \times 10^3\right)$ b) $\left(3 \times 10^9\right) + \left(2 \times 10^{10}\right)$ c) $\left(6 \times 10^{-9}\right) - \left(6 \times 10^{-11}\right)$

What did you get? Find the solution below to double-check your answer.

Solution to Example 15

a) $\left(8 \times 10^3\right) + \left(7 \times 10^3\right)$
Solution
For this, we will take out the common power of 10. Thus

$$\left(8 \times 10^3\right) + \left(7 \times 10^3\right) = 10^3 \times (8 + 7) = 10^3 \times (15) = 15 \times 10^3$$

Hence,

    i)   Scientific notation $15 \times 10^3 = 1.5 \times 10^4$
    ii)   Engineering notation $15 \times 10^3 = 15 \times 10^3$

b) $\left(3 \times 10^9\right) + \left(2 \times 10^{10}\right)$

Solution

For this, we will take out the common power of 10, preferably 10^9, which is the smaller of the two. Thus

$$\left(3 \times 10^9\right) + \left(2 \times 10^{10}\right) = \left(3 \times 10^9\right) + \left(2 \times 10^9 \times 10^1\right)$$
$$= \left(3 \times 10^9\right) + \left(20 \times 10^9\right)$$
$$= 10^9 \times (3 + 20) = 10^9 \times (23) = 23 \times 10^9$$

Hence,

    i)   Scientific notation $23 \times 10^9 = \mathbf{2.3} \times \mathbf{10^{10}}$

    ii)   Engineering notation $23 \times 10^9 = \mathbf{23} \times \mathbf{10^9}$

---

c) $\left(6 \times 10^{-9}\right) - \left(6 \times 10^{-11}\right)$

Solution

For this, we will take out the common power of 10, preferably 10^{-11}, which is the smaller of the two. Thus

$$\left(6 \times 10^{-9}\right) - \left(6 \times 10^{-11}\right) = \left(6 \times 10^2 \times 10^{-11}\right) - \left(6 \times 10^{-11}\right)$$
$$= \left(600 \times 10^{-11}\right) - \left(6 \times 10^{-11}\right)$$
$$= 10^{-11} \times (600 - 6)$$
$$= 10^{-11} \times (594) = 594 \times 10^{-11}$$

Hence,

    i)   Scientific notation $594 \times 10^{-11} = 5.94 \times 10^2 \times 10^{-11} = \mathbf{5.94} \times \mathbf{10^{-9}}$

    ii)   Engineering notation $594 \times 10^{-11} = 5.94 \times 10^2 \times 10^{-11} = \mathbf{5.94} \times \mathbf{10^{-9}}$

1.8 POWER AND ROOT

Power (or index or exponent) is the superscript on a number (or term) called **base**, as illustrated in (Figure 1.3).

y (in y^x in Figure 1.3) is the **base** and x is the **power**. We read this as '**y to the power of x**' or '**y to the x**' or simply '**y power x**'.

In general, the power can be any real number, rational or irrational. We will however restrict our discussion to positive integers in this section. This subject will be revisited and extensively covered when we discuss indices in Chapter 11.

Power is a shorthand way of saying that a number is being multiplied by itself the number of times dictated by the value of the power. As such, $\mathbf{3^2}$ means that 3 is multiplied by itself once, while $\mathbf{2^5}$

FIGURE 1.3 Base and power of a number illustrated.

implies that 2 is multiplied by itself four times. When the power is 2, 3, and 4, we specifically say that the number is squared, cubed, and quadrupled respectively.

The square of any number (positive and negative) is always positive, while the cube of a number is positive if the number is positive and negative if the number is negative. In general, if the number is positive, the answer is always positive, but if the number is negative, the answer is positive only if the power is even, otherwise the answer is negative (i.e., for odd powers).

On the other hand, the root is a direct inverse of power and is denoted by the symbol $\sqrt{}$, $\sqrt[3]{}$, and $\sqrt[4]{}$ for square root, cube root, and fourth root respectively. The number next to the radical $\sqrt{}$ sign can be any positive integer and therefore $\sqrt[n]{..}$ is termed **nth root**. For example, when the number is 5 or $n = 5$, this will be called the fifth root and denoted as $\sqrt[5]{}$.

You will notice that in square root we do not write the '2' beside the operator $\sqrt{}$; it is customary to leave this out, though you may find it written as $\sqrt[2]{}$. If so, do not be alarmed, they mean the same. In general, when a root is mentioned, it is understood as the square root. The square root is also known to give two answers, positive and negative, though the negative answer is sometimes ignored.

So, what does root represent? Well, it may not be straightforward to articulate, but let's take a particular case to explain it. Given that:

$$2^3 = 8$$

implies that:

$$\sqrt[3]{8} = 2$$

Do you get the link? This relationship between power and root is further illustrated in Figure 1.4.

So, given that:

$$3^4 = 81$$

implies that:

$$\sqrt[4]{81} = 3$$

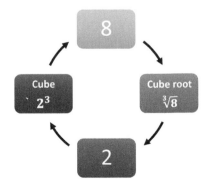

FIGURE 1.4 Cube root of a number as an inverse of the cube of the same number illustrated.

How do we derive this? Take the last example, if I know a number that can be multiplied by itself three times to get 81, then that number is the fourth root of 81. Mathematically expressed as:

$$\sqrt[4]{3 \times 3 \times 3 \times 3} = \sqrt[4]{3^4} = 3$$

It is as though the power cancels the number outside the radical sign $\sqrt{}$. **We can therefore say that the nth root of a number x is the number which when it multiplies itself $(n-1)$ times will be equal to the number x.**

When a whole number's square root is also a whole number, this is called a **perfect square**. 4, 9, 25 and 36 are examples of perfect squares.

For roots given by $\sqrt[n]{x}$, the answer is a positive real number if x is positive for all valid values of n. If however x is negative, then the result will be:

- negative real number if n is odd (e.g., 3, 5, 7, etc.).
- complex number if n is even (e.g., 2, 4, 6, etc.). This is beyond the scope of this book but duly covered in *Advanced Mathematics for Engineers and Scientists with Worked Examples* by the same author.

Great! Let's attempt a few examples.

Example 16

Simplify the following:

a) 5^2 b) 3^4 c) $(-10)^5$ d) $\sqrt[3]{27}$ e) $\sqrt[4]{16}$ f) $\sqrt[5]{(-32)}$

What did you get? Find the solution below to double-check your answer.

Solution to Example 15

a) $5^2 = 5 \times 5 = 25$

b) $3^4 = 3 \times 3 \times 3 \times 3$
$= 81$

c) $(-10)^5 = -10 \times -10 \times$
$-10 \times -10 \times -10 = -100\ 000$

d) $\sqrt[3]{27} = \sqrt[3]{3 \times 3 \times 3}$
$= \sqrt[3]{3^3} = 3$

e) $\sqrt[4]{16}$
$= \sqrt[4]{2 \times 2 \times 2 \times 2}$
$= \sqrt[4]{2^4} = 2$

f) $\sqrt[5]{-32}$
$= \sqrt[5]{-2 \times -2 \times -2 \times -2 \times -2}$
$= \sqrt[5]{(-2)^5} = -2$

1.9   CHAPTER SUMMARY

1) Arithmetic is derived from the Greek word '**arithmos**', which means 'number'. It is a branch of mathematics that deals with numbers and the relationships that exist between them.

2) Numbers can be broadly divided into two:

   • **Common numbers** include natural numbers, integers, real numbers, complex numbers, rational numbers, and irrational numbers.

   • **Special numbers** include negative numbers, even numbers, odd numbers, and prime numbers.

3) A factor (or factors) f of a number N are numbers that can divide an original number without leaving any remainder. Informally, we say that the number 'goes into it' without a remainder. A factor that is also a prime number is called a **prime factor**.

4) Multiples of a number M are numbers obtained when the original number is multiplied by natural numbers 1, 2, 3, etc.

5) A **perfect number** is a number that is exactly half of the sum of its factors.

6) A mathematical operator is generally a symbol that is used to perform a task on a number or numbers. Mathematical operations are carried out in the order captured in a single mnemonic **BIDMAS**, **BEDMAS**, or **BODMAS**.

7) The term **decimal place** is shortened to **d.p.** 1 decimal place means one digit (or more precisely one non-zero digit) after a decimal point, 2 decimal places means two digits after the decimal point, and so on.

8) The decimal place(s) of a number can be changed from a higher decimal place to a lower one; say, from 3 decimal places to 2 d.p. or 1 d.p. In this case, all excess digits to the right are discarded. It is also possible to increase the number of decimal places, say from 1 decimal place to 2, 3, etc., decimal places.

9) When a decimal number does not end, this is called a **recurring decimal** (as against a **terminating decimal**), and three dots after the last number to the right are used to indicate recurrence.

10) A significant figure implies the number of 'significant' digits.

11) A number (N), small or large, written in scientific form takes the form:

$$\boxed{N = A \times 10^{x}}$$

where A is a number between 1 and 10 and x is any integer, positive or negative.

12) Engineering notation is a special form of scientific form such that a number (M), can also be written as:

$$\boxed{M = B \times 10^{y}}$$

where B is a number between 1 and 1000 and y is any integer that is a multiple of 3.

13) Power, also known as index or exponent, refers to the superscript attached to a number or term called the **base**. It serves as a shorthand, indicating that the base is multiplied by itself a specific number of times determined by the value of the power.

****

1.10 FURTHER PRACTICE

To access complementary contents, including additional exercises, please go to www.dszak.com.

2 Fraction, Decimal, and Percentage

2.1 INTRODUCTION

In the preceding chapter, we looked at numbers and their various types and carried out basic operations involving them. It is now time to deal with the second part of rational numbers, i.e., fractions, the first part being the integers. This chapter covers the principles of fractions, decimals, and percentages and the relationships between them. The skills here will be the backbone for working with algebraic fractions, which will be covered in Chapter 3.

2.2 FRACTIONS

In general, a fraction is a number that can be expressed as $\pm \frac{a}{b}$ (in its simplest or lowest form) such that a and b are integers, excluding 0 for both and ± 1 for b only. The number or variable that is denoted by a and placed above the line is called the **numerator** while b (the number or variable below the line) is the **denominator** of the fraction, as illustrated in Figure 2.1.

Both $\frac{2}{1}$ and $-\frac{3}{1}$ are not fractions because their denominators are 1. Similarly, $\frac{15}{3}$ and $-\frac{20}{5}$ are also not fractions, although their denominators are apparently not 1 (but in fact they are 1). This is because when these numbers are expressed in their simplified forms, the denominators will become 1. In other words, $\frac{15}{3}$ is equal to $\frac{5}{1}$ and $-\frac{20}{5}$ equals $-\frac{4}{1}$.

A fraction is said to be in its simplest or lowest form when it is expressed such that no other common integer can divide both the numerator and denominator. To achieve this, we need to divide the numerator and denominator by a common factor and repeat this process until all common factors are exhausted.

DOI:10.1201/9781003027928-2

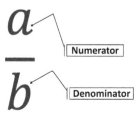

FIGURE 2.1 Position of numerator and denominator in a fraction illustrated.

The term '**complex fraction**' is sometimes used when the numerator a or the denominator b or both in $\pm\frac{a}{b}$ are also fractions. For example, $\dfrac{\left(\frac{2}{3}\right)}{5}$, $\dfrac{1}{\left(\frac{5}{3}\right)}$, and $\dfrac{\left(\frac{7}{12}\right)}{\left(\frac{5}{6}\right)}$ are regarded as complex fractions. However, as will be shown later in this chapter, the above fractions can be simplified such that the numerators and denominators are no longer fractions. In fact, the simplified form of the above fractions are $\frac{2}{15}$, $\frac{3}{5}$, and $\frac{7}{10}$ respectively.

Fractions can be one of three types, namely:

1) Proper fraction	This is when the denominator is greater than the numerator. This is indeed the true fraction, such that the value of the fraction is greater than -1 and less than **1**. $-\frac{2}{3}$, $\frac{5}{7}$, and $\frac{17}{205}$ are all examples of proper fractions.
2) Improper fraction	This is when the denominator is less than the numerator. In this case, the value of the fraction is less than -1 or greater than **1**. $\frac{4}{3}$, $-\frac{9}{5}$, and $\frac{21}{5}$ are all examples of improper fractions.
3) Mixed fraction	This is when a fraction is a combination of an integer and a fraction. This takes the form $\pm a\frac{b}{c}$ where a, b, and c are all integers and $\frac{b}{c}$ should be a proper fraction ideally. Like an improper fraction, the value of a mixed fraction is generally less than -1 or greater than 1. Examples of mixed fractions include $3\frac{1}{2}$, $5\frac{3}{4}$, and $-2\frac{6}{7}$.

When both the numerator and the denominator of a fraction are multiplied (or divided) by a constant number, a new fraction is formed. These two fractions are said to be **equivalent** as they are numerically the same. In fact, a fraction has an unlimited number of its equivalent fractions. For example, given a fraction $\frac{1}{2}$, if we multiply the numerator and denominator by 2, 3, 4, and 5, we obtain four equivalent fractions, namely: $\frac{2}{4}$, $\frac{3}{6}$, $\frac{4}{8}$, and $\frac{5}{10}$ respectively. This said, $\frac{1}{2}$ is said to be the simplest form of all its equivalents.

Note that a fraction does not change the value if both the numerator and denominator are multiplied or divided by the same number, but the value changes if the same number is added to or subtracted from the numerator and denominator. Let's illustrate this using the four main operations as:

Multiplying with a constant	$\dfrac{1}{2} = \dfrac{1 \times 2}{2 \times 2} = \dfrac{2}{4}$	$\dfrac{1}{2} = \dfrac{1 \times 3}{2 \times 3} = \dfrac{3}{6}$	$\dfrac{1}{2} = \dfrac{1 \times 4}{2 \times 4} = \dfrac{4}{8}$	$\dfrac{1}{2} = \dfrac{1 \times 5}{2 \times 5} = \dfrac{5}{10}$
Dividing by a constant	$\dfrac{30}{60} = \dfrac{30 \div 2}{60 \div 2} = \dfrac{15}{30}$	$\dfrac{30 \div 6}{60 \div 6} = \dfrac{5}{10}$	$\dfrac{30 \div 10}{60 \div 10} = \dfrac{3}{6}$	$\dfrac{30 \div 30}{60 \div 30} = \dfrac{1}{2}$
Adding a constant	$\dfrac{1}{2} \neq \dfrac{1+2}{2+2} \neq \dfrac{3}{4}$	$\dfrac{1}{2} \neq \dfrac{1+3}{2+3} \neq \dfrac{4}{5}$	$\dfrac{1}{2} \neq \dfrac{1+4}{2+4} \neq \dfrac{5}{6}$	$\dfrac{1}{2} \neq \dfrac{1+5}{2+5} \neq \dfrac{6}{7}$
Subtracting a constant	$\dfrac{7}{9} \neq \dfrac{7-2}{9-2} \neq \dfrac{5}{7}$	$\dfrac{7}{9} \neq \dfrac{7-3}{9-3} \neq \dfrac{4}{6}$	$\dfrac{7}{9} \neq \dfrac{7-4}{9-4} \neq \dfrac{3}{5}$	$\dfrac{7}{9} \neq \dfrac{7-5}{9-5} \neq \dfrac{2}{4}$

Note the use of $\neq$ in addition and subtraction. The operator $\neq$ is read as '**not equal to**' and it implies that the LHS is not equal to the RHS.

Let's try some examples.

Example 1

Express each of the following fractions in its simplest form.

a) $\dfrac{36}{100}$ b) $\dfrac{30}{75}$ c) $\dfrac{210}{360}$ d) $\dfrac{168}{98}$

What did you get? Find the solution below to double-check your answer.

Solution to Example 1

a) $\dfrac{36}{100} = \dfrac{18}{50} = \dfrac{9}{25}$ b) $\dfrac{30}{75} = \dfrac{6}{15} = \dfrac{2}{5}$

c) $\dfrac{210}{360} = \dfrac{21}{36} = \dfrac{7}{12}$ d) $\dfrac{168}{98} = \dfrac{84}{49} = \dfrac{12}{7}$

2.2.1 Conversion between Various Types of Fractions

Once in its simplified form, a proper fraction does not require any further processing. We may however need to change from an improper to a mixed fraction or vice versa as:

Mixed to improper fraction Given a mixed fraction $a\frac{b}{c}$, the improper fraction equivalent is:

$$\frac{a \times c + b}{c}$$

Follow these steps to obtain an improper fraction as illustrated in the above expression:

Step 1: Multiply the integer (whole number a) with the denominator.

Step 2: Add the numerator to the result.

Step 3: Divide overall with the original denominator.

Improper to mixed fraction The relationship below can be used to convert an improper fraction to a mixed fraction.

$$\frac{u}{v} = m\frac{r}{v}$$

We can see from the above that the denominator is the same in both formats. When u is divided by v, the answer is m and the remainder is r, where r is a positive integer excluding zero.

Let's put this into practice.

Example 2

Convert the following mixed fractions to improper fractions.

a) $2\frac{1}{7}$ b) $1\frac{4}{15}$ c) $3\frac{11}{12}$ d) $6\frac{2}{5}$ e) $7\frac{3}{4}$

What did you get? Find the solution below to double-check your answer.

Solution to Example 2

a) $2\frac{1}{7} = \frac{(2\times7)+1}{7} = \frac{14+1}{7} = \frac{15}{7}$ b) $1\frac{4}{15} = \frac{(1\times15)+4}{15} = \frac{15+4}{15} = \frac{19}{15}$

c) $3\frac{11}{12} = \frac{(3\times12)+11}{12} = \frac{36+11}{12} = \frac{47}{12}$ d) $6\frac{2}{5} = \frac{(6\times5)+2}{5} = \frac{30+2}{5} = \frac{32}{5}$

e) $7\frac{3}{4} = \frac{(7\times4)+3}{4} = \frac{28+3}{4} = \frac{31}{4}$

Another set of examples to try.

Example 3

Convert the following improper fractions to mixed fractions.

a) $\dfrac{7}{2}$ b) $\dfrac{13}{5}$ c) $\dfrac{35}{6}$ d) $\dfrac{22}{3}$ e) $\dfrac{75}{7}$

What did you get? Find the solution below to double-check your answer.

Solution to Example 3

HINT

In this case, we want to write the fraction as $a\dfrac{b}{c}$. Follow the stated approach.

a) $\dfrac{7}{2} = 3\dfrac{1}{2}$ b) $\dfrac{13}{5} = 2\dfrac{3}{5}$ c) $\dfrac{35}{6} = 5\dfrac{5}{6}$ d) $\dfrac{22}{3} = 7\dfrac{1}{3}$ e) $\dfrac{75}{7} = 10\dfrac{5}{7}$

2.2.2 OPERATION OF FRACTIONS

We have demonstrated basic operations with numbers in Chapter 1. In the same way, we will show the same with fractions.

2.2.2.1 Addition and Subtraction

There are two instances that we need to discuss in relation to addition and subtraction of fractions, namely:

Case 1 When the denominator is the same for all

To add two or more fractions with the same value of denominator, simply add the numerators and divide the result with the common denominator. To subtract, carry out the subtraction of the numerators accordingly and divide the resulting answer with the common denominator.

In general, we have:

$$\boxed{\dfrac{n_1}{d} \pm \dfrac{n_2}{d} \pm \dfrac{n_3}{d} \pm \cdots \pm \dfrac{n_n}{d} = \dfrac{n_1 \pm n_2 \pm n_3 \pm \cdots \pm n_n}{d}}$$ (2.1)

where $n_1, n_2, n_3, \ldots, n_n$ are the numerators and d is the denominator, which is the same for all the fractions.

Case 2 When the denominator is not the same for all

When the denominators are different, we need to take the following extra steps to perform addition and subtraction of two or more fractions:

Step 1: Find the LCM of the denominators $d_1, d_2, d_3, \dots, d_n$. Let this be denoted as d_{LCM}.
Step 2: Divide d_{LCM} by d_1 and denote the answer by m_1.
Step 3: Multiply m_1 with n_1.
Step 4: Repeat steps 2 and 3 for the remaining fractions using their $d_2, d_3, \dots, d_n$.
Step 5: Find the algebraic sum of $m_n n_n$ or $\sum_i^n m_i n_i$ and divide this by d_{LCM}.

The above steps are summarised in this expression:

$$\boxed{\frac{n_1}{d_1} \pm \frac{n_2}{d_2} \pm \frac{n_3}{d_3} \pm \cdots \pm \frac{n_n}{d_n} = \frac{m_1 n_1 \pm m_2 n_2 \pm m_3 n_3 \pm \dots m_n n_n}{d_{\text{LCM}}}} \qquad (2.2)$$

If there are only two fractions involved, we use what is generally called '**cross multiplication**'. Given two fractions $\frac{n_1}{d_1}$ and $\frac{n_2}{d_2}$, we can apply cross multiplication to add or subtract them. This is written as:

$$\boxed{\frac{n_1}{d_1} + \frac{n_2}{d_2} = \frac{d_2 n_1 + d_1 n_2}{d_1 d_2}} \ \text{OR} \ \boxed{\frac{n_1}{d_1} - \frac{n_2}{d_2} = \frac{d_2 n_1 - d_1 n_2}{d_1 d_2}} \qquad (2.3)$$

Cross multiplication is further illustrated in Figure 2.2.

Notice that the common denominator d in case 1 is the same as the LCM of the denominators d_{LCM} in case 2 expressed above. Also, what is important in this operation (addition and subtraction of fractions) is the ability to ensure that the denominators of all the fractions are the same. Once this is achieved – either through the LCM as explained above or any other method – we are left with adding/subtracting the numerators.

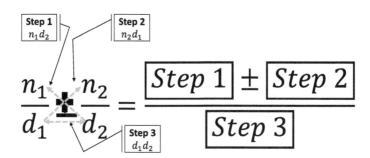

FIGURE 2.2 Addition and subtraction of two fractions using cross multiplication.

Let's try some examples.

Example 4

Work out the following fractions and present the answers in their simplest forms.

a) $\dfrac{2}{5} + \dfrac{1}{5}$ b) $\dfrac{1}{8} + \dfrac{3}{8}$ c) $\dfrac{7}{9} - \dfrac{5}{9}$ d) $\dfrac{11}{12} - \dfrac{5}{12}$ e) $\dfrac{4}{15} + \dfrac{2}{15} + \dfrac{1}{15}$ f) $\dfrac{11}{20} - \dfrac{3}{20} + \dfrac{7}{20}$

What did you get? Find the solution below to double-check your answer.

Solution to Example 4

HINT

Since the fractions here have the same denominators, we just need to add/subtract the numerator.

a) $\dfrac{2}{5} + \dfrac{1}{5}$

Solution

$$\frac{2}{5} + \frac{1}{5} = \frac{2+1}{5} = \frac{3}{5}$$

b) $\dfrac{1}{8} + \dfrac{3}{8}$

Solution

$$\frac{1}{8} + \frac{3}{8} = \frac{1+3}{8} = \frac{4}{8} = \frac{1}{2}$$

c) $\dfrac{7}{9} - \dfrac{5}{9}$

Solution

$$\frac{7}{9} - \frac{5}{9} = \frac{7-5}{9} = \frac{2}{9}$$

d) $\dfrac{11}{12} - \dfrac{5}{12}$

Solution

$$\frac{11}{12} - \frac{5}{12} = \frac{11-5}{12} = \frac{6}{12} = \frac{1}{2}$$

e) $\dfrac{4}{15} + \dfrac{2}{15} + \dfrac{1}{15}$

Solution

$$\frac{4}{15} + \frac{2}{15} + \frac{1}{15} = \frac{4+2+1}{15} = \frac{7}{15}$$

f) $\frac{11}{20} - \frac{3}{20} + \frac{7}{20}$

Solution

$$\frac{11}{20} - \frac{3}{20} + \frac{7}{20} = \frac{11 - 3 + 7}{20} = \frac{15}{20} = \frac{3}{4}$$

Let's try examples, where the denominators are not the same.

Example 5

Work out the following fractions by using cross multiplication. Present the answers in their simplest forms.

a) $\frac{1}{3} + \frac{1}{2}$ **b)** $\frac{2}{5} + \frac{1}{6}$ **c)** $\frac{3}{4} - \frac{5}{8}$ **d)** $\frac{2}{3} - \frac{3}{7}$ **e)** $1\frac{1}{3} + 2\frac{1}{4}$ **f)** $5\frac{1}{2} - 3\frac{1}{3}$

What did you get? Find the solution below to double-check your answer.

Solution to Example 5

HINT

Here the denominators of the fractions are not the same. We therefore need to make them the same or use cross multiplication. Where the fractions are mixed, we first need to convert them to improper fractions and then proceed.

a) $\frac{1}{3} + \frac{1}{2}$

Solution

$$\frac{1}{3} + \frac{1}{2} = \frac{(1 \times 2) + (1 \times 3)}{3 \times 2} = \frac{2 + 3}{3 \times 2} = \frac{5}{6}$$

b) $\frac{2}{5} + \frac{1}{6}$

Solution

$$\frac{2}{5} + \frac{1}{6} = \frac{(2 \times 6) + (1 \times 5)}{5 \times 6} = \frac{12 + 5}{5 \times 6} = \frac{17}{30}$$

c) $\frac{3}{4} - \frac{5}{8}$

Solution

$$\frac{3}{4} - \frac{5}{8} = \frac{(3 \times 8) - (5 \times 4)}{4 \times 8} = \frac{24 - 20}{4 \times 8} = \frac{4}{32} = \frac{1}{8}$$

d) $\frac{2}{3} - \frac{3}{7}$

Solution

$$\frac{2}{3} - \frac{3}{7} = \frac{(2 \times 7) - (3 \times 3)}{3 \times 7} = \frac{14 - 9}{21} = \frac{5}{21}$$

---

e) $1\frac{1}{3} + 2\frac{1}{4}$

Solution

$$1\frac{1}{3} + 2\frac{1}{4} = \frac{4}{3} + \frac{9}{4} = \frac{(4 \times 4) + (9 \times 3)}{3 \times 4} = \frac{16 + 27}{12} = \frac{43}{12} \text{ OR } 3\frac{7}{12}$$

ALTERNATIVE METHOD

$$1\frac{1}{3} + 2\frac{1}{4} = 1 + \frac{1}{3} + 2 + \frac{1}{4} = 3 + \frac{1}{3} + \frac{1}{4}$$

Now let's simplify the fraction part as:

$$\frac{1}{3} + \frac{1}{4} = \frac{(1 \times 4) + (1 \times 3)}{3 \times 4} = \frac{4 + 3}{12} = \frac{7}{12}$$

Hence,

$$1\frac{1}{3} + 2\frac{1}{4} = 3 + \frac{1}{3} + \frac{1}{4} = 3 + \frac{7}{12} = 3\frac{7}{12} \text{ OR } \frac{43}{12}$$

---

f) $5\frac{1}{2} - 3\frac{1}{3}$

Solution

$$5\frac{1}{2} - 3\frac{1}{3} = \frac{11}{2} - \frac{10}{3} = \frac{(11 \times 3) - (10 \times 2)}{2 \times 3} = \frac{33 - 20}{6} = \frac{13}{6} \text{ OR } 2\frac{1}{6}$$

ALTERNATIVE METHOD

$$5\frac{1}{2} - 3\frac{1}{3} = 5 + \frac{1}{2} - 3 - \frac{1}{3} = 2 + \frac{1}{2} - \frac{1}{3}$$

Now let's simplify the fraction part as:

$$\frac{1}{2} - \frac{1}{3} = \frac{(1 \times 3) - (1 \times 2)}{2 \times 3} = \frac{3 - 2}{6} = \frac{1}{6}$$

Hence,

$$5\frac{1}{2} - 3\frac{1}{3} = 2 + \frac{1}{2} - \frac{1}{3} = 2 + \frac{1}{6} = 2\frac{1}{6} \text{ OR } \frac{13}{6}$$

Let's try some examples, involving three fractions.

Example 6

Work out the following fractions by first expressing them in the same denominator. Present the answers in their simplest forms.

a) $\frac{1}{2} + \frac{1}{3} + \frac{1}{5}$ **b)** $\frac{3}{4} + \frac{2}{5} - \frac{5}{6}$ **c)** $\frac{7}{8} - \frac{1}{9} - \frac{5}{12}$

What did you get? Find the solution below to double-check your answer.

Solution to Example 6

a) $\frac{1}{2} + \frac{1}{3} + \frac{1}{5}$
Solution
The LCM of 2, 3, and 5 is 30, so we have:

$$\frac{1}{2} + \frac{1}{3} + \frac{1}{5} = \left(\frac{1}{2} \times \frac{15}{15}\right) + \left(\frac{1}{3} \times \frac{10}{10}\right) + \left(\frac{1}{5} \times \frac{6}{6}\right)$$
$$= \left(\frac{1 \times 15}{2 \times 15}\right) + \left(\frac{1 \times 10}{3 \times 10}\right) + \left(\frac{1 \times 6}{5 \times 6}\right) = \left(\frac{15}{30}\right) + \left(\frac{10}{30}\right) + \left(\frac{6}{30}\right)$$
$$= \frac{15 + 10 + 6}{30} = \frac{31}{30}$$
$$\therefore \frac{1}{2} + \frac{1}{3} + \frac{1}{5} = \frac{31}{30}$$

The answer can also be given as $1\frac{1}{30}$.

b) $\frac{3}{4} + \frac{2}{5} - \frac{5}{6}$
Solution
The LCM of 4, 5, and 6 is 60, so we have:

$$\frac{3}{4} + \frac{2}{5} - \frac{5}{6} = \left(\frac{3}{4} \times \frac{15}{15}\right) + \left(\frac{2}{5} \times \frac{12}{12}\right) - \left(\frac{5}{6} \times \frac{10}{10}\right)$$
$$= \left(\frac{3 \times 15}{4 \times 15}\right) + \left(\frac{2 \times 12}{5 \times 12}\right) - \left(\frac{5 \times 10}{6 \times 10}\right) = \left(\frac{45}{60}\right) + \left(\frac{24}{60}\right) - \left(\frac{50}{60}\right)$$
$$= \frac{45 + 24 - 50}{60} = \frac{19}{60}$$
$$\therefore \frac{3}{4} + \frac{2}{5} - \frac{5}{6} = \frac{19}{60}$$

c) $\dfrac{7}{8} - \dfrac{1}{9} - \dfrac{5}{12}$

Solution

The LCM of 8, 9, and 12 is 72, so we have:

$$\frac{7}{8} - \frac{1}{9} - \frac{5}{12} = \left(\frac{7}{8} \times \frac{9}{9}\right) - \left(\frac{1}{9} \times \frac{8}{8}\right) - \left(\frac{5}{12} \times \frac{6}{6}\right)$$

$$= \left(\frac{7 \times 9}{8 \times 9}\right) - \left(\frac{1 \times 8}{9 \times 8}\right) - \left(\frac{5 \times 6}{12 \times 6}\right) = \left(\frac{63}{72}\right) - \left(\frac{8}{72}\right) - \left(\frac{30}{72}\right)$$

$$= \frac{63 - 8 - 30}{72} = \frac{25}{72}$$

$$\therefore \frac{7}{8} - \frac{1}{9} - \frac{5}{12} = \frac{25}{72}$$

2.2.2.2 Multiplication

To multiply two or more fractions, carry out the following:

Step 1: Multiply all numerators to obtain the overall numerator.
Step 2: Multiply all denominators to obtain the overall denominator.

The above steps are summarised in this expression:

- For two fractions

$$\boxed{\frac{n_1}{d_1} \times \frac{n_2}{d_2} = \frac{n_1 \times n_2}{d_1 \times d_2}}$$ (2.4)

- For any number of fractions

$$\boxed{\frac{n_1}{d_1} \times \frac{n_2}{d_2} \times \frac{n_3}{d_3} \times \ldots \times \frac{n_n}{d_n} = \frac{n_1 \times n_2 \times n_3 \times \ldots \times n_n}{d_1 \times d_2 \times d_3 \times \ldots \times d_n}}$$ (2.5)

Easy one! Let's try some examples.

Example 7

Work out the following fractions. Present the answers in their simplest forms.

a) $\dfrac{2}{3} \times \dfrac{1}{4}$ **b)** $\dfrac{1}{2} \times \dfrac{2}{3} \times \dfrac{5}{7}$ **c)** $7\dfrac{1}{3} \times \dfrac{2}{5}$ **d)** $1\dfrac{2}{3} \times 2\dfrac{2}{5} \times \left(-\dfrac{5}{7}\right)$

What did you get? Find the solution below to double-check your answer.

Solution to Example 7

HINT

If the fractions are mixed, we first need to convert them to improper fractions and then proceed.

---

a) $\frac{2}{3} \times \frac{1}{4}$

Solution

$$\frac{2}{3} \times \frac{1}{4} = \frac{2 \times 1}{3 \times 4} = \frac{2}{12} = \frac{1}{6}$$

---

b) $\frac{1}{2} \times \frac{2}{3} \times \frac{5}{7}$

Solution

$$\frac{1}{2} \times \frac{2}{3} \times \frac{5}{7} = \frac{1 \times 2 \times 5}{2 \times 3 \times 7} = \frac{1 \times 5}{3 \times 7} = \frac{5}{21}$$

---

c) $7\frac{1}{3} \times \frac{2}{5}$

Solution

$$7\frac{1}{3} \times \frac{2}{5} = \frac{22}{3} \times \frac{2}{5} = \frac{22 \times 2}{3 \times 5} = \frac{44}{15} \text{ OR } 2\frac{14}{15}$$

---

d) $1\frac{2}{3} \times 2\frac{2}{5} \times \left(-\frac{5}{7}\right)$

Solution

$$1\frac{2}{3} \times 2\frac{2}{5} \times \left(-\frac{5}{7}\right) = \frac{5}{3} \times \frac{12}{5} \times \left(-\frac{5}{7}\right) = \frac{5 \times 12 \times (-5)}{3 \times 5 \times 7}$$
$$= \frac{12 \times (-5)}{3 \times 7} = \frac{4 \times (-5)}{7} = \frac{-20}{7} \text{ OR } -2\frac{6}{7}$$

2.2.2.3 Division

Before proceeding, let's take a moment to highlight something. Remember that a fraction is a division, the sign '−', called 'over', implies division. The numerator (or the number above the sign) is called the **dividend** while the denominator (or the number below the sign) is the **divisor**. Now to carry out the division of two fractions, follow these steps:

Step 1: Keep the left fraction as it is.
Step 2: Flip the second fraction such that the denominator becomes the numerator and vice versa.
Step 3: Change the division sign '÷' to multiplication '×'.
Step 4: Multiply the two new fractions together.

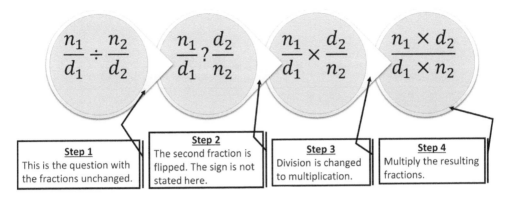

FIGURE 2.3 Procedure for carrying out the division of two fractions.

The above steps are summarised in this expression:

$$\frac{n_1}{d_1} \div \frac{n_2}{d_2} = \frac{n_1}{d_1} \times \frac{d_2}{n_2} = \frac{n_1 \times d_2}{d_1 \times n_2}$$

(2.6)

It is further illustrated in Figure 2.3.

You may also find the division of two fractions in this style:

$$\left. \frac{n_1}{d_1} \middle/ \frac{n_2}{d_2} \right. \quad \text{OR} \quad \frac{\left(\frac{n_1}{d_1} \right)}{\left(\frac{n_2}{d_2} \right)}$$

(2.7)

It is all the same. You will encounter the above style in algebraic fractions instead of the common division ÷ sign; it is simply a style, and the approach is the same.

Let's try some examples.

Example 8

Work out the following fractions. Present the answers in their simplest forms.

a) $4 \div \frac{2}{3}$ b) $\frac{12}{13} \div 6$ c) $\frac{2}{15} \div \frac{4}{5}$ d) $2\frac{1}{5} \div \frac{11}{20}$ e) $3\frac{4}{7} \div 1\frac{3}{7}$

What did you get? Find the solution below to double-check your answer.

Solution to Example 8

a) $4 \div \frac{2}{3}$

Solution

$$4 \div \frac{2}{3} = \frac{4}{1} \div \frac{2}{3} = \frac{4}{1} \times \frac{3}{2} = \frac{4 \times 3}{1 \times 2} = \frac{12}{2} = 6$$

b) $\frac{12}{13} \div 6$

Solution

$$\frac{12}{13} \div 6 = \frac{12}{13} \div \frac{6}{1} = \frac{12}{13} \times \frac{1}{6} = \frac{12 \times 1}{13 \times 6} = \frac{2}{13 \times 1} = \frac{2}{13}$$

c) $\frac{2}{15} \div \frac{4}{5}$

Solution

$$\frac{2}{15} \div \frac{4}{5} = \frac{2}{15} \times \frac{5}{4} = \frac{2 \times 5}{15 \times 4} = \frac{10}{60} = \frac{1}{6}$$

d) $2\frac{1}{5} \div \frac{11}{20}$

Solution

$$2\frac{1}{5} \div \frac{11}{20} = \frac{11}{5} \div \frac{11}{20} = \frac{11}{5} \times \frac{20}{11} = \frac{11 \times 20}{5 \times 11} = \frac{20}{5} = 4$$

e) $3\frac{4}{7} \div 1\frac{3}{7}$

Solution

$$3\frac{4}{7} \div 1\frac{3}{7} = \frac{25}{7} \div \frac{10}{7} = \frac{25}{7} \times \frac{7}{10} = \frac{25 \times 7}{7 \times 10} = \frac{25}{10} = \frac{5}{2} \text{ OR } 2\frac{1}{2}$$

2.2.3 Decimals

We attach place value to numbers. Counting from the right to the left of any number, we have units, tens, hundreds, thousands, etc., in the place value. Thus, **222** is a number which consists of three different '**2s**' because each represents a different value. Starting with the far right, we have 2, 20, and 200. This is straightforward.

In the last section, we looked at fractions and concluded that, in general, a positive fraction is a number that is less than 1 except for improper and mixed fractions. Additionally, there are ten different digits, from 0 to 9, and a combination of these with fractions provides us with unlimited real numbers, as there are countless fractions between 0 and 1.

$\frac{2}{3}$ means 2 out of 3 portions and so $\frac{4}{7}$ means 4 out of 7 portions. In both cases, the portions are less than a whole. It is not always straightforward to identify the bigger (or biggest fraction) when comparing fractions. This problem is solved if the fractions are represented as decimals.

A decimal number has three parts, namely:

1) Whole number part	This is a number that is equal or greater than **1** and is placed to the left of the decimal point. It starts with units, then tens, then hundreds, and so on, starting from the decimal point to the left.
2) Decimal point	It is a dot (.) that is used to separate the whole and the decimal parts.
3) Decimal number part	This is a number that is less than 1 and placed to the right of the decimal point. It begins with tenth, then hundredth, then thousandth, and so on, starting from the decimal point and moving to the right.

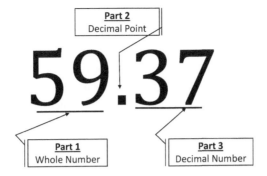

FIGURE 2.4 Decimal number and its parts.

Figure 2.4 illustrates these three parts for the decimal number 59.37.

The parts are further explained below.

Number	5	9	.	3	7
	↓	↓		↓	↓
Value	5×10	9×1	.	$\dfrac{3}{10}$	$\dfrac{7}{100}$
	↓	↓		↓	↓
	Tens	Units		Tenths	Hundredths
	↓	↓		↓	↓
	Fifty	Nine	Decimal point	Three tenths	Seven hundredths

The above number should be read as '**fifty-nine point three seven**'. Note that anything after the decimal point is less than one, so the digits must be read one after the other for all the numbers after the decimal point. It is incorrect to read 1.29 as '**one point twenty-nine**', because 0.29 represents 29 parts in 100 and is thus less than 1.

Some useful points about decimal numbers are:

Note 1 A decimal number that continues forever can be truncated using a sign or symbol to indicate the recurrence. For example, **5/3** can be written in decimal form as **1.66...** The three dots indicate the recurrence or continuity of **6**. It is also possible to write it as **1.6̄6̄, 1.6̇6̇,** or **1.66ʳ**.

The number has been given correct to 2 d.p., but it is possible to extend this and write 5/3 as 1.666̇ or 1.6666̇ for example. Alternatively, we can write the same number as **1.67** correct to 2 decimal places.

Note that recurring decimals are still rational as they can be written as a fraction, as seen in **5/3**, though we use the three dots for them as we do for irrational numbers such as $\sqrt{5} = \mathbf{2.236067...}$

Note 2 When a decimal number has no whole number, then it is possible to write zero or leave it empty. For example, 0.359 can be written as .359.

Note 3 Carrying out addition and subtraction of decimal numbers is the same as with whole numbers, but one must ensure that the decimal point is aligned. Multiplication and division are better carried out if the number can be converted to a standard form before proceeding with operations. This has been covered in Chapter 1.

Note 4 Recall that the word 'decimal' is generally related to ten. Thus, a decimal number can be viewed as a special fraction with a denominator that is a multiple of 10. This is then written in such a way that only the numerator is represented while the denominator, which is a multiple of 10, is shown by the position of the decimal point. For example, the decimal number 0.34 is the same as the fraction $\frac{34}{100}$ while the decimal number 12.05 is the same as the fraction $\frac{1205}{100}$, both fractions are not in their simplest forms.

Let's try some examples.

Example 9

Work out the following additions and present the answers correct to 1 decimal place.

a) 15.45 m + 11.39 m   **b)** 12.34 kN + 7.69 kN   **c)** 327 mm + 0.38 cm

---

What did you get? Find the solution below to double-check your answer.

Solution to Example 9

HINT

> - Align the numbers according to the position of their decimal point.
> - Watch out when the units are not the same, as there is a need to convert to the other; that is, express them in the same unit.

a) 15.45 m + 11.39 m
Solution

$$
\begin{array}{r r c c c c l}
  & 1 & 5 & . & 4 & 5 & \text{m} \\
+ & 1 & 1 & . & 3 & 9 & \text{m} \\
\hline
= & 2 & 6 & . & 8 & 4 & \text{m} \\
\hline
\end{array}
$$

$$\therefore \mathbf{15.45 \ m + 11.39 \ m = 26.8 \ m \ (1 \ d.p.)}$$

b) 12.34 kN + 7.69 kN
Solution

		1	2	.	3	4	kN
+			7	.	6	9	kN
=		2	0	.	0	3	kN

$$\therefore \textbf{12.34 kN} + \textbf{7.69 kN} = \textbf{20.0 kN (1 d.p.)}$$

---

c) 327 mm + 0.38 cm
Solution
To proceed, we need to present the numbers in the same unit. Here we have two options. In the first case, we will use mm as our reference unit and convert 0.38 cm to 3.8 mm by multiplying the former by 10. Hence, we have:

$$327 \text{ mm} + 0.38 \text{ cm} = 327 \text{ mm} + 3.8 \text{ mm}$$

Let's set the working out as:

	3	2	7	.		mm
+			3	.	8	mm
=	3	3	0	.	8	mm

$$\therefore \textbf{327 mm} + \textbf{0.38 cm} = \textbf{330.8 mm (1 d.p.)}$$

ALTERNATIVE METHOD

Alternatively, we can use cm as our reference unit and convert 327 mm to 32.7 cm by dividing the former by 10. Hence, we have

$$327 \text{ mm} + 0.38 \text{ cm} = 32.7 \text{ cm} + 0.38 \text{ cm}$$

Let's set the working out as:

	3	2	.	7		cm
+			.	3	8	cm
=	3	3	.	0	8	cm

$$\therefore \textbf{327 mm} + \textbf{0.38 cm} = \textbf{33.1 cm (1 d.p.)}$$

NOTE

The answers look a bit different but remember that if the first answer (**330.8 mm**) is changed to cm we will have **33.08 cm** and when expressed correct to 1 d.p., we will have **33.1 cm**. This is now the same as the second answer we obtained using the alternative method.

Another set of examples for subtraction.

Example 10

Work out the following subtractions and present the answers correct to 2 decimal places.

a) 1.105 kg – 0.430 kg   **b)** 2.06 kΩ – 301 Ω   **c)** 320 ms – 0.142 s

---

What did you get? Find the solution below to double-check your answer.

Solution to Example 10

a) 1.105 kg – 0.430 kg
Solution

	1	.	1	0	5		kg
−	0	.	4	3	0		kg
=	**0**	.	**6**	**7**	**5**		**kg**

$$\therefore \textbf{1.105 kg} - \textbf{0.430 kg} = \textbf{0.68 kg (2 d.p.)}$$

---

b) 2.06 kΩ – 301 Ω
Solution

For this example, question we will need to work in one unit. The choice here is kΩ. We will convert 301 Ω to 0.301 kΩ by dividing the former by 1000. (Remember to move the decimal point three steps to the left as discussed in the previous chapter). Hence, we have:

$$2.06 \text{ k}\Omega - 301 \ \Omega = 2.06 \text{ k}\Omega - 0.301 \text{ k}\Omega$$

Let's set the working out as:

	2	.	0	6		kΩ
−	0	.	3	0	1	kΩ
=	**1**	.	**7**	**5**	**9**	**kΩ**

$$\therefore \textbf{2.06 k}\Omega - \textbf{301 }\Omega = \textbf{1.76 k}\Omega \textbf{ (2 d.p.)}$$

c) 320 ms − 0.142 s

Solution

Again, for this question we need to work in one unit. The choice here is s (or seconds). We will convert 320 ms to 0.320 s by dividing the former by 1000. (Remember to move the decimal point three steps to the left as discussed in the previous chapter). Hence, we have

$$320 \text{ ms} - 0.142 \text{ s} = 0.320 \text{ s} - 0.142 \text{ s}$$

Let's set the working out as:

	0	.	3	2	0	s
−	0	.	1	4	2	s
=	**0**	.	**1**	**7**	**8**	s

$$\therefore \textbf{320 ms} - \textbf{0.142 s} = \textbf{0.18 s (2 d.p.)}$$

2.2.4 PERCENTAGE

Percentage or simply **per cent** (from the Latin phrase "per centu") implies one hundred parts and is indicated using the symbol '%'. It is another way of expressing fractions, where the denominator is set at 100 by default. Hence, $x\%$ means x part(s) in **100** or $x/100$ if expressed in fraction.

In science, we also have another similar notation called **ppm (part per million)**. Thus, 15 ppm means 15 parts in a million or $\frac{15}{1\,000\,000}$ or $\left(\frac{15}{1\,000\,000}\right)$ 100% or 0.0015%. This ppm is a special form of per cent, we should say.

Back to %, let's put it in context, 30% discount on an item valued at £100 translates to a £30 reduction for every £100. Generally, per cent is typically between 0 and 100, which translates to a proper fraction. It is however possible that a percentage is more than 100, and this does not change the definition. For example, 150% profit means that for every £100 invested one gets £150 profit, excluding the capital. This last example can be likened to our improper fraction, as it is more than 1.

Let's try some examples.

Example 11

Without using a calculator, calculate each of the following:

a) 2% of £38   **b)** 5% of 80 kg   **c)** 12.5% of 40 cm   **d)** 0.4% of 50 kΩ

What did you get? Find the solution below to double-check your answer.

Solution to Example 11

HINT

Note that 'of' implies multiplication here.

a) 2% of £38
Solution

$$2\% \text{ of } £38 = \frac{2}{100} \times £38 = \frac{2 \times £38}{100} = \frac{£76}{100} = £0.76$$

b) 5% of 80 kg
Solution

$$5\% \text{ of } 80 \text{ kg} = \frac{5}{100} \times 80 \text{ kg} = \frac{5 \times 80 \text{ kg}}{100} = \frac{400 \text{ kg}}{100} = 4 \text{ kg}$$

c) 12.5% of 40 cm
Solution

$$12.5\% \text{ of } 40 \text{ cm} = \frac{12.5}{100} \times 40 \text{ cm} = \frac{12.5 \times 40 \text{ cm}}{100} = \frac{125 \times 4 \text{ cm}}{100} = \frac{500 \text{ cm}}{100} = 5 \text{ cm}$$

d) 0.4% of 50 kΩ
Solution

$$0.4\% \text{ of } 50 \text{ k}\Omega = \frac{0.4}{100} \times 50 \text{ k}\Omega = \frac{0.4 \times 50 \text{ k}\Omega}{100} = \frac{4 \times 5 \text{ k}\Omega}{100} = \frac{20 \text{ k}\Omega}{100} = 0.2 \text{ k}\Omega$$

Let's try another set of examples.

Example 12

Work out the following addition and present the answers correct to 2 significant figures.

a) 15% + 23%   **b)** 62% − 8%   **c)** 30% × 10%   **d)** 13% × 13%

What did you get? Find the solution below to double-check your answer.

Solution to Example 12

HINT

There is nothing special about these examples. Just consider the numbers without the per cent sign (%) and bring it back once you've completed the operation.

a) 15% + 23%
Solution

$$15\% + 23\% = 38\% \text{ (2 s.f.)}$$

b) 62% − 8%
Solution

$$62\% - 8\% = 54\% \text{ (2 s.f.)}$$

c) $30\% \times 10\%$
Solution
For multiplication, we will need to note that 30% means 30 in 100. In other words, $30\% = \frac{30}{100}$. This
is the gist for working with multiplication. Now let's go back to the question and work this out as:

$$30\% \times 10\% = \frac{30}{100} \times \frac{10}{100} = \frac{300}{10\,000}$$

Before we turn this back to per cent, we need to pause and think about the denominator, which is
currently 10 000. We need to make sure that it is 100 as $\frac{300}{10\,000}$ means 300 in 10 000. There are a few
ways to do this, but here is one of them.

$$\frac{300}{10\,000} = \frac{3 \times 100}{100 \times 100} = \frac{3}{100} = 3\%$$

Hence,

$$30\% \times 10\% = \textbf{3.0\% (2 s.f.)}$$

---

d) $13\% \times 13\%$
Solution

$$13\% \times 13\% = \frac{13}{100} \times \frac{13}{100}$$
$$= \frac{169}{10\,000} = \frac{1.69 \times 100}{100 \times 100}$$
$$= \frac{1.69}{100} = 1.69\%$$

Hence,

$$13\% \times 13\% = \textbf{1.7\% (2 s.f.)}$$

NOTE

In this case, we have mentioned previously that one should move the decimal point in 169 to the left
twice when dividing by 100. Two questions might have come to mind.

(1) The first is that 'where is the decimal point in the first place?' The answer is that it is to the
     right of 9 (i.e., $169 = 169$).
(2) The second question is why do we need to move twice? The answer to this is because there
     are two zeros to be deleted in 10 000 to make it 100. Thus, 1.69×100, an equivalent of 169,
     is arrived at because we have divided 169 by 100.

Let's try more examples.

Example 13

Without using a calculator, calculate the new value when:

   a) a load of 500 N is increased by 10%.
   b) an electromotive force (emf) of 120 V is reduced by 20%.

What did you get? Find the solution below to double-check your answer.

Solution to Example 13

a) A load of 500 N is increased by 10%
Solution

$$10\% \text{ of } 500 \text{ N} = \frac{10}{100} \times 500 \text{ N}$$
$$= \frac{10 \times 500 \text{ N}}{100} = \frac{5000 \text{ N}}{100}$$
$$= \mathbf{50 \text{ N}}$$

Therefore, the new value is

$$500 \text{ N} + 50 \text{ N} = \mathbf{550 \text{ N}}$$

ALTERNATIVE METHOD

500 N implies 100% plus an increase of 10% we have the total percentage of

$$100\% + 10\% = 110\%$$

Thus, the new value is

$$110\% \text{ of } 500 \text{ N} = \frac{110}{100} \times 500 \text{ N}$$
$$= \frac{110 \times 500 \text{ N}}{100} = 110 \times 5 \text{ N}$$
$$= \mathbf{550 \text{ N}}$$

b) An emf of 120 V is reduced by 20%
Solution

$$20\% \text{ of } 120 \text{ V} = \frac{20}{100} \times 120 \text{ V}$$
$$= \frac{20 \times 120 \text{ V}}{100} = 2 \times 12$$
$$= \mathbf{24 \text{ V}}$$

Therefore, the new value is

$$120 \text{ V} - 24 \text{ V} = \mathbf{96 \text{ V}}$$

ALTERNATIVE METHOD

120 V implies 100%, but a reduction of 20% means that the total percentage is

$$100\% - 20\% = 80\%$$

Thus, the new value is

$$
\begin{aligned}
80\% \text{ of } 120 \text{ V} &= \frac{80}{100} \times 120 \text{ V} \\
&= \frac{80 \times 120 \text{ V}}{100} = 8 \times 12 \text{ V} \\
&= \mathbf{96\,V}
\end{aligned}
$$

2.2.5 CONVERSION BETWEEN FRACTION, DECIMAL, AND PERCENTAGE

Any of the above three forms (i.e., fraction, decimal, and percentage) of number representation can be converted to the others. For example, $\frac{1}{2}$ can also be represented as 50% and 0.50. Out of these, changing from fraction to the other two forms (i.e., decimal and percentage) may sometimes present a bit of challenge, the others are straightforward. Let's look at the three possible cases.

(1) Percentage to others

    a) To change a percentage to a decimal, simply move the decimal point two steps to the left. If there is no decimal point in the number, assume one to be at the far right of the given number, and if the number is a single digit, add a zero to its left side. This is the same as dividing the number by 100. For example, $5\% = 05\% = 0.05$. In the same way, we have that $25\% = 0.25$ and $63.5\% = 0.635$.

    b) To change a percentage to its fraction equivalent, simply divide the number by 100 and simplify if possible. For example, $12\% = \frac{12}{100} = \frac{3}{25}$ and $2.5\% = \frac{2.5}{100} = \frac{25}{1000} = \frac{1}{40}$.

(2) Decimal to others

    a) To change a decimal number to a percentage, move the decimal point two steps to the right and add the per cent sign (%). For example, $0.03 = 3\%$, $0.145 = 14.5\%$, and $1.65 = 165\%$.

    b) To change a decimal number to its fraction equivalent, simply move the point to the right of the last number and divide the number by 10 to a power corresponding to the number of steps of movement. In other words, if you move the point once, dividing by 10, moving the point twice means that the number should be divided by 100 (i.e., 10^2) and so on. Simplify the resulting fraction. For example, $0.36 = \frac{36}{100} = \frac{9}{25}$ and $2.5 = \frac{25}{10} = \frac{5}{2}$.

(3) Fraction to others

For this case, it suffices to look at the conversion of either of the other two forms, since changing from decimal to percentage and vice versa is easy and has been covered. Therefore, we will look at converting a fraction to a percentage. To do this, we have the following two options:

Option 1 Given a fraction, multiply (or divide) both the numerator and denominator with the same number such that the denominator becomes 100. The numerator is the percentage once the sign is added. For example,

$$\frac{2}{5} = \frac{2 \times 20}{5 \times 20} = \frac{40}{100} = 40\%$$

Option 2 If option 1 is not possible, use the long division method (or a calculator) to evaluate it, move the decimal point two places to the right, and add the per cent sign. Sometimes, we say multiply the answer by 100 and add the percent sign. Both are the same because, multiplying a number by 100 is the same as moving the decimal point two steps to the right.

As a way of quick reference and summary, Table 2.1 provides some common percentages and their decimal and fraction equivalents.

Let's try some examples.

Example 14

Without using a calculator, write the following fractions as decimals.

a) $\frac{3}{5}$ b) $\frac{11}{25}$ c) $\frac{3}{40}$ d) $\frac{81}{250}$ e) $3\frac{1}{2}$

TABLE 2.1
Common Percentage and Their Respective Decimal and Fraction Equivalents

Percentage	Decimal	Fraction	Percentage	Decimal	Fraction
1%	$\frac{1}{100}$	0.01	60%	$\frac{60}{100} = \frac{3}{5}$	0.6
5%	$\frac{5}{100} = \frac{1}{20}$	0.05	70%	$\frac{70}{100} = \frac{7}{10}$	0.7
10%	$\frac{10}{100} = \frac{1}{10}$	0.1	75%	$\frac{75}{100} = \frac{3}{4}$	0.75
25%	$\frac{25}{100} = \frac{1}{4}$	0.25	80%	$\frac{80}{100} = \frac{4}{5}$	0.8
40%	$\frac{40}{100} = \frac{2}{5}$	0.4	90%	$\frac{90}{100} = \frac{9}{10}$	0.9
50%	$\frac{50}{100} = \frac{1}{2}$	0.5	100%	$\frac{100}{100} = 1$	1

What did you get? Find the solution below to double-check your answer.

Solution to Example 14

HINT

> The key to solving these questions is to ensure that the denominator is **10, 100, 1000**, etc. This can be done by multiplying/dividing the numerator and denominator by the same number.

a) $\frac{3}{5}$

Solution

$$\frac{3}{5} = \frac{3}{5} \times \frac{2}{2} = \frac{3 \times 2}{5 \times 2} = \frac{6}{10} = \textbf{0.6}$$

b) $\frac{11}{25}$

Solution

$$\frac{11}{25} = \frac{11}{25} \times \frac{4}{4} = \frac{11 \times 4}{25 \times 4} = \frac{44}{100} = \textbf{0.44}$$

c) $\frac{3}{40}$

Solution

$$\frac{3}{40} = \frac{3}{40} \times \frac{\left(\frac{5}{2}\right)}{\left(\frac{5}{2}\right)} = \frac{3 \times \left(\frac{5}{2}\right)}{40 \times \left(\frac{5}{2}\right)} = \frac{\left(\frac{15}{2}\right)}{100} = \frac{7.5}{100} = \textbf{0.075}$$

d) $\frac{81}{250}$

Solution

$$\frac{81}{250} = \frac{81}{250} \times \frac{4}{4} = \frac{81 \times 4}{250 \times 4} = \frac{324}{1000} = \textbf{0.324}$$

e) $3\frac{1}{2}$

Solution

$$3\frac{1}{2} = \frac{7}{2} = \frac{7}{2} \times \frac{5}{5} = \frac{7 \times 5}{2 \times 5} = \frac{35}{10} = \textbf{3.5}$$

Let's try more examples.

Example 15

Without using a calculator, write the following decimals as fractions. Present your answers in their simplest form.

a) 0.24   b) 3.5   c) 0.048   d) 1.25

What did you get? Find the solution below to double-check your answer.

Solution to Example 15

a) 0.24
Solution

$$0.24 = \frac{24}{100}$$
$$= \frac{12}{50} = \frac{6}{25}$$

b) 3.5
Solution

$$3.5 = \frac{35}{10}$$
$$= \frac{7}{2}$$

c) 0.048
Solution

$$0.048 = \frac{48}{1000} = \frac{24}{500}$$
$$= \frac{12}{250} = \frac{6}{125}$$

d) 1.25
Solution

$$1.25 = \frac{125}{100}$$
$$= \frac{25}{20} = \frac{5}{4}$$

Let's try more examples.

Example 16

Without using a calculator, write each of the following percentages as a:

    i) Decimal

    ii) Fraction

Present the answers in their simplest form.

a) 76%   **b)** 1.4%   **c)** 6.25%   **d)** 30.8%

What did you get? Find the solution below to double-check your answer.

Solution to Example 16

a) 76%
Solution

    i) Decimal $76\% = \mathbf{0.76}$

    ii) Fraction $76\% = \dfrac{76}{100} = \dfrac{38}{50} = \mathbf{\dfrac{19}{25}}$

b) 1.4%
Solution

    i) Decimal $1.4\% = \mathbf{0.014}$

    ii) Fraction $1.4\% = \dfrac{1.4}{100} = \dfrac{14}{1000} = \dfrac{7}{500} = \mathbf{\dfrac{7}{500}}$

c) 6.25%
Solution

    i) Decimal $6.25\% = \mathbf{0.0625}$

    ii) Fraction $6.25\% = \dfrac{6.25}{100} = \dfrac{\left(\dfrac{625}{100}\right)}{100} = \dfrac{625}{10000}$

                           $= \dfrac{25}{400} = \dfrac{1}{16}$

                           $= \mathbf{\dfrac{1}{16}}$

d) 30.8%
Solution

i) Decimal $\quad$ 30.8% = **0.308**

ii) Fraction $\quad$ $30.8\% = \dfrac{30.8}{100} = \dfrac{308}{1000} = \dfrac{154}{500} = \dfrac{\mathbf{77}}{\mathbf{250}}$

2.3 RATIO AND PROPORTION

Ratio is the proportion of one quantity (sample) compared to another similar quantity (population). It is a fraction or percentage of the sample in the population. It is given as $x:y$ if there are two samples, $x:y:z$ if there are three, and so on.

In general, if we have n different samples $s_1, s_2, s_3, \ldots, s_n$ to share a quantity Q in the ratio $r_1:r_2:r_3: \ldots :r_n$, the share and quantity for each are given in Table 2.2.

We use $\sum r_i$ to imply addition or summation of r values, where r is any physical quantity (e.g., money, age, mass). The subscript i is for referencing the member of the set of the physical quantity. To be precise about the members to be added, we use $\sum_i^n r_i$ to imply that we add from the ith member to the nth member of the set. For example, given that r is a data set of weights of students in a class, then $\sum_2^{10} r_i$ means to add the weights of the second, third, etc., up to the tenth student in the class.

Note that we simplify ratios similar to fractions as $2:3$ is $\dfrac{2}{5}$ and $\dfrac{3}{5}$ of the quantity.

TABLE 2.2
Illustrating Ratio and Proportion

Sample	Ratio	Fraction	Percentage	Share (or Amount)
s_1	r_1	$\dfrac{r_1}{r_1+r_2+\cdots+r_n}$	$\dfrac{r_1}{r_1+r_2+\cdots+r_n} \times 100\%$	$\dfrac{r_1}{\sum_i^n r_i} \times Q$ $= \dfrac{r_1}{r_1+r_2+\cdots+r_n} \times Q$
s_2	r_2	$\dfrac{r_2}{r_1+r_2+\cdots+r_n}$	$\dfrac{r_2}{r_1+r_2+\cdots+r_n} \times 100\%$	$\dfrac{r_2}{\sum_i^n r_i} \times Q$ $= \dfrac{r_2}{r_1+r_2+\cdots+r_n} \times Q$
s_3	r_3	$\dfrac{r_3}{r_1+r_2+\cdots+r_n}$	$\dfrac{r_3}{r_1+r_2+\cdots+r_n} \times 100\%$	$\dfrac{r_3}{\sum_i^n r_i} \times Q$ $= \dfrac{r_3}{r_1+r_2+\cdots+r_n} \times Q$
s_n	r_n	$\dfrac{r_n}{r_1+r_2+\cdots+r_n}$	$\dfrac{r_n}{r_1+r_2+\cdots+r_n} \times 100\%$	$\dfrac{r_n}{\sum_i^n r_i} \times Q$ $= \dfrac{r_n}{r_1+r_2+\cdots+r_n} \times Q$

Let's try some examples.

Example 17

Write each of the following ratios in their simplest form.

a) $2:4$ **b)** $7:21$ **c)** $25:10$ **d)** $2:6:12$ **e)** $24:15:9$ **f)** $22:55:33$

What did you get? Find the solution below to double-check your answer.

Solution to Example 17

a) $2:4$
Solution

$$2:4 = 1:2$$

b) $7:21$
Solution

$$7:21 = 1:3$$

c) $25:10$
Solution

$$25:10 = 5:2$$

d) $2:6:12$
Solution

$$2:6:12 = 1:3:6$$

e) $24:15:9$
Solution

$$24:15:9 = 8:5:3$$

f) $22:55:33$
Solution

$$22:55:33 = 2:5:3$$

Let's try another set of examples.

Example 18

Determine the missing ratio, denoted by a letter, in the following. Re-write the ratios using the value obtained.

a) $1:4 = x:20$ **b)** $2:3 = 26:y$ **c)** $a:5 = 42:35$ **d)** $72:b = 12:7$

What did you get? Find the solution below to double-check your answer.

Solution to Example 18

a) $1:4 = x:20$
Solution
Given that

$$1:4 = x:20$$

we can write that

$$\frac{1}{4} = \frac{x}{20}$$

$$\frac{1}{4} \times 20 = x$$

$$x = 5$$

$$\therefore 1:4 = 5:20$$

b) $2:3 = 26:y$
Solution
Given that

$$2:3 = 26:y$$

we can write that

$$\frac{2}{3} = \frac{26}{y}$$

$$2 \times y = 26 \times 3$$

$$y = \frac{26 \times 3}{2} = 39$$

$$\therefore 2:3 = 26:39$$

c) $a:5 = 42:35$
Solution
Given that

$$a:5 = 42:35$$

we can write that

$$\frac{a}{5} = \frac{42}{35}$$

$$a = \frac{42}{35} \times 5$$

$$a = \frac{42}{7} = 6$$

$$\therefore 6:5 = 42:35$$

---

d) $72:b = 12:7$
Solution
Given that

$$72:b = 12:7$$

we can write that

$$\frac{72}{b} = \frac{12}{7}$$

$$72 \times 7 = 12 \times b$$

$$b = \frac{72 \times 7}{12} = 6 \times 7 = 42$$

$$\therefore 72:42 = 12:7$$

Let's try more examples.

Example 19

In a class of 24 girls and 20 boys, determine the ratio of boys to girls in this class. Express your answer in the simplest form.

What did you get? Find the solution below to double-check your answer.

Solution to Example 19

Ratio of boys to girls implies:

$$\text{boys}:\text{girls} = 20:24$$

To express this in simplified form, divide both sides by 4:

$$\therefore 20:24 = 5:6$$

Let's try more examples.

Example 20

Solve the following problems.

a) Share £30 in the ratio $1:4$.   **b)** Divide $360°$ in the ratio $1:3:5$.

What did you get? Find the solution below to double-check your answer.

Solution to Example 20

a) Share £30 in the ratio $1:4$.
Solution
The total share is

$$1 + 4 = 5$$

That is five parts of £30 or £30 in five parts.

If Q_n is the amount received because of n share, then we have:

$$Q_1 = \frac{1}{5} \times £30 = £6$$
$$Q_4 = \frac{4}{5} \times £30 = £24$$

Alternatively, since we know the amount representing a share, we simply multiply this by 4 to get the four shares. In other words:

$$Q_4 = 4Q_1 = 4\,(£6) = £24$$

---

b) Divide $360°$ in the ratio $1:3:5$.
Solution
The total share is

$$1 + 3 + 5 = 9$$

If Q_n is the amount in n share, then we have:

$$Q_1 = \frac{1}{9} \times 360° = 40°$$

Thus,

$$Q_3 = 3Q_1 = 3\,(40°) = 120°$$

and

$$Q_5 = 5Q_1 = 5\,(40°) = 200°$$

Let's try another example.

Example 21

The lengths l_1, l_2, and l_3 of a scalene triangle are in the ratio $3:5:7$. Calculate the length of each side if the perimeter is 45 cm.

What did you get? Find the solution below to double-check your answer.

Solution to Example 21

The total ratio is

$$3 + 5 + 7 = 15$$

If Q_n is the length in n share, then we have:

$$Q_3 = \frac{3}{15} \times 45 \text{ cm} = \textbf{9 cm}$$

$$Q_5 = \frac{5}{15} \times 45 \text{ cm} = \textbf{15 cm}$$

and

$$Q_7 = \frac{7}{15} \times 45 \text{ cm} = \textbf{21 cm}$$

Alternatively, we can obtain the third length by using

$$Q_{\text{total}} = Q_3 + Q_5 + Q_7$$
$$Q_7 = Q_{\text{total}} - Q_3 - Q_5 = 45 - 9 - 15 = \textbf{21 cm}$$

2.4 CHAPTER SUMMARY

1) A fraction is a number that can be expressed as $\pm\frac{a}{b}$ (in its simplest or lowest form) such that a and b are integers, excluding 0 for both and ± 1 for b. a is called the **numerator** while b is the **denominator** of the fraction.

2) Fractions can either be a proper, an improper, or a mixed fraction.

3) An improper fraction can be changed to a mixed fraction and vice versa.

4) 'Cross multiplication is a method of carrying out addition and subtraction of two fractions, which can be represented as:

$$\boxed{\frac{n_1}{d_1} + \frac{n_2}{d_2} = \frac{d_2 n_1 + d_1 n_2}{d_1 d_2}} \text{ OR } \boxed{\frac{n_1}{d_1} - \frac{n_2}{d_2} = \frac{d_2 n_1 - d_1 n_2}{d_1 d_2}}$$

5) Multiplication of two or more fractions is carried out using:

$$\frac{n_1}{d_1} \times \frac{n_2}{d_2} = \frac{n_1 \times n_2}{d_1 \times d_2} \quad \text{OR} \quad \frac{n_1}{d_1} \times \frac{n_2}{d_2} \times \frac{n_3}{d_3} \times \dots \times \frac{n_n}{d_n} = \frac{n_1 \times n_2 \times n_3 \times \dots \times n_n}{d_1 \times d_2 \times d_3 \times \dots \times d_n}$$

6) Division of two fractions is carried out by changing it to multiplication using:

$$\frac{n_1}{d_1} \div \frac{n_2}{d_2} = \frac{n_1}{d_1} \times \frac{d_2}{n_2} = \frac{n_1 \times d_2}{d_1 \times n_2}$$

7) A number can be expressed in fraction, decimal, and percentage. For example, $\frac{1}{2}$ = 0.5 = 50%.

8) Ratio is the proportion of one quantity (sample) compared to another similar quantity (population).

****

2.5 FURTHER PRACTICE

To access complementary contents, including additional exercises, please go to www.dszak.com.

3 Algebraic Expression

Learning Outcomes

Once you have studied the content of this chapter, you should be able to:

- Explain the term variable and its different types
- Carry out addition, subtraction, multiplication, and division of algebraic expressions
- Open or expand brackets involving algebraic expressions
- Factorise algebraic expressions
- Simplify algebraic expressions
- Change the subject of a formula

3.1 INTRODUCTION

We earlier discussed numbers and carried out basic arithmetic operations involving them. We will now extend our application of this concept to the domain of algebra – a branch of mathematics that deals with arithmetical operations of variables or changing numbers, which are represented by letters or symbols. This chapter covers the fundamentals of algebraic expressions, including opening brackets, factorisation, algebraic simplification, and transposition.

3.2 ALGEBRA

In a simplistic form, algebra is a branch of mathematics which involves the manipulation of variables and numbers. As such, basic operations ($+$, $-$, $\times$, and $\div$) as well as other operators can be applied to this branch of mathematics.

3.2.1 VARIABLE

The term **variable** or something that varies (also called the '**unknown**') is a quantity that we cannot assign a fixed numerical value to and generally represents changing quantities (time, amount, force, height, etc.). Although it is very common to use x and y or a, b, and c for arbitrary quantities, in theory any symbol or letter (Arabic, English, Greek, etc.) can be used to denote a variable. We have two types of variables.

---

1) Independent variable   These are variables that we control or change. They are usually found on the right-hand side (RHS) of the equation or formula.

DOI:10.1201/9781003027928-3

2) Dependent variables   These are the resulting variables due to the changes made in the control variables. They are usually found on the left-hand side (LHS) of the equation and are called the subjects of the formula.

There are certain letters or symbols that have fixed numerical values and are referred to as **constants** as opposed to variables. These include *pi* (π), acceleration due to gravity (g), and Euler's number (e) among others.

3.2.2 ALGEBRAIC EXPRESSION

Algebra can be used to represent a situation; this is called an **algebraic expression**. It can also be used to show a relationship between equalities or solve a problem, and this is called an **algebraic equation**. $3x + 12$, $a + 2b - 5$, and $\frac{3}{x+5}$ are examples of expressions whilst $3x + 12 = 0$, $a + 2b - 5 = y$, and $\frac{3}{x+5} = 1$ are examples of equations. It is apparent that the distinguishing factor between an expression and an equation is the presence (or absence) of the equal sign '='.

Let's put this into context. If I bought three apples and two bananas for £2.70, we can write an algebraic expression and an algebraic equation from this. To do this, we need to assign letter/symbol to the quantities. It is a common practice to use the first letter of a quantity to denote its variable, as a result, let apple be represented by a and banana by b. Therefore, $3a$ means 3 × apple, i.e., 3 apples (1 apple + 1 apple + 1 apple) while $2b$ means 2 × banana, i.e., 2 bananas. Hence, for this, we can write:

- the algebraic expression as: $3a + 2b$
- the algebraic equation as: $3a + 2b = £2.70$

Furthermore, the above are examples of a **binomial expression** and **equation** because they contain only two terms.

Another difference between the two is that an algebraic expression does not require an answer, but an equation demands that we find the numerical value for the unknown, provided we have sufficient information. Unfortunately, for the equation above, that is $3a + 2b = £2.70$, we cannot say the price for both apple and banana as we do not have enough information to determine this. We will discuss this further in Chapter 7.

One last point to emphasise before we move on is that each inseparable or undetachable 'part' is called a **term**. This can either be a number (**2**, **3.5**, $\frac{1}{4}$, etc.), a variable (**x**, **y**, **m**, etc.) or variables (**xy**, **xyz**, **bc**, etc.), or both (**3x**, **2xy**, etc.). We have illustrated this in Figure 3.1 for an expression consisting of three terms, which are added together.

Now let's try some examples.

Example 1

State the number of terms in the following expressions and list them.

a) $a - 3b$ **b)** $2uv + y^2$ **c)** $3 - xyz$ **d)** $ax^2 + bx + c$ **e)** $m^2 + 3xy - 5z^2n + 10$

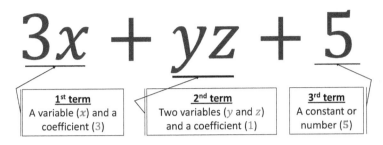

FIGURE 3.1 Explaining the constituents of an expression.

TABLE 3.1
Solution to Example 1

	Expression	Number of Terms	List
a)	$a - 3b$	2	a and $-3b$
b)	$2uv + y^2$	2	$2uv$ and y^2
c)	$3 - xyz$	2	3 and $-xyz$
d)	$ax^2 + bx + c$	3	ax^2, bx, and c
e)	$m^2 + 3xy - 5z^2n + 10$	4	m^2, $3xy$, $-5z^2n$, and 10

What did you get? Find the solution to double-check your answer.

Solution to Example 1

Solution
See Table 3.1.

Let's finish this section by summarising some key tips to note when working with an expression:

Note 1 Between each element of a term there is an unwritten multiplication sign, omitted for brevity (to save time and space) and clarity (to remove any confusion between ×, as a multiplication sign, and x representing a variable). In other words, $3a$ is the same as $3 \times a$, $5y^2$ is the same as $5 \times y \times y$, and $5uv$ is the same as $5 \times u \times v$. Note that this does not apply to numbers, thus 47 is not the same as 4×7.

Note 2 ab is the same as ba, but the former is a preferred option. Similarly, abc has five other variants: acb, bac, bca, cab, and cba. Again abc is the preferred choice. This is simply based on the English alphabetical order.

Note 3 When a term consists of a number and variable, the number is called the numerical coefficient (or simply coefficient) of the term. For example, in the expression $3a + 2b$, there are two terms. The coefficient of the first term is 3 while the coefficient of the second term is 2. Also, the coefficient of $\frac{x}{3}$ is $\frac{1}{3}$ because $\frac{x}{3} = \frac{1}{3} \times x$ just like $\frac{7}{3} = \frac{1}{3} \times 7$.

Note 4 A coefficient must always be the first element. For this reason, $3a$ cannot be written as $a3$ and $5uv$ cannot be written as $u5v$ or $uv5$, though technically they are equivalent, i.e., $3a = a3$ and $5uv = u5v = uv5$.

Note 5 A coefficient can have more than one element. For example, 2π is the coefficient of the formula for angular velocity $\omega = 2\pi f$.

Note 6 When a term does not (apparently) have a coefficient, this is taken to be 1. Thus, the coefficients of ab and x are both 1. In other words, $ab = 1ab$ and $x = 1x$. Also, when a variable does not appear in an expression, then it can be assumed that the coefficient of the variable is zero. More examples on this later.

Note 7 The position of the letter also plays a crucial role and as such the 'b' in $2b$, 2^b, and 2_b are not the same. In $2b$, 2 is multiplied by b, while in 2^b and 2_b, the b is a superscript and a subscript respectively.

Note 8 The case of the letter is important, and as such, t is not the same as T. Similarly, F is not equal to f.

Note 9 We usually italicise a variable, especially when it's a capital letter. Thus, it is better to write $2t$ instead of **2t** and $2N$ instead of **2N**. This is because a capital letter (without italic) is reserved as a unit whilst a capital letter (with italic) is a quantity (or variable). For example, the capital letter C is used for capacitance (a physical quantity or a variable) and **C** denotes a unit, known as Coulomb. Based on this, $2C$ is 2 × capacitance and 2 C is 2 Coulomb. In a more technical style, we leave a space between a number and unit. The number in the latter case is not regarded as a coefficient but a magnitude.

3.3 BASIC OPERATIONS

The four basic operations can be carried out on algebraic expressions as will be shown shortly.

3.3.1 ADDITION (+) AND SUBTRACTION (−)

We can only carry out addition and subtraction if the terms are identical. Two or more terms are identical if they have the same type and number of variables, and the variables have the same powers. For example, $3p$, $-6p$, and $p/3$ are identical terms as they are all multiples of p and can be added; the same can be said about $7q^2$, $-0.5q^2$, and $\frac{2q^2}{3}$.

The terms $7p$, $2pq$, and $-2p^2$ are not alike and cannot be added. This is because the number of variables in them are not the same. The first and the third have only one variable, which is p while the second has two variables (p and q). In addition, the power of the variable is not the same; the power of p in the first and second terms is 1, but it is 2 in the third term.

Consequently, we say that when we have like (or identical) terms, we can add or subtract the coefficients and keep the variable. As previously mentioned, the **coefficient** of a term is a numerical value or a multiplier that appears in front of a variable, and when the coefficient is 1, we generally omit writing it. Hence, 1, 2, and 3 are the coefficients of xy, $2x^2$, and $3xy^2$ respectively.

Given two variables x and y, an addition of the two is represented as $x + y$ while the subtraction of y from x is written as $x - y$. Note that $x + y = y + x$ but $x - y \neq y - x$. In other words, addition is considered **commutative** because the sequence of x and y does not affect the result, whereas subtraction is **non-commutative** since the order of x and y is important.

Also, if we are given three variables p, q, and r, we can write that $(p + q) + r = p + (q + r)$. In other words, when given three variables to be added, any two can be added first and the result obtained will be added to the third variable. Technically, we say that addition is **associative**. These two properties, **commutativity** and **associativity**, are also valid for numbers.

Before we try examples, it is important to introduce the term '**simplify**' or '**simplification**'. When either of these terms is used, it implies that the expression should be evaluated using all suitable rules until the simplest form is obtained.

Now it's time to try some examples.

Example 2

Simplify the following expressions.

a) $2x + 3y + 5x + 6y$ **b)** $10m - 13n + 8n + m$ **c)** $2p - 7 - 8q + 15 + q$

What did you get? Find the solution below to double-check your answer.

Solution to Example 2

HINT

Watch out for when we apply the commutative and associative rules.

a) $2x + 3y + 5x + 6y$
Solution

$$2x + 3y + 5x + 6y = (2x + 5x) + (3y + 6y)$$
$$= (7x) + (9y)$$
$$= \mathbf{7x + 9y}$$

b) $10m - 13n + 8n + m$
Solution

$$10m - 13n + 8n + m = (10m + m) + (-13n + 8n)$$
$$= (11m) + (-5n)$$
$$= \mathbf{11m - 5n}$$

c) $2p - 7 - 8q + 15 + q$
Solution

$$2p - 7 - 8q + 15 + q = (2p) + (15 - 7) + (q - 8q)$$
$$= (2p) + (8) + (-7q)$$
$$= \mathbf{2p + 8 - 7q} \textbf{ OR } \mathbf{2p - 7q + 8}$$

Let's try more examples.

Example 3

Add the following pairs of expressions together.

a) $3a^2 + x^2 - 4$ **and** $a^2 - 7x^2 + 16$ **b)** $5mn - y^3 + 10$ **and** $6mn - xy^3 - 5$

What did you get? Find the solution below to double-check your answer.

Solution to Example 3

a) $3a^2 + x^2 - 4$ **and** $a^2 - 7x^2 + 16$
Solution

$$(3a^2 + x^2 - 4) + (a^2 - 7x^2 + 16) = (3a^2 + a^2) + (x^2 - 7x^2) + (-4 + 16)$$
$$= (4a^2) + (-6x^2) + (12)$$
$$= \mathbf{4a^2 - 6x^2 + 12}$$

NOTE
We can write the final answer as $\mathbf{2(2a^2 - 3x^2 + 6)}$. This alternative answer is derived through factorisation, which will be covered shortly.

b) $5mn - y^3 + 10$ **and** $6mn - xy^3 - 5$
Solution
Some terms in the first and second expressions are not completely identical so watch out.

$$(5mn - y^3 + 10) + (6mn - xy^3 - 5) = (5mn + 6mn) + (-y^3) + (-xy^3) + (10 - 5)$$
$$= (11mn) + (-y^3) + (-xy^3) + (5)$$
$$= \mathbf{11mn - xy^3 - y^3 + 5}$$

NOTE

- In this example, you may wish to consider the absence of an identical term with a term having zero as its coefficient. Therefore, we can write the above as $5mn - y^3 + 0xy^3 + 10$ and $6mn + 0y^3 - xy^3 - 5$. It is now obvious that each expression has four distinct terms, but one term in each expression is irrelevant since its coefficient is zero.

- We can factorise and write the final answer as $\mathbf{11mn - (xy^3 + y^3) + 5}$. Again, we will cover factorisation shortly.

Let's try more examples.

Example 4

Subtract the second expression from the first expression in each of the following pairs.

a) $5k^4 + j^2 + 3l$ and $k^4 + 3j^2 + 13l$ **b)** $5 - 3ab + 4bc + ac$ and $7ab - bc + ac$

What did you get? Find the solution below to double-check your answer.

Solution to Example 4

a) $5k^4 + j^2 + 3l$ and $k^4 + 3j^2 + 13l$
Solution
When we subtract, the key focus is to change the sign (from negative to positive and vice versa) of each term in the subtrahend (the number or term that appears after the minus sign). Look carefully at how this is carried out; it becomes easier once we get the 'trick'.

$$\begin{aligned}
\left(5k^4 + j^2 + 3l\right) - \left(k^4 + 3j^2 + 13l\right) &= \left(5k^4 + j^2 + 3l\right) - k^4 - 3j^2 - 13l \\
&= \left(5k^4 + j^2 + 3l\right) + \left(-k^4 - 3j^2 - 13l\right) \\
&= \left(5k^4 - k^4\right) + \left(j^2 - 3j^2\right) + \left(3l - 13l\right) \\
&= \left(4k^4\right) + \left(-2j^2\right) + \left(-10l\right) \\
&= \mathbf{4k^4 - 2j^2 - 10l}
\end{aligned}$$

NOTE
We can write the final answer as $\mathbf{2\left(2k^4 - j^2 - 5l\right)}$.

b) $5 - 3ab + 4bc + ac$ and $7ab - bc + ac$
Solution

$$\begin{aligned}
(5 - 3ab + 4bc + ac) - (7ab - bc + ac) &= (5 - 3ab + 4bc + ac) + (-7ab + bc - ac) \\
&= (5) + (-3ab - 7ab) + (4bc + bc) + (ac - ac) \\
&= (5) + (-10ab) + (5bc) + (0) \\
&= 5 - 10ab + 5bc \\
&= \mathbf{5 - 10ab + 5bc}
\end{aligned}$$

NOTE
Let's make the following observation about this example.

- A term (ac) disappears; that's fine, and in fact more could disappear. We can even end up with a number when we carry out simplification.

- The final answer can also be written as $\mathbf{5\left(1 - 2ab + bc\right)}$ to make it simpler. Recall that we're 'simplifying' here.

3.3.2 MULTIPLICATION (×) AND DIVISION (÷)

We can multiply any number of terms together and they do not have to be identical. This operation results in a term, which is written without any multiplication sign (×, ., or ∗). For example, $5 \times y = 5y$ and $x \times y = xy$.

We handle divisions by employing factors (or multiples) or by cancelling out. Note that $\frac{5}{5} = 1$ so $\frac{x}{x} = 1$ and $\frac{5^2}{5} = \frac{5 \times 5}{5} = 5$ so $\frac{y^2}{y} = \frac{y \times y}{y} = y$. Again, we can say that multiplication is both commutative and associative because $xy = yx$ and $(pq)\,r = p\,(qr)$ respectively. Division is not commutative because $\frac{x}{y} \neq \frac{y}{x}$.

Table 3.2 has been previously produced when we discussed numbers in Chapter 1, it is also relevant here thus its re-production but with algebraic examples.

Let's now try multiplication and division, but at a very basic level. Multiplication will be covered further when opening brackets and division will be shown in algebraic fractions later in this chapter.

Example 5

Multiply the following pairs of terms.

a) $5x$ **and** x b) ab **and** $6c$ c) $2pq^2$ **and** $-3p^2q$

TABLE 3.2
Rules of Multiplying and Dividing Negative Numbers

	Operation	Result	Example
	$(positive) \times (positive)$	positive	$(3) \times (a) = 3a$
	$(positive) \times (negative)$	negative	$(3) \times (-a) = -3a$
Multiplication	$(negative) \times (positive)$	negative	$(-5a) \times (b) = -5ab$
	$(negative) \times (negative)$	positive	$(-5a) \times (-2b) = 10ab$
	$(positive) \div (positive)$	positive	$(30a) \div (6b) = \frac{5a}{b}$
	$(positive) \div (negative)$	negative	$(30a) \div (-6b) = -\frac{5a}{b}$
Division	$(negative) \div (positive)$	negative	$(-30a) \div (6b) = -\frac{5a}{b}$
	$(negative) \div (negative)$	positive	$(-30a) \div (-6b) = \frac{5a}{b}$

What did you get? Find the solution below to double-check your answer.

Solution to Example 5

a) $5x$ **and** x
Solution

$$(5x) \times (x) = 5 \times x \times x$$
$$= 5x^2$$

b) ab **and** $6c$
Solution

$$(ab) \times (6c) = a \times b \times 6 \times c$$
$$= 6 \times a \times b \times c$$
$$= 6abc$$

c) $2pq^2$ **and** $-3p^2q$
Solution
The minus sign in the second term can be attached to any of the constituents of the term. Here, we will give it to the coefficient (3) of the term.

$$(2pq^2) \times (-3p^2q) = 2 \times p \times q^2 \times -3 \times p^2 \times q$$
$$= 2 \times -3 \times p \times p^2 \times q^2 \times q$$
$$= -6 \times p^3 \times q^3$$
$$= -6p^3q^3$$

Let's try examples on division.

Example 6

Divide the first term (or expression) by the second term (or expression) in the following pairs.

a) $10g$ **and** $2g$ **b)** $18uv^2$ **and** $-3v$ **c)** $-4ab^2$ **and** $-8a^2c$

What did you get? Find the solution below to double-check your answer.

Solution to Example 6

a) $10g$ **and** $2g$
Solution

$$(10g) \div (2g) = \frac{10g}{2g} = \frac{10 \times g}{2 \times g}$$
$$= \frac{10}{2} \times \frac{g}{g} = 5 \times 1$$
$$= 5$$

NOTE
Here, the final answer is a number or constant.

b) $18uv^2$ **and** $-3v$
Solution

$$\left(18uv^2\right) \div (-3v) = \frac{18uv^2}{-3v} = \frac{18}{-3} \times \frac{uv^2}{v}$$
$$= -6 \times \frac{u \times v \times v}{v} = -6 \times uv$$
$$= -6uv$$

c) $-4ab^2$ **and** $-8a^2c$
Solution

$$\left(-4ab^2\right) \div \left(-8a^2c\right) = \frac{-4ab^2}{-8a^2c} = \frac{-4}{-8} \times \frac{ab^2}{a^2c}$$
$$= \frac{1}{2} \times \frac{a \times b \times b}{a \times a \times c}$$
$$= \frac{1}{2} \times \frac{b \times b}{a \times c} = \frac{1}{2} \times \frac{b^2}{ac}$$
$$= \frac{b^2}{2ac}$$

3.4 ALGEBRA – OPENING THE BRACKETS

Opening the brackets is a key part of algebra; it is sometimes referred to as '**expanding the brackets**' or '**multiplying out**' or '**removing the brackets**'. Note that a pair of brackets, such as (), indicates multiplication. The procedure for opening brackets is generally determined by the number and type of brackets. This is usually one or two, but it is possible to have more than two brackets as will be shown shortly.

3.4.1 SINGLE BRACKETS

This is where there is a single pair of brackets and a term (number, letter, or both) outside it, usually before the brackets, e.g., $2(x + 3)$ and $3b\,(a - 2c)$. The general expression for this is:

$$\boxed{x\,(y + z) = xy + xz} \quad \text{OR} \quad \boxed{x\,(y - z) = xy - xz} \tag{3.1}$$

where x, y, and z are variables with their associated coefficients.

Equation 3.1 implies that we multiply the term outside the brackets with each term in the brackets one at a time. Let's quickly illustrate this.

$$2(x + 3) = (2 \times x) + (2 \times 3)$$
$$= 2x + 6$$

or

$$3b(a - 2c) = (3b \times a) - (3b \times 2c)$$
$$= 3ab - 6bc$$

It is possible that one of the terms in the brackets is a number, but not both, otherwise the contents of the brackets can be simplified easily. Also, we have illustrated with two terms in the brackets, but the principle is applicable to any number of terms in the brackets as will be demonstrated in the worked examples below. Let's attempt some examples.

Example 7

Expand the brackets in the following expressions.

a) $5(x + 3)$ **b)** $2x(y - 1)$ **c)** $-3(s + 2t - 5)$

d) $-2pq(3qr - 5pr^3)$ **e)** $2b(a^2c - 3bd + 7)$

What did you get? Find the solution below to double-check your answer.

Solution to Example 7

a) $5(x + 3)$
Solution

$$5(x + 3) = (5 \times x) + (5 \times 3)$$
$$= 5x + 15$$

b) $2x(y - 1)$
Solution

$$2x(y - 1) = (2x \times y) + (2x \times -1)$$
$$= 2xy - 2x$$

c) $-3(s + 2t - 5)$
Solution

$$-3(s + 2t - 5) = (-3 \times s) + (-3 \times 2t) + (-3 \times -5)$$
$$= (-3s) + (-6t) + (15)$$
$$= -3s - 6t + 15$$

NOTE

- The above final answer is not a preferred style, as we tend to start with a positive sign in algebra. We can therefore write this as:

$$15 - 3s - 6t$$

- We can also factorise and write the final answer as $3(5 - s - 2t)$, though this is the reverse of what the question asks for, i.e. expand.

---

d) $-2pq(3qr - 5pr^3)$
Solution

$$\begin{aligned}
-2pq(3qr - 5pr^3) &= (-2pq \times 3qr) + (-2pq \times -5pr^3) \\
&= (-6pq^2r) + (10p^2qr^3) \\
&= -6pq^2r + 10p^2qr^3 \\
&= 10p^2qr^3 - 6pq^2r
\end{aligned}$$

---

e) $2b(a^2c - 3bd + 7)$
Solution

$$\begin{aligned}
2b(a^2c - 3bd + 7) &= (2b \times a^2c) + (2b \times -3bd) + (2b \times 7) \\
&= (2ba^2c) + (-6b^2d) + (14b) \\
&= 2a^2bc - 6b^2d + 14b
\end{aligned}$$

NOTE

Notice that we try to re-arrange the order of the variables a, b, and c. This is a good skill, and it helps in quickly identifying like terms. Imagine if we were presented with b^2ac and $5ab^2c$, it may take a bit of time before we realise that they are like terms. But if they are presented as ab^2c and $5ab^2c$, their 'likeness' is instantly visible.

3.4.2 DOUBLE BRACKETS

This is where there are two separate pairs of brackets. The general expression is:

$$\boxed{(p + q)(r + s) = pr + ps + qr + qs} \tag{3.2}$$

where $p, q, r,$ and s are variables with their associated coefficients.

The procedure for opening double brackets is illustrated in Figure 3.2.

We have numbered them for reference purposes, but the order is irrelevant. What is essential is to complete the four different multiplications and simplify if necessary; recall that addition and multiplication are commutative.

Another way to open the brackets is by multiplying each term in one pair of brackets (often the first) with the terms in the second. This will result in two pairs of brackets if there are only two terms in

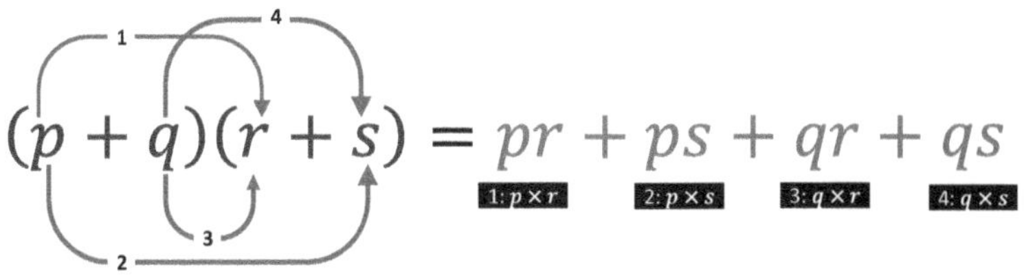

FIGURE 3.2 Opening double brackets explained.

the chosen brackets. This can then be simplified as shown for single brackets above. In fact, we can treat this case as though there are two pairs of brackets as illustrated below.

$$(p+q)(r+s) = p(r+s) + q(r+s)$$
$$= pr + ps + qr + qs$$

There is yet another way to open double brackets based on a tabular form (Table 3.3). In this method, the terms in the first pair of brackets are distributed in the first row and the terms in the second brackets are arranged vertically in the far-left column. Let's call the cells or boxes occupied by these terms as reference boxes. We then fill each of the remaining boxes by multiplying the corresponding terms in the horizontal and vertical reference boxes. Table 3.3 illustrates the use of this method for $(p+q)(r+s)$, where there are two terms in each of the two brackets. It can be used for three or more terms in each bracket; we simply create more rows and/or columns to accommodate the terms.

Whichever approach we fancy or use, they all follow the same procedure: multiply out (or open the brackets) and simplify. That's all! Let's try some examples.

Example 8

Expand the brackets in each of the following expressions.

a) $(a+3)(a+5)$ $\qquad$ **b)** $(j-1)(j+2)$ $\qquad$ **c)** $(2x-7)(3x-5)$

d) $(x-2y)(5x+y)$ $\qquad$ **e)** $(m-1)(m+2)(2m-3)$ $\qquad$ **f)** $(2q+5)(q^2-2q+3)$

TABLE 3.3
Alternative Method of Opening Double Brackets

	p	$+q$
r	pr	$+qr$
$+s$	$+ps$	$+qs$

What did you get? Find the solution below to double-check your answer.

Solution to Example 8

a) $(a + 3)(a + 5)$
Solution

$$(a + 3)(a + 5) = a(a + 5) + 3(a + 5)$$
$$= a^2 + 5a + 3a + 15 = a^2 + 8a + 15$$
$$\therefore (a + 3)(a + 5) = a^2 + 8a + 15$$

b) $(j - 1)(j + 2)$
Solution

$$(j - 1)(j + 2) = j(j + 2) - 1(j + 2)$$
$$= j^2 + 2j - j - 2 = j^2 + j - 2$$
$$\therefore (j - 1)(j + 2) = j^2 + j - 2$$

NOTE
$j^2 + j - 2$ can be simplified further to have $j - 3$ as j is more than just a letter or variable. It is an operator that represents a 90-degree rotation in an anticlockwise direction. This is beyond the scope of this book, but it's covered in *Advanced Mathematics for Engineers and Scientists with Worked Examples* by the same author.

c) $(2x - 7)(3x - 5)$
Solution

$$(2x - 7)(3x - 5) = 2x(3x - 5) - 7(3x - 5)$$
$$= 6x^2 - 10x - 21x + 35 = 6x^2 - 31x + 35$$
$$\therefore (2x - 7)(3x - 5) = 6x^2 - 31x + 35$$

d) $(x - 2y)(5x + y)$
Solution

$$(x - 2y)(5x + y) = x(5x + y) - 2y(5x + y)$$
$$= 5x^2 + xy - 10xy - 2y^2 = 5x^2 - 9xy - 2y^2$$
$$\therefore (x - 2y)(5x + y) = 5x^2 - 9xy - 2y^2$$

e) $(m - 1)(m + 2)(2m - 3)$
Solution
This is a case of multiple brackets, which will require opening the double brackets twice. As multiplication is associative, we can start with any pair. Let's do this in steps.

Step 1: Expand $(m + 2)(2m - 3)$

$$(m - 1)(m + 2)(2m - 3) = (m - 1)[m(2m - 3) + 2(2m - 3)]$$
$$= (m - 1)[2m^2 - 3m + 4m - 6]$$
$$= (m - 1)(2m^2 + m - 6)$$

Step 2: Now expand $(m-1)\left(2m^2 + m - 6\right)$

$$= m\left(2m^2 + m - 6\right) - \left(2m^2 + m - 6\right)$$
$$= \left(2m^3 + m^2 - 6m\right) + \left(-2m^2 - m + 6\right)$$

Step 3: Collect the like terms and simplify

$$= \left(2m^3\right) + \left(m^2 - 2m^2\right) + \left(-6m - m\right) + (6)$$
$$= 2m^3 - m^2 - 7m + 6$$
$$\therefore (m-1)(m+2)(2m-3) = 2m^3 - m^2 - 7m + 6$$

---

f) $(2q+5)(q^2 - 2q + 3)$

This is not in any way different from others, except that we have three terms in the second brackets.

Step 1: Now expand $(2q+5)(q^2 - 2q + 3)$

$$(2q+5)(q^2 - 2q + 3) = 2q(q^2 - 2q + 3) + 5(q^2 - 2q + 3)$$
$$= \left(2q^3 - 4q^2 + 6q\right) + \left(5q^2 - 10q + 15\right)$$

Step 2: Collect the like terms and simplify

$$= \left(2q^3\right) + \left(5q^2 - 4q^2\right) + (6q - 10q) + (15)$$
$$= 2q^3 + q^2 - 4q + 15$$
$$\therefore (2q+5)(q^2 - 2q + 3) = 2q^3 + q^2 - 4q + 15$$

3.4.3 LONG BRACKETS

Here we will consider when one or both brackets have three or more terms. The procedure is the same, and you will have noticed this in Example 8(e). Let's try some examples.

Example 9

Expand the brackets in the following expressions.

a) $(2x - 3y)(x - 2y + 3z)$ **b)** $(2q+5)\left(q^2 - 2q + 3\right)$

c) $\left(x^2 - 2x + 3\right)\left(3x^2 - x + 5\right)$ **d)** $\left(x^2 + x - 1\right)\left(2x^2 - 3x + 7\right)$

What did you get? Find the solution below to double-check your answer.

Solution to Example 9

a) $(2x - 3y)(x - 2y + 3z)$

Solution

Step 1: Multiply each term of the first pair of brackets with the entire second pair of brackets.

$$(2x - 3y)(x - 2y + 3z) = 2x(x - 2y + 3z) - 3y(x - 2y + 3z)$$

Step 2: Open the brackets.

$$= \left(2x^2 - 4xy + 6xz\right) + \left(-3xy + 6y^2 - 9yz\right)$$

Step 3: Let's collect like terms.

$$= \left(2x^2\right) + \left(-4xy - 3xy\right) + \left(6xz\right) + \left(6y^2\right) + \left(-9yz\right)$$

Only one like term.

Step 4: Let's simplify.

$$= \left(2x^2\right) + \left(-7xy\right) + \left(6xz\right) + \left(6y^2\right) + \left(-9yz\right)$$
$$= 2x^2 + 6y^2 - 7xy + 6xz - 9yz$$
$$\therefore \left(2x - 3y\right)\left(x - 2y + 3z\right) = 2x^2 + 6y^2 - 7xy + 6xz - 9yz$$

ALTERNATIVE METHOD

Step 1: Let's create the table and fill the cells (see Table 3.4).
Step 2: Write the terms. We've done this by row.

$$= \left(2x^2 - 4xy + 6xz\right) + \left(-3xy + 6y^2 - 9yz\right)$$

NOTE

We could have done this by column and have the answer below, but we will arrive at the same result.

$$= \left(2x^2 - 3xy\right) + \left(-4xy + 6y^2\right) + \left(+6xz - 9yz\right)$$

We will continue with the first option, obtained by row.

Step 3: Let's collect like terms.

$$= \left(2x^2\right) + \left(-4xy - 3xy\right) + \left(6xz\right) + \left(6y^2\right) + \left(-9yz\right)$$

There is only one like term.

Step 4: Let's simplify.

$$= \left(2x^2\right) + \left(-7xy\right) + \left(6xz\right) + \left(6y^2\right) + \left(-9yz\right)$$
$$= 2x^2 + 6y^2 - 7xy + 6xz - 9yz$$
$$\therefore \left(2x - 3y\right)\left(x - 2y + 3z\right) = 2x^2 + 6y^2 - 7xy + 6xz - 9yz$$

TABLE 3.4
Solution to Example 9(a)

	x	$-2y$	$3z$
$2x$	$2x^2$	$-4xy$	$+6xz$
$-3y$	$-3xy$	$+6y^2$	$-9yz$

b) $(2q + 5)(q^2 - 2q + 3)$

Solution

Let's take a different approach for this and also be brief.

$$(2q + 5)(q^2 - 2q + 3) = 2q(q^2 - 2q + 3) + 5(q^2 - 2q + 3)$$
$$= 2q^3 - 4q^2 + 6q + 5q^2 - 10q + 15 = 2q^3 + q^2 - 4q + 15$$
$$\therefore (2q + 5)(q^2 - 2q + 3) = 2q^3 + q^2 - 4q + 15$$

---

c) $(x^2 - 2x + 3)(3x^2 - x + 5)$

Solution

We have one quadratic expression multiplying another one. This may be a long one.

Step 1: Multiply each term of the first pair of brackets with the entire second pair of brackets.

$$(x^2 - 2x + 3)(3x^2 - x + 5) = x^2(3x^2 - x + 5) - 2x(3x^2 - x + 5) + 3(3x^2 - x + 5)$$

Step 2: Open the brackets.

$$= (3x^4 - x^3 + 5x^2) + (-6x^3 + 2x^2 - 10x) + (9x^2 - 3x + 15)$$

Step 3: Let's collect like terms.

$$= (3x^4) + (-x^3 - 6x^3) + (5x^2 + 2x^2 + 9x^2) + (-10x - 3x) + (15)$$

Unlike in Example 9(a), there are several like terms here. This is because there is only one variable in this expression, but the former has x, y, and z as its variables.

Step 4: Let's simplify.

$$= (3x^4) + (-7x^3) + (16x^2) + (-13x) + (15)$$
$$= 3x^4 - 7x^3 + 16x^2 - 13x + 15$$
$$\therefore (x^2 - 2x + 3)(3x^2 - x + 5) = 3x^4 - 7x^3 + 16x^2 - 13x + 15$$

ALTERNATIVE METHOD

Step 1: Let's create the table and fill the cells (see Table 3.5).

Step 2: Write the terms. We've done this by row.

$$= (3x^4 - x^3 + 5x^2) + (-6x^3 + 2x^2 - 10x) + (9x^2 - 3x + 15)$$

Let's collect like terms.

$$= (3x^4) + (-x^3 - 6x^3) + (5x^2 + 2x^2 + 9x^2) + (-10x - 3x) + (15)$$

Step 4: Let's simplify.

$$= (3x^4) + (-7x^3) + (16x^2) + (-13x) + (15)$$
$$= 3x^4 - 7x^3 + 16x^2 - 13x + 15$$
$$\therefore (x^2 - 2x + 3)(3x^2 - x + 5) = 3x^4 - 7x^3 + 16x^2 - 13x + 15$$

TABLE 3.5
Solution to Example 9(b)

	$3x^2$	$-x$	5
x^2	$3x^4$	$-x^3$	$+5x^2$
$-2x$	$-6x^3$	$+2x^2$	$-10x$
3	$+9x^2$	$-3x$	$+15$

d) $(x^2 + x - 1)(2x^2 - 3x + 7)$
Solution

$$(x^2 + x - 1)(2x^2 - 3x + 7) = x^2(2x^2 - 3x + 7) + x(2x^2 - 3x + 7) - 1(2x^2 - 3x + 7)$$
$$= 2x^4 - 3x^3 + 7x^2 + 2x^3 - 3x^2 + 7x - 2x^2 + 3x - 7$$
$$= 2x^4 - 3x^3 + 2x^3 + 7x^2 - 3x^2 - 2x^2 + 7x + 3x - 7$$
$$= 2x^4 - x^3 + 2x^2 + 10x - 7$$
$$\therefore (x^2 + x - 1)(2x^2 - 3x + 7) = 2x^4 - x^3 + 2x^2 + 10x - 7$$

From Examples 9(a) and 9(c) above, we can see that the two methods are nearly identical and only differ in step 1. The tabular formula method additionally provides a visual insight and is less prone to error.

3.4.4 NESTED BRACKETS

When we have layers of brackets, the procedure is to simplify the innermost first and work our way out. It is also customary to use different types of brackets for different layers to avoid confusion. Let's try a couple of examples on this.

Example 10

Expand the brackets in the following expressions.

a) $3(2x - x[x^2 - 5x - 7])$ **b)** $xy(3y - 2xy - y\{3x + 2y + 1\} - 3y^2)$

What did you get? Find the solution below to double-check your answer.

Solution to Example 10

a) $3\left(2x - x\left[x^2 - 5x - 7\right]\right)$

Solution

Step 1: Open the brackets of the inner layer contained in square brackets [] and simplify.

$$3\left(2x - x\left[x^2 - 5x - 7\right]\right) = 3\left(2x - x^3 + 5x^2 + 7x\right)$$
$$= 3\left(-x^3 + 5x^2 + 9x\right)$$

Step 2: Open the brackets of the outer layer contained in parenthesis () and simplify if necessary.

$$= -3x^3 + 15x^2 + 27x$$
$$\therefore 3\left(2x - x\left[x^2 - 5x - 7\right]\right) = 27x + 15x^2 - 3x^3$$

We have re-arranged the expressions so that it is in ascending power of x., i.e. we started with power of 1, then 2 and followed by 3. This is just to avoid starting the expression with a negative term.

---

b) $xy\left(3y - 2xy - y\{3x + 2y + 1\} - 3y^2\right)$

Solution

Step 1: Open the brackets of the inner layer contained in curly brackets { } and simplify.

$$xy\left(3y - 2xy - y\{3x + 2y + 1\} - 3y^2\right) = xy\left(3y - 2xy - 3xy - 2y^2 - y - 3y^2\right)$$
$$= xy\left(2y - 5xy - 5y^2\right)$$

Step 2: Open the brackets of the outer layer contained in () brackets and simplify if necessary.

$$= 2xy^2 - 5x^2y^2 - 5xy^3$$
$$\therefore xy\left(3y - 2xy - y\{3x + 2y + 1\} - 3y^2\right) = 2xy^2 - 5x^2y^2 - 5xy^3$$

3.4.5 SQUARED BRACKETS

Squared brackets are where we have double brackets that are exactly the same, and they should be treated in a similar way as double brackets. For example, let us multiply $(p + q)$ by itself. Note that $5^2 = 5 \times 5$, so multiplying $(p + q)$ by itself equals $(p + q)^2$.

$$(p + q)^2 = (p + q)(p + q)$$
$$= p^2 + pq + pq + q^2$$
$$= p^2 + 2pq + q^2$$

Let's try to multiply $(p - q)$ by itself.

$$(p - q)^2 = (p - q)(p - q)$$
$$= p^2 - pq - pq + q^2$$
$$= p^2 - 2pq + q^2$$

From the above examples, we have two primary insights to note.

1) It is clear from the above examples that:

$$(p+q)^2 \neq p^2 + q^2 \quad \textbf{AND} \quad (p-q)^2 \neq p^2 - q^2$$

The above is a common mistake, so do not fall into it.

2) The efficient way of evaluating a binomial squared brackets $(p \pm q)^2$ is as follows:

Step 1: Square the first term to have p^2.
Step 2: Square the second/last term to have q^2.
Step 3: Multiply the first and second terms together to obtain $\pm pq$ and double the answer to obtain $\pm 2pq$. If there is a positive sign between the two terms, you will obtain $2pq$, otherwise you will have $-2pq$.
Step 4: Add the answers from 1 to 3 to obtain the three terms you need for this expansion as: $p^2 \pm 2pq + q^2$.

Take note of the above steps, as it is an efficient way of expanding squared brackets. We'll show and apply these in the worked examples.

Let's now try a couple of examples.

Example 11

Expand the brackets in the following expressions.

a) $(3x + 2y)^2$ **b)** $\left(ab^2 - 5c\right)^2$

What did you get? Find the solution below to double-check your answer.

Solution to Example 11

a) $(3x + 2y)^2$
Solution

Step 1: Square the first term.

$$(3x)^2 = 3^2 \times x^2 = 9x^2$$

Step 2: Square the second term.

$$(2y)^2 = 2^2 \times y^2 = \mathbf{4y^2}$$

Step 3: Find the product of the first and second terms and double the answer.

$$2\,(3x)\,(2y) = \mathbf{12xy}$$

Step 4: Add the answers from the above three steps.

$$= 9x^2 + 4y^2 + 12xy$$

That's everything.

$$\therefore (3x + 2y)^2 = \mathbf{9x^2 + 12xy + 4y^2}$$

---

b) $\left(ab^2 - 5c\right)^2$

Solution

Step 1: Square the first term.

$$\left(ab^2\right)^2 = a^2 \times \left(b^2\right)^2 = \mathbf{a^2b^4}$$

Step 2: Square the second term.

$$(-5c)^2 = (-5)^2 \times c^2 = \mathbf{25c^2}$$

We know that the square of a negative number is positive. For this step (assuming you don't have to demonstrate the working out), you can leave out the negative sign as:

$$(5c)^2 = (5)^2 \times c^2 = \mathbf{25c^2}$$

Step 3: Double the product of the first and second terms.

$$2\left(ab^2\right)(-5c) = \mathbf{-10ab^2c}$$

For this step, do NOT ignore the sign.

Step 4: Add the answers from the above three steps.

$$= a^2b^4 + 25c^2 - 10ab^2c$$

That's all.

$$\therefore \left(ab^2 - 5c\right)^2 = \mathbf{a^2b^4 + 25c^2 - 10ab^2c}$$

3.5 FACTORISATION

Factorisation is a process of writing an algebraic expression as a product of two or more simple algebraic expressions and numbers. The multiplier or multiplicand is contained in a pair of brackets unless they consist of a term only. The algebraic term $3ab$ can be written in a factored form as $(3)(a)(b)$. This is fine. However, when we talk about factorisation, we usually intend two or more terms in the brackets.

In fact, factorisation is the inverse of opening the brackets. The procedure in factorising an expression is primarily determined by the nature of the expression. Here, we will look at a few cases.

3.5.1 FACTORISING A LINEAR EXPRESSION

This is an expression in which the power of any unknown is 1. If there is only one type of variable involved, we will have the general expression $ax + b$ or $ax + bx$ where a and b are constants (or numerical values) and x is the variable.

We do not however have a specific expression for where there are two or more types of variables involved, as this can be a binomial (two terms) or trinomial (three terms) expression and can be even more.

Whatever the case, we will need to look at the HCF of the terms in the expression and take it 'outside the brackets'. If there is one variable involved, the HCF of the coefficients will be the focus, otherwise the HCF of the entire terms will be considered.

Let's try some examples.

Example 12

Factorise the following expressions.

a) $6a + 9$ **b)** $12x - 30$ **c)** $5 + 35y$ **d)** $42 - 21v$

What did you get? Find the solution below to double-check your answer.

Solution to Example 12

HINT

For this case, look at the coefficient of the two terms and determine their HCF.

a) $6a + 9$
Solution
The two terms are $6a$ and 9. So, we need the HCF of 6 and 9, which is 3. Thus, we have:

$$6a + 9 = 3 \times 2a + 3 \times 3$$
$$= 3 \times (2a) + 3 \times (3)$$
$$= 3(2a) + 3(3)$$

It is obvious that 3 is common to both terms and can be taken outside the brackets as:

$$= 3\,(2a + 3)$$

Did you get the gist there? Great!

$$\therefore 6a + 9 = 3\,(2a + 3)$$

---

b) $12x - 30$

Solution

The two terms are $12x$ and -30. So, we need the HCF of 12 and 30, which is 6. Thus, we have

$$12x - 30 = 6 \times 2x - 6 \times 5$$
$$= 6 \times (2x) - 6 \times (5)$$
$$= 6\,(2x) - 6\,(5)$$

It is obvious that 6 is common to both terms and can be taken outside the brackets as:

$$= 6\,(2x - 5)$$

Do take note how we handled the negative sign.

$$\therefore 12x - 30 = 6\,(2x - 5)$$

ALTERNATIVE METHOD

In case we find it difficult or time consuming to start working out the HCF, we may decide to take the common factor one at a time. Start with the least of the common factors, often 2, and continue until nothing is common to them.

A factor may be repeated, but with experience, you will be able to spot when a factor can be used twice. For example, instead of taking factor 2 twice, you can quickly take 4 and instead of 3 twice, 9 can be used.

Let's set this out for this question as:

Step 1: 2 is common to both 12 and 30.

$$12x - 30 = 2 \times 6x + 2 \times -15 = 2\,(6x - 15)$$

Notice that -30 is split as 2×-15 and not -2×15. This is a choice we've taken because we need 2 and not -2.

Step 2: The expression in the brackets $(6x - 15)$ can still be factorised, as 3 is common to both 6 and 15.

$$2\,(6x - 15) = 2\,(3 \times 2x + 3 \times -5)$$
$$= 2\,[3 \times (2x - 5)]$$
$$= 2 \times 3\,(2x - 5)$$
$$= 6\,(2x - 5)$$

Notice that we do not have any integer that can go in both 2 and 5. Hence, this is the simplest form. All done!

c) $5 + 35y$

Solution

The two terms are 5 and $35y$. So, we need the HCF of 5 and 35, which is 5. Thus, we have

$$
\begin{aligned}
5 + 35y &= 5 \times 1 + 5 \times 7y \\
&= 5 \times (1) + 5 \times (7y) \\
&= 5(1) + 5(7y) \\
&= 5(1 + 7y) \\
\therefore 5 + 35y &= \mathbf{5(1 + 7y)}
\end{aligned}
$$

d) $42 - 21v$

Solution

The two terms are 42 and $-21v$. So, we need the HCF of 42 and 21, which is 21. Thus, we have

$$
\begin{aligned}
42 - 21v &= 21 \times 2 + 21 \times -v \\
&= 21 \times (2) + 21 \times (-v) \\
&= 21(2) + 21(-v) \\
&= 21(2 - v) \\
\therefore 42 - 21v &= \mathbf{21(2 - v)}
\end{aligned}
$$

The examples above are linear with one variable. Let's try examples with more than one variable.

Example 13

Factorise the following expressions.

a) $4m - 6n$ **b)** $28 - 21x - 7y$

What did you get? Find the solution below to double-check your answer.

Solution to Example 13

a) $4m - 6n$

Solution

The two terms are $4m$ and $6n$. So, we need the HCF of 4 and 6, which is 2. Thus, we have

$$
\begin{aligned}
4m - 6n &= 2 \times 2m - 2 \times 3n \\
&= 2 \times (2m) - 2 \times (3n) \\
&= 2(2m) - 2(3n)
\end{aligned}
$$

It is obvious that 2 is common to both terms and can be taken outside the brackets as:

$$= 2\,(2m - 3n)$$
$$\therefore 4m - 6n = 2\,(2m - 3n)$$

---

b) $28 - 21x - 7y$

Solution

The three terms are 28, $-21x$, and $7y$. So, we need the HCF of 28, 21, and 7, which is 7. Thus, we have

$$28 - 21x - 7y = 7 \times 4 - 7 \times 3x - 7 \times y$$
$$= 7 \times (4) - 7 \times (3x) - 7 \times (y)$$
$$= 7\,(4) - 7\,(3x) - 7\,(y)$$

It is obvious that 7 is common to both terms and can be taken outside the brackets as:

$$= 7\,(4 - 3x - 7y)$$

Did you get the gist there? Great!

$$\therefore 28 - 21x - 7y = 7\,(4 - 3x - 7y)$$

3.5.2 FACTORISING DIFFERENCE OF TWO SQUARES

Difference of two squares (sometimes shortened as DOTS) implies '**a squared term minus another squared term**'. The minus implies '**difference**' so $x^2 - y^2$, $a^2b^2 - 2^2$, and $9d^2 - 25$ are examples of the difference of two squares.

You may wonder about the last example being a difference of two squares. We will need to modify the expression to give us the desired format, so we can write $9d^2 - 25$ as $3^2d^2 - 5^2$, and then the latter as $(3d)^2 - 5^2$. It is now obvious that there is a difference of two squares.

How to factorise a difference of two squares? We will use $x^2 - y^2$ to illustrate the procedures as:

Step 1: Establish that there are two squared terms with a minus between them.
Step 2: Copy the expression and remove their squares to have $(x - y)$.
Step 3: Copy the answer in step 2 and change the sign to obtain $(x + y)$.
Step 4: Multiply the answers from steps 2 and 3 to obtain $(x - y)(x + y)$. This is the factorised form of $x^2 - y^2$.

We can therefore say that

$$x^2 - y^2 = (x + y)(x - y)$$

Notice that we swapped the factors or signs. This is not a problem since multiplication is commutative, so either can come first or last. That's all! Let's try some examples.

Example 14

Factorise the following expressions completely.

a) $a^2 - b^2$ **b)** $x^2y^2 - d^2$ **c)** $4u^2 - 9v^2$ **d)** $3x^2 - 48$

e) $xy^3 - 4xy$ **f)** $p^2q^4 - 16r^2$ **g)** $8m^2x - 18x$ **h)** $\left(x^2 + x - 1\right)^2 - \left(x^2 - x + 1\right)^2$

What did you get? Find the solution below to double-check your answer.

Solution to Example 14

a) $a^2 - b^2$

Solution

This is already in the right format so let's proceed.

Step 1: Open the two brackets and put the first term without the square.

$$a^2 - b^2 = (a + ?)(a - ?)$$

Step 2: Put the second term without the square in both brackets.

$$a^2 - b^2 = (a + b)(a - b)$$

That's all really.

$$\therefore a^2 - b^2 = (a + b)(a - b)$$

b) $x^2y^2 - d^2$

Solution

Step 1: This is already in the right format, but let's modify it a bit for clarity.

$$x^2y^2 - d^2 = (xy)^2 - d^2$$

Step 2: Open the two brackets and put the first term without the square.

$$(xy)^2 - d^2 = (xy + ?)(xy - ?)$$

Step 3: Put the second term without the square in both brackets.

$$(xy)^2 - d^2 = (xy + d)(xy - d)$$

All done!

$$\therefore x^2y^2 - d^2 = (xy + d)(xy - d)$$

c) $4u^2 - 9v^2$

Solution

Step 1: This is not in the right format; let's modify it.

$$4u^2 - 9v^2 = 2^2u^2 - 3^2v^2$$
$$= (2u)^2 - (3v)^2$$

Step 2: Open the two brackets and put the first term without the square.

$$(2u)^2 - (3v)^2 = (2u \ + \ ?)(2u \ - \ ?)$$

Step 3: Put the second term without the square in both brackets.

$$(2u)^2 - (3v)^2 = (2u + 3v)(2u - 3v)$$

All done!

$$\therefore \ \mathbf{4u^2 - 9v^2 = (2u + 3v)(2u - 3v)}$$

---

d) $3x^2 - 48$

Solution

Step 1: This is not in the right format; let's modify it.

The two terms are $3x^2$ and -48, so the HCF here is 3. Thus, we have:

$$3x^2 - 48 = 3 \times x^2 - 3 \times 16$$
$$= 3(x^2 - 16)$$
$$= 3(x^2 - 4^2)$$

Our difference of two squares is $x^2 - 4^2$, so we can proceed.

Step 2: Open the two brackets and put the first term without the square.

$$3(x^2 - 4^2) = 3(x \ + \ ?)(x \ - \ ?)$$

Step 3: Put the second term without the square in both brackets.

$$3(x^2 - 4^2) = 3(x + 4)(x - 4)$$

All done!

$$\therefore \ \mathbf{3x^2 - 48 = 3(x + 4)(x - 4)}$$

---

e) $xy^3 - 4xy$

Solution

Step 1: Again, this is not in the right format; let's modify it.

The two terms are xy^3 and $-4xy$, so the HCF here is xy. Thus, we have

$$xy^3 - 4xy = xy \times y^2 - xy \times 4$$
$$= xy(y^2 - 4)$$
$$= xy(y^2 - 2^2)$$

Our difference of two squares is $y^2 - 2^2$, so we can proceed

Step 2: Open the two brackets and put the first term without the square.

$$xy(y^2 - 2^2) = xy(y + ?)(y - ?)$$

Step 3: Put the second term without the square in both brackets.

$$xy(y^2 - 2^2) = xy(y + 2)(y - 2)$$

All done!

$$\therefore xy^3 - 4xy = xy(y + 2)(y - 2)$$

---

f) $p^2q^4 - 16r^2$

Solution

Step 1: This is not in the right format; let's modify it.

$$p^2q^4 - 16r^2 = p^2q^4 - 4^2r^2$$
$$= (pq^2)^2 - (4r)^2$$

Step 2: Open the two brackets and put the first term without the square.

$$(pq^2)^2 - (4r)^2 = (pq^2 + ?)(pq^2 - ?)$$

Step 3: Put the second term without the square in both brackets.

$$(pq^2)^2 - (4r)^2 = (pq^2 + 4r)(pq^2 - 4r)$$

All done!

$$\therefore p^2q^4 - 16r^2 = (pq^2 + 4r)(pq^2 - 4r)$$

---

g) $8m^2x - 18x$

Solution

Step 1: This is not in the right format and looks a bit unusual. Let's modify it.

$$8m^2x - 18x = 2x(4m^2 - 9)$$
$$= 2x[(2m)^2 - 3^2]$$

We will now need to apply the difference of two squares to the expression within the square brackets [].

Step 2: Open the two brackets and put the first term without the square.

$$2x[(2m)^2 - 3^2] = 2x[(2m + ?)(2m - ?)]$$

Step 3: Put the second term without the square in both brackets.

$$2x[(2m)^2 - 3^2] = 2x[(2m + 3)(2m - 3)]$$

All done!

$$\therefore 8m^2x - 18x = 2x(2m + 3)(2m - 3)$$

---

h) $(x^2 + x - 1)^2 - (x^2 - x + 1)^2$

Solution

This is already in the right format though a bit complicated, but let's see.

Step 1: Open the two brackets and put the first term without the square.

$$(x^2 + x - 1)^2 - (x^2 - x + 1)^2 = [(x^2 + x - 1) + ?][(x^2 + x - 1) - ?]$$

Step 2: Put the second term without the square in both brackets.

$$(x^2 + x - 1)^2 - (x^2 - x + 1)^2 = [(x^2 + x - 1) + (x^2 - x + 1)][(x^2 + x - 1) - (x^2 - x + 1)]$$

Step 3: We will need to simplify here.

$$(x^2 + x - 1)^2 - (x^2 - x + 1)^2 = [x^2 + x - 1 + x^2 - x + 1][x^2 + x - 1 - x^2 + x - 1]$$
$$= [2x^2][2x - 2] = 2x^2(2x - 2)$$

2 is common to the two terms in $(2x - 2)$, so let's take it out as:

$$= 2x^2 \times 2(x - 1) = 4x^2(x - 1)$$

That's all really.

$$\therefore (x^2 + x - 1)^2 - (x^2 - x + 1)^2 = 4x^2(x - 1)$$

Look at how this complicated expression has been simplified.

3.5.3 FACTORISING A QUADRATIC EXPRESSION

A quadratic expression involving one variable is an expression of the form $ax^2 + bx + c$, where a, b, and c are any rational number. The value of b or c can be zero, meaning that a term or two may not exist however, a must not be zero. In other words, a term with a squared variable must be present in the expression. Consequently, the following are examples of quadratic expressions: $m^2 + m - 2$, $2a^2 - 5$, $x^2 + 2x$, and $7w^2$.

In general, a quadratic expression in one variable, when factorised, should take the format:

$$\boxed{ax^2 + bx + c = (mx + u)(nx + v)}$$
(3.3)

where m, n, u, and v are the numerical values to be determined.

Factorising quadratic expressions can fall under two categories, determined by the value of the coefficient of x^2.

Case 1 When $a = 1$

When the coefficient of x^2 is equal to 1, i.e., $a = 1$, the quadratic expression becomes $x^2 + bx + c$, thus:

$$\boxed{x^2 + bx + c = (mx + u)(nx + v)}$$

For this to be reversible, $m = n = 1$, hence we have:

$$\boxed{x^2 + bx + c = (x + u)(x + v)}$$

We are now left with guessing u and v. A guess is made easy if we can establish that:

$$\boxed{u + v = b, \ uv = c}$$

The above implies that the **sum** of u and v is b and the **product** of u and v is c.

Case 2 When $a \neq 1$

When the coefficient of x^2 in $ax^2 + bx + c$ is NOT equal to 1, follow these steps to factorise the expression.

Step 1: Find the product ac.
Step 2: Find two numbers whose product is ac and whose sum is b. Let this be b_1 and b_2.
Step 3: Remove bx and replace it with $b_1x + b_2x$. Thus, the expression becomes $ax^2 + b_1x + b_2x + c$.
Step 4: Factorise the expression using any suitable method.

Note that the steps above can also be applied when $a = 1$, but in step 1 ac is the same as c since $a = 1$. Both cases will be further covered in Chapter 6 when we study quadratic equations. We will also introduce another method of factorisation.

Let's take some examples here.

Example 15

Factorise the following expressions.

a) $x^2 + 3x + 2$ b) $d^2 + 2d - 15$ c) $v^2 - 11v + 28$ d) $w^2 - 10w + 25$

e) $2 + x - 3x^2$ f) $6y^2 + 11y + 4$ g) $4x^2 - 16x + 15$ h) $16u^2 + 8u + 1$

What did you get? Find the solution below to double-check your answer.

Solution to Example 15

a) $x^2 + 3x + 2$
Solution

Step 1: Let's open the two brackets and put the variable x within them.

$$x^2 + 3x + 2 = (x + u)(x + v)$$

Notice that we placed u and v too; we will need to replace them shortly.

Step 2: From the expression, 3 and 2 are our target. So, think about two numbers that will be 2 when multiplied together and 3 when added. What did you get?

I obtained 1 and 2 and we will therefore replace u and v with them. It does not matter which one you plug in first.

$$x^2 + 3x + 2 = (x + 1)(x + 2)$$

It is after all a multiplication, so the order is not essential (as per the commutative rule). Thus,

$$x^2 + 3x + 2 = (x + 2)(x + 1)$$

Job done! Let's go for another one.

b) $d^2 + 2d - 15$
Solution

Step 1: Let's open the two brackets and put the variable d within them.

$$d^2 + 2d - 15 = (d + u)(d + v)$$

We've placed u and v too; we will need to replace them shortly.

Step 2: From the expression, 2 and -15 are our target. So, think about two numbers that will be -15 when multiplied together and 2 when added. What did you get?

I obtained 5 and -3 and we will therefore replace u and v with them.

$$d^2 + 2d - 15 = (d + 5)(d - 3)$$

c) $v^2 - 11v + 28$
Solution

Step 1: Let's open the two brackets and put the variable v within them.

$$v^2 - 11v + 28 = (v + u)(v + v)$$

We've placed u and v too; we will need to replace them shortly.

Step 2: From the expression, -11 and 28 are our target. So, think about two numbers that will be 28 when multiplied together and -11 when added. What did you get?

I obtained -4 and -7 and we will therefore replace u and v with them.

$$\therefore v^2 - 11v + 28 = (v - 4)(v - 7)$$

---

d) $w^2 - 10w + 25$
Solution

NOTE

We've chosen to use a different approach for this expression though it is generally used when $a \neq 1$. This is to show the universality of the approach.

Here we have $a = 1$, $b = -10$, and $c = 25$.

Step 1: Product of ac

$$ac = 1 \times 25 = 25$$

Step 2: Finding b_1 and b_2

The two numbers whose product is 25 and sum is 25 are

$$-5, \ -5$$

Because

$$-5 \times -5 = 25 \text{ and } -5 - 5 = -10$$

Step 1: Re-write the expression using b_1 and b_2

$$w^2 - 5w - 5w + 25$$

Step 2: Factorise

$$w(w - 5) - 5(w - 5)$$

Notice that $(w - 5)$ is common to both terms, i.e., $w(w - 5)$ and $-5(w - 5)$.
So, let's factorise again to have:

$$(w - 5)(w - 5)$$

Note that the factor to be taken out can be an expression with two or more terms.

$$\therefore w^2 - 10w + 25 = (w - 5)(w - 5) = (w - 5)^2$$

---

e) $2 + x - 3x^2$

Solution

Let's first write it in standard form $ax^2 + bx + c$, i.e., $-3x^2 + x + 2$. Therefore, we have $a = -3$, $b = 1$, and $c = 2$.

Step 1: Product of ac

$$ac = -3 \times 2 = -6$$

Step 2: Finding b_1 and b_2

The two numbers whose product is -6 and sum is 1 are

$$-2, 3$$

Because

$$-3 \times 2 = -6 \text{ and } 3 - 2 = 1$$

Step 3: Re-write the expression using b_1 and b_2

$$-3x^2 - 2x + 3x + 2$$

Note that in this case, if you were to write $3x$ before $-2x$ to obtain $-3x^2 + 3x - 2x + 2$ it will still work. This is the same in all cases.

Step 4: Factorise

$$-x(3x + 2) + 1(3x + 2)$$

Notice that $(3x + 2)$ is common to both terms, i.e., $-x(3x + 2)$ and $1(3x + 2)$
So, let's factorise again to have

$$(3x + 2)(-x + 1)$$

Note that $-x + 1$ is the same as $1 - x$, so let's re-write this

$$(3x + 2)(1 - x)$$

To keep the format of the factors the same, we will swap the positions of the terms in the first factor from $(3x + 2)$ to $(2 + 3x)$, thus we have:

$$(2 + 3x)(1 - x)$$
$$\therefore 2 + x - 3x^2 = (2 + 3x)(1 - x)$$

---

f) $6y^2 + 11y + 4$

Solution

Here, we have $a = 6$, $b = 11$, and $c = 4$.

Step 1: Product of ac

$$ac = 6 \times 4 = 24$$

Step 2: Finding b_1 and b_2

The two numbers whose product is 24 and sum is 11 are

$$\textbf{3, 8}$$

Because
$3 \times 8 = \textbf{24}$ and $3 + 8 = \textbf{11}$

Step 3: Re-write the expression using b_1 and b_2

$$6y^2 + 8y + 3y + 4$$

Step 4: Factorise

$$2y(3y + 4) + 1(3y + 4)$$

We've placed 1 outside the brackets because, in theory, that is what is common to both $3y$ and 4. More importantly, we need it to complete our next step. Now we can see that $(3y + 4)$ is common to both terms, i.e., $2y(3y + 4)$ and $1(3y + 4)$, so let's factorise again.

$$(3y + 4)(2y + 1)$$
$$\therefore \textbf{6} \textbf{\textit{y}}^2 + \textbf{11} \textbf{\textit{y}} + \textbf{4} = (\textbf{3} \textbf{\textit{y}} + \textbf{4})(\textbf{2} \textbf{\textit{y}} + \textbf{1})$$

NOTE
It's good to see the changes that occur if we rewrite using the second option as:

$$6y^2 + 3y + 8y + 4 = 3y(2y + 1) + 4(2y + 1)$$
$$= (\textbf{2} \textbf{\textit{y}} + \textbf{1})(\textbf{3} \textbf{\textit{y}} + \textbf{4})$$

This is the same as we obtained previously except that the order is reversed, but this is not a problem since multiplication is commutative. That is to say $(\textbf{3} \textbf{\textit{y}} + \textbf{4})(\textbf{2} \textbf{\textit{y}} + \textbf{1})$ is the same as $(\textbf{2} \textbf{\textit{y}} + \textbf{1})(\textbf{3} \textbf{\textit{y}} + \textbf{4})$.

---

g) $4x^2 - 16x + 15$
Solution
Here, we have $a = 4$, $b = -16$, and $c = 15$.

Step 1: Product of ac

$$ac = 4 \times 15 = 60$$

Step 2: Finding b_1 and b_2

The two numbers whose product is 60 and sum is -16 are

$$\textbf{-6, -10}$$

Because

$$-6 \times -10 = \textbf{60} \text{ and } -6 - 10 = \textbf{-16}$$

Step 3: Re-write the expression using b_1 and b_2

$$4x^2 - 10x - 6x + 15$$

Step 4: Factorise

$$2x(2x-5) - 3(2x-5)$$

Notice that $(2x-5)$ is common to both terms, i.e., $2x(2x-5)$ and $-3(2x-5)$, so let's factorise again:

$$(2x-5)(2x-3)$$
$$\therefore 4x^2 - 16x + 15 = (2x-5)(2x-3)$$

---

h) $16u^2 + 8u + 1$
Solution
Here we have $a = 16$, $b = 8$, and $c = 1$.

Step 1: Product of ac

$$ac = 16 \times 1 = 16$$

Step 2: Finding b_1 and b_2

The two numbers whose product is 16 and sum is 8 are

$$\mathbf{4, 4}$$

Because

$$4 \times 4 = \mathbf{16} \text{ and } 4 + 4 = \mathbf{8}$$

Step 3: Re-write the expression using b_1 and b_2

$$16u^2 + 4u + 4u + 1$$

Step 4: Factorise

$$4u(4u+1) + 1(4u+1)$$

Notice that $(4u+1)$ is common to both terms, i.e., $4u(4u+1)$ and $1(4u+1)$, so let's factorise again:

$$(4u+1)(4u+1)$$
$$\therefore \mathbf{16u^2 + 8u + 1 = (4u+1)(4u+1) = (4u+1)^2}$$

In Examples 15(d) and 15(h), you will notice that both $w^2 - 10w + 25$ and $16u^2 + 8u + 1$ have factors that are repeated such that we present them in squared forms as $w^2 - 10w + 25 = (w-5)^2$ and $16u^2 + 8u + 1 = (4u+1)^2$ respectively. Such quadratic expressions are termed **perfect squares**, and the resulting factors – for this case, $(w-5)^2$ and $(4u+1)^2$ – are termed **exact square** or **complete square**.

While perfect squares are not frequently encountered in quadratic expressions, we will find it necessary to express them in squared form. This is the basis of solving quadratic equations by a method known as **completing the square** and it will be covered in Chapter 7.

3.5.4 Miscellaneous

There are several other expressions that do not fit into any of those cases that we've covered and are required to be factorised. The approach we took in the above cases might also apply to them. Let's try some examples to illustrate.

Example 16

Factorise the following expressions.

a) $x^2 y^2 - xy - 12$　　　　　　**b)** $a^2 + 5ab - 14b^2$　　　　　　**c)** $3u^2 v^2 + 13uv - 10$

What did you get? Find the solution below to double-check your answer.

Solution to Example 16

HINT

If we look at these examples carefully, it will become apparent that they are similar to our quadratic expressions, and indeed they are, except that they have two variables. We will evaluate them in the same way.

a) $x^2 y^2 - xy - 12$
Solution
Let's re-write $x^2 y^2 - xy - 12$ to look like the standard form $x^2 + bx + c$:

$$(xy)^2 - (xy) - 12$$

Step 1: Let's open the two brackets and place the variable xy within them.

$$(xy)^2 - (xy) - 12 = (xy + u)(xy + v)$$

We've placed u and v too; we will need to replace them shortly.

Step 2: From the expression, -1 and -12 are our target. So, think about two numbers that will be -12 when multiplied together and -1 when added. What did you get?

I obtained -4 and 3 and we will therefore replace u and v with them.

$$\therefore x^2 y^2 - xy - 12 = (xy - 4)(xy + 3)$$

ALTERNATIVE METHOD

Recall that $x^2 y^2 - xy - 12 = (xy)^2 - (xy) - 12$

Step 1: Product of ac

$$ac = 1 \times -12 = -12$$

Step 2: Finding b_1 and b_2

The two numbers whose product is -12 and sum is -1 are

$$-4, 3$$

Because

$$-4 \times 3 = -12 \text{ and } -4 + 3 = -1$$

Step 3: Re-write the expression using b_1 and b_2

$$x^2y^2 - 4xy + 3xy - 12$$

Step 4: Factorise

$$xy(xy - 4) + 3(xy - 4)$$

Notice that $(xy - 4)$ is common to both terms, i.e., $xy(xy - 4)$ and $3(xy - 4)$.
So, let's factorise again to have:

$$(xy - 4)(xy + 3)$$

$$\therefore x^2y^2 - xy - 12 = (xy - 4)(xy + 3)$$

---

b) $a^2 + 5ab - 14b^2$
Solution
We will approach this and the next question differently, but still similar others.

Step 1: Product of the first and last terms

$$(a^2)(-14b^2) = -14a^2b^2$$

Step 2: Finding b_1 and b_2

The two terms whose product is $-14a^2b^2$ and sum is $5ab$ are

$$7ab, -2ab$$

Because

$$7ab \times -2ab = -14a^2b^2 \text{ and } 7ab - 2ab = 5ab$$

Step 3: Re-write the expression using b_1 and b_2

$$a^2 + 7ab - 2ab - 14b^2$$

Step 4: Factorise

$$a\,(a+7b)-2b\,(a+7b)$$

Notice that $(a+7b)$ is common to both terms, i.e., $a\,(a+7b)$ and $-2b\,(a+7b)$.
So, let's factorise again to have:

$$(a+7b)\,(a-2b)$$
$$\therefore a^2+5ab-14b^2=(a+7b)\,(a-2b)$$

c) $3u^2v^2+13uv-10$
Solution

Step 1: Product of the first and last terms

$$\left(3u^2v^2\right)(-10)=-30u^2v^2$$

Step 2: Finding b_1 and b_2

The two terms whose product is $-30u^2v^2$ and sum is $13uv$ are

$$\mathbf{15uv,-2uv}$$

Because

$$15uv\times-2uv=\mathbf{-30u^2v^2}\text{ and }15uv-2uv=\mathbf{13uv}$$

Step 3: Re-write the expression using b_1 and b_2

$$3u^2v^2+15uv-2uv-10$$

Step 4: Factorise

$$3uv\,(uv+5)-2(uv+5)$$

Notice that $(uv+5)$ is common to both terms, i.e., $3uv\,(uv+5)$ and $-2(uv+5)$. So, let's factorise again to have:

$$(uv+5)\,(3uv-2)$$
$$\therefore 3u^2v^2+13uv-10=(uv+5)\,(3uv-2)$$

Let's try another set of examples.

Example 17

Factorise the following expressions.

a) y^4-y^2-20 **b)** $10y^4-11y^2+3$ **c)** $3y^6+y^3-10$

What did you get? Find the solution below to double-check your answer.

Solution to Example 17

HINT

Apparently, these examples are not quadratic as the highest power is greater than two. However, they can be modelled as though they are.

a) $y^4 - y^2 - 20$

Solution

Step 1: Re-write the expression to look like the standard form $x^2 + bx + c$

$$y^4 - y^2 - 20 = (y^2)^2 - (y^2) - 20$$

Step 2: Let's open the two brackets and put the variable y^2 within them.

$$(y^2)^2 - (y^2) - 20 = (y^2 + u)(y^2 + v)$$

We've placed u and v too; we will need to replace them shortly.

Step 3: From the expression, -1 and -20 are our target. So, think about two numbers that will be -20 when multiplied together and -1 when added. What did you get?

I obtained -5 and 4 and we will therefore replace u and v with them.

$$\therefore y^4 - y^2 - 20 = (y^2 - 5)(y^2 + 4)$$

b) $10y^4 - 11y^2 + 3$

Solution

Step 1: Re-write the expression to look like the standard form $x^2 + bx + c$

$$10y^4 - 11y^2 + 3 = 10(y^2)^2 - 11(y^2) + 3$$

Step 2: Product of ac

$$ac = 10 \times 3 = 30$$

Step 3: Finding b_1 and b_2

The two terms whose product is 30 and the sum is -11 are

$$-5, \; -6$$

Step 4: Re-write the expression using b_1 and b_2

$$10y^4 - 5y^2 - 6y^2 + 3$$

Step 5: Factorise

$$5y^2\left(2y^2-1\right)-3\left(2y^2-1\right)$$

Notice that $\left(2y^2-1\right)$ is common to both terms, i.e., $5y^2\left(2y^2-1\right)$ and $-3\left(2y^2-1\right)$.
So, let's factorise again to have

$$\left(5y^2-3\right)\left(2y^2-1\right)$$
$$\therefore 10y^4-11y^2+3=\left(5y^2-3\right)\left(2y^2-1\right)$$

---

c) $3y^6+y^3-10$
Solution

Step 1: Re-write the expression to look like the standard form x^2+bx+c

$$3y^6+y^3-10=3\left(y^3\right)^2+\left(y^3\right)-10$$

Step 2: Product of ac

$$ac=3\times-10=-30$$

Step 3: Finding b_1 and b_2

The two terms whose product is -30 and the sum is 1 are

$$\mathbf{-5,\ 6}$$

Step 4: Re-write the expression using b_1 and b_2

$$3y^6-5y^3+6y^3-10$$

Step 5: Factorise

$$y^3\left(3y^3-5\right)+2\left(3y^3-5\right)$$

Notice that $\left(3y^3-5\right)$ is common to both terms, i.e., $y^3\left(3y^3-5\right)$ and $2\left(3y^3-5\right)$.
So, let's factorise again to have

$$\left(y^3+2\right)\left(3y^3-5\right)$$
$$\therefore 3y^6+y^3-10=\left(y^3+2\right)\left(3y^3-5\right)$$

Let's take a final set of examples.

Example 18

Factorise the following expressions.

a) $2x^2 + 3x$ **b)** $6u^2 - 9u$ **c)** $21p^2 + 15p^3$

d) $15ab^2 - 12a^2b$ **e)** $2x^2yz - xy^2z + 5xyz^2$

f) $2(x+1)^3 - 3x(x+1)^2$ **g)** $4v(v^2-1) + 2(v^2-1)^2 - 4(v^2-1)$

What did you get? Find the solution below to double-check your answer.

Solution to Example 18

HINT

For this case, determine the HCF of the terms involved. You can split them into two: one HCF for the coefficients (i.e., the number in front of the terms) and another HCF for the variables. Multiply these together to obtain the overall HCF.

a) $2x^2 + 3x$
Solution
The two terms are $2x^2$ and $3x$, so the HCF here is x. Thus, we have:

$$2x^2 + 3x = x \times 2x + x \times 3$$
$$= x(2x) + x(3)$$
$$= x(2x + 3)$$
$$\therefore 2x^2 + 3x = x(2x + 3)$$

b) $6u^2 - 9u$
Solution
The two terms are $6u^2$ and $-9u$, so the HCF here is $3u$. Thus, we have:

$$6u^2 - 9u = 3u \times 2u + 3u \times -3$$
$$= 3u(2u) + 3u(-3)$$
$$= 3u(2u - 3)$$
$$\therefore 6u^2 - 9u = 3u(2u - 3)$$

c) $21p^2 + 15p^3$

Solution

The two terms are $21p^2$ and $15p^3$, so the HCF here is $3p^2$. Thus, we have:

$$21p^2 + 15p^3 = 3p^2 \times 7 + 3p^2 \times 5p$$
$$= 3p^2\,(7) + 3p^2\,(5p)$$
$$= 3p^2\,(7 + 5p)$$
$$\therefore 21p^2 + 15p^3 = 3p^2\,(7 + 5p)$$

d) $15ab^2 - 12a^2b$

Solution

The two terms are $15ab^2$ and $-12a^2b$, so the HCF here is $3ab$. Thus, we have:

$$15ab^2 - 12a^2b = 3ab \times 5b + 3ab \times -4a$$
$$= 3ab\,(5b) + 3ab\,(-4a)$$
$$= 3ab\,(5b - 4a)$$
$$\therefore 15ab^2 - 12a^2b = 3ab\,(5b - 4a)$$

e) $2x^2yz - xy^2z + 5xyz^2$

Solution

The three terms are $2x^2yz$, $-xy^2z$ and $5xyz^2$, so the HCF here is xyz. Thus, we have:

$$2x^2yz - xy^2z + 5xyz^2 = xyz \times 2x + xyz \times -y + xyz \times 5z$$
$$= xyz\,(2x) + xyz\,(-y) + xyz\,(5z)$$
$$= xyz\,(2x - y + 5z)$$
$$\therefore 2x^2yz - xy^2z + 5xyz^2 = xyz\,(2x - y + 5z)$$

f) $2(x + 1)^3 - 3x(x + 1)^2$

Solution

The case here is a little bit different, but we can say that we have two terms: $2(x + 1)^3$ and $-3x(x + 1)^2$, so the HCF here is $(x + 1)^2$. Thus, we have:

$$2(x + 1)^3 - 3x(x + 1)^2 = (x + 1)^2\,[2\,(x + 1) - 3x]$$
$$= (x + 1)^2\,[2x + 2 - 3x]$$
$$= (x + 1)^2\,(2 - x)$$
$$\therefore 2(x + 1)^3 - 3x(x + 1)^2 = (x + 1)^2\,(2 - x)$$

g) $4v\,(v^2 - 1) + 2\,(v^2 - 1)^2 - 4\,(v^2 - 1)$

Solution

Again, this is a different case but we can say that we have three terms: $4v\,(v^2 - 1)$, $2\,(v^2 - 1)^2$, and $-4\,(v^2 - 1)$. So, the HCF here is $2\,(v^2 - 1)$. Thus, we have:

Step 1: Let's take out the HCF.

$$4v\,(v^2 - 1) + 2\,(v^2 - 1)^2 - 4\,(v^2 - 1) = 2\,(v^2 - 1)\,[2v + (v^2 - 1) - 2]$$

Step 2: Simplify the right brackets.

$$= 2\left(v^2 - 1\right)\left[2v + v^2 - 1 - 2\right]$$
$$= 2\left(v^2 - 1\right)\left(v^2 + 2v - 3\right)$$

Step 3: Simplify $\left(v^2 - 1\right)$ using the difference of two squares.

$$= 2\left(v + 1\right)\left(v - 1\right)\left(v^2 + 2v - 3\right)$$

Step 4: Simplify $\left(v^2 + 2v - 3\right)$, as it is a quadratic expression that can be factorised:

$$2\left(v + 1\right)\left(v - 1\right)\left(v + 3\right)\left(v - 1\right) = 2\left(v + 1\right)\left(v + 3\right)\left(v - 1\right)^2$$

All done now!

$$\therefore 4v\left(v^2 - 1\right) + 2\left(v^2 - 1\right)^2 - 4\left(v^2 - 1\right) = 2\left(v + 1\right)\left(v + 3\right)\left(v - 1\right)^2$$

3.6 ALGEBRAIC FRACTIONS

Algebraic fractions are very similar to numerical fractions except that either the numerator, the denominator or both are algebraic expressions. Examples include $\frac{x}{2y}$, $\frac{5}{xy}$, $\frac{x^2 - 2x + 1}{13}$, $\frac{x}{2y - 3}$, and $\frac{x - 4}{2y^2 - 9}$. Addition, subtraction, multiplication, and division of algebraic fractions will follow the same rules and procedures as we did for numeric fractions.

3.6.1 ADDITION AND SUBTSRACTION

In this section we will focus on addition and subtraction only.

Let's try some examples where the denominators are the same.

Example 19

Simplify the following algebraic expressions.

a) $\dfrac{1}{x} + \dfrac{2}{x}$ b) $\dfrac{7}{2xy} - \dfrac{3}{2xy}$ c) $\dfrac{5c}{ab} - \dfrac{11}{ab}$ d) $\dfrac{3}{v^2} + \dfrac{u}{v^2} - \dfrac{6}{v^2}$ e) $\dfrac{d^2}{u+1} - \dfrac{d}{u+1} - \dfrac{3}{u+1}$

What did you get? Find the solution below to double-check your answer.

Solution to Example 19

HINT

Since the denominators are the same, just take one of the denominators and add/subtract the numerators.

a) $\dfrac{1}{x} + \dfrac{2}{x}$

Solution

$$\frac{1}{x} + \frac{2}{x} = \frac{1+2}{x} = \frac{3}{x}$$

$$\therefore \frac{1}{x} + \frac{2}{x} = \frac{3}{x}$$

---

b) $\dfrac{7}{2xy} - \dfrac{3}{2xy}$

Solution

$$\frac{7}{2xy} - \frac{3}{2xy} = \frac{7-3}{2xy}$$

$$= \frac{4}{2xy} = \frac{2}{xy}$$

$$\therefore \frac{7}{2xy} - \frac{3}{2xy} = \frac{2}{xy}$$

---

c) $\dfrac{5c}{ab} - \dfrac{11}{ab}$

Solution

$$\frac{5c}{ab} - \frac{11}{ab} = \frac{5c-11}{ab}$$

$$\therefore \frac{5c}{ab} - \frac{11}{ab} = \frac{5c-11}{ab}$$

---

d) $\dfrac{3}{v^2} + \dfrac{u}{v^2} - \dfrac{6}{v^2}$

Solution

$$\frac{3}{v^2} + \frac{u}{v^2} - \frac{6}{v^2} = \frac{3+u-6}{v^2}$$

$$= \frac{u-3}{v^2}$$

$$\therefore \frac{3}{v^2} + \frac{u}{v^2} - \frac{6}{v^2} = \frac{u-3}{v^2}$$

---

e) $\dfrac{d^2}{u+1} - \dfrac{d}{u+1} - \dfrac{3}{u+1}$

Solution

$$\frac{d^2}{u+1} - \frac{d}{u+1} - \frac{3}{u+1} = \frac{d^2-d-3}{u+1}$$

Let's try more examples.

Example 20

Simplify the following algebraic expressions.

a) $\dfrac{7}{3x} + \dfrac{5}{2x}$ **b)** $\dfrac{3}{x} + \dfrac{2}{y} - \dfrac{1}{z}$ **c)** $\dfrac{13}{v} - \dfrac{5}{v^2}$ **d)** $\dfrac{10}{x^2 y} - \dfrac{9}{xy^2}$ **e)** $\dfrac{b}{a} - \dfrac{2}{ab} - \dfrac{a}{c^2}$

What did you get? Find the solution below to double-check your answer.

Solution to Example 20

a) $\dfrac{7}{3x} + \dfrac{5}{2x}$

Solution

The two denominators are $3x$ and $2x$ and their LCM is $3 \times 2 \times x = 6x$.

Step 1: Divide the LCM by the denominator of the first fraction $\dfrac{6x}{3x} = 2$ and multiply the numerator and denominator of the first fraction by 2. Do the same for the second fraction as:

$$\frac{7}{3x} + \frac{5}{2x} = \frac{7 \times 2}{3x \times 2} + \frac{5 \times 3}{2x \times 3}$$
$$= \frac{14}{6x} + \frac{15}{6x}$$

Step 2: Now the denominators are the same, add the numerators:

$$= \frac{14 + 15}{6x} = \frac{29}{6x}$$

The job is done!

$$\therefore \frac{7}{3x} + \frac{5}{2x} = \frac{29}{6x}$$

ALTERNATIVE METHOD

Step 1: Carry out cross-multiplication.

$$\frac{7}{3x} + \frac{5}{2x} = \frac{7(2x) + 5(3x)}{(3x)(2x)}$$

Step 2: Open the brackets in the numerator and the denominator.

$$= \frac{7(2x) + 5(3x)}{(3x)(2x)} = \frac{14x + 15x}{6x^2}$$

Step 3: Simplify

$$= \frac{29x}{6x^2}$$

Notice that this is different from the answer above. This is because we can still simplify further as:

$$= \frac{29x}{6x^2} = \frac{29 \times x}{6x \times x}$$

Cancel the x

$$\frac{29x}{6x^2} = \frac{29}{6x}$$

$$\therefore \frac{7}{3x} + \frac{5}{2x} = \frac{29}{6x}$$

---

b) $\frac{3}{x} + \frac{2}{y} - \frac{1}{z}$

Solution

The denominators are x, y, and z and their LCM is $x \times y \times z = xyz$.

Step 1: Divide the LCM by the denominator of the first fraction $\frac{xyz}{x} = yz$ and multiply the numerator and denominator of the first fraction by yz. Do the same for the second and third fractions as:

$$\frac{3}{x} + \frac{2}{y} - \frac{1}{z} = \frac{3 \times yz}{x \times yz} + \frac{2 \times xz}{y \times xz} - \frac{1 \times xy}{z \times xy}$$

$$= \frac{3yz}{xyz} + \frac{2xz}{yxz} - \frac{xy}{zxy}$$

Step 2: Now the denominators are the same, add the numerators:

$$= \frac{3yz + 2xz - xy}{xyz}$$

Nothing more to simplify, so job done!

$$\therefore \frac{3}{x} + \frac{2}{y} - \frac{1}{z} = \frac{3yz + 2xz - xy}{xyz}$$

---

c) $\frac{13}{v} - \frac{5}{v^2}$

Solution

The denominators are v and v^2 and their LCM is $v \times v = v^2$.

Step 1: Divide the LCM by the denominator of the first fraction $\frac{v^2}{v} = v$ and multiply the numerator and denominator of the first fraction by v. Do the same for the second fraction as:

$$\frac{13}{v} - \frac{5}{v^2} = \frac{13 \times v}{v \times v} - \frac{5 \times 1}{v^2 \times 1} = \frac{13v}{v^2} - \frac{5}{v^2}$$

Step 2: Now the denominators are the same, subtract the numerators:

$$= \frac{13v - 5}{v^2}$$

Nothing more to simplify, so job done!

$$\therefore \frac{13}{v} - \frac{5}{v^2} = \frac{13v - 5}{v^2}$$

d) $\dfrac{10}{x^2y} - \dfrac{9}{xy^2}$

Solution

The denominators are x^2y and xy^2 and their LCM is $x \times x \times y \times y = x^2y^2$.

Step 1: Divide the LCM by the denominator of the first fraction $\dfrac{x^2y^2}{x^2y} = y$ and multiply the numerator and denominator of the first fraction by y. Do the same for the second fraction as:

$$\frac{10}{x^2y} - \frac{9}{xy^2} = \frac{10 \times y}{x^2y \times y} - \frac{9 \times x}{xy^2 \times x}$$

$$= \frac{10y}{x^2y^2} - \frac{9x}{x^2y^2}$$

Step 2: Now the denominators are the same, add the numerators:

$$= \frac{10y - 9x}{x^2y^2}$$

Nothing more to simplify, so job done!

$$\therefore \frac{10}{x^2y} - \frac{9}{xy^2} = \frac{10y - 9x}{x^2y^2}$$

---

e) $\dfrac{b}{a} - \dfrac{2}{ab} - \dfrac{a}{c^2}$

Solution

The denominators are a, ab, and c^2 and their LCM is $a \times b \times c^2 = abc^2$.

Step 1: Divide the LCM by the denominator of the first fraction $\dfrac{abc^2}{a} = bc^2$ and multiply the numerator and denominator of the first fraction by bc^2. Do the same for the second and third fractions as:

$$\frac{b}{a} - \frac{2}{ab} - \frac{a}{c^2} = \frac{b \times bc^2}{a \times bc^2} - \frac{2 \times c^2}{ab \times c^2} - \frac{a \times ab}{c^2 \times ab}$$

$$= \frac{b^2c^2}{abc^2} - \frac{2c^2}{abc^2} - \frac{a^2b}{abc^2}$$

Step 2: Now the denominators are the same, add the numerators:

$$= \frac{b^2c^2 - 2c^2 - a^2b}{abc^2}$$

Nothing more to simplify, so job done!

$$\therefore \frac{b}{a} - \frac{2}{ab} - \frac{a}{c^2} = \frac{b^2c^2 - 2c^2 - a^2b}{abc^2}$$

Let's try more examples.

Example 21

Simplify the following algebraic expressions.

a) $\dfrac{1}{m-13} - \dfrac{1}{2m}$ b) $\dfrac{1}{x-3} + \dfrac{3}{x+2}$ c) $\dfrac{x}{x+5} - \dfrac{7}{x-1}$ d) $\dfrac{y-1}{y+2} - \dfrac{y-2}{y+1}$

e) $3 - \dfrac{q}{p+q} - \dfrac{p}{p-q}$ f) $\dfrac{5}{x} + \dfrac{1}{x+5} - \dfrac{2}{x-1}$ g) $\dfrac{1}{w+1} - \dfrac{2w}{(w+1)^2} + \dfrac{w^2-2}{(w+1)^3}$

What did you get? Find the solution below to double-check your answer.

Solution to Example 21

HINT

In this case, we will use cross-multiplication where there are only two fractions.

a) $\dfrac{1}{m-13} - \dfrac{1}{2m}$
Solution

Step 1: Cross-multiply.

$$\frac{1}{m-13} - \frac{1}{2m} = \frac{1 \times 2m - 1 \times (m-13)}{2m(m-13)}$$

Step 2: Open the brackets in the numerator and simplify.

$$= \frac{2m - m + 13}{2m(m-13)} = \frac{m+13}{2m(m-13)}$$

Job done!

$$\therefore \frac{1}{m-13} - \frac{1}{2m} = \frac{m+13}{2m(m-13)}$$

b) $\dfrac{1}{x-3} + \dfrac{3}{x+2}$
Solution

Step 1: Cross-multiply.

$$\frac{1}{x-3} + \frac{3}{x+2} = \frac{(x+2) + 3(x-3)}{(x-3)(x+2)}$$

Step 2: Open the brackets in the numerator and simplify.

$$= \frac{x + 2 + 3x - 9}{(x-3)(x+2)} = \frac{4x - 7}{(x-3)(x+2)}$$

$$\therefore \frac{1}{x-3} + \frac{3}{x+2} = \frac{4x-7}{(x-3)(x+2)}$$

---

c) $\frac{x}{x+5} - \frac{7}{x-1}$

Solution

Step 1: Cross-multiply.

$$\frac{x}{x+5} - \frac{7}{x-1} = \frac{x(x-1) - 7(x+5)}{(x+5)(x-1)}$$

Step 2: Open the brackets in the numerator and simplify.

$$= \frac{x^2 - x - 7x - 35}{(x+5)(x-1)} = \frac{x^2 - 8x - 35}{(x+5)(x-1)}$$

$$\therefore \frac{x}{x+5} - \frac{7}{x-1} = \frac{x^2 - 8x - 35}{(x+5)(x-1)}$$

NOTE

Although the numerator is a quadratic expression, but it cannot be factorised. The reason will be clear when we get to Chapter 7.

---

d) $\frac{y-1}{y+2} - \frac{y-2}{y+1}$

Solution

Step 1: Cross-multiply.

$$\frac{y-1}{y+2} - \frac{y-2}{y+1} = \frac{(y-1)(y+1) - (y-2)(y+2)}{(y+2)(y+1)}$$

Step 2: Open the brackets in the numerator. You should notice the difference of squares here.

$$= \frac{(y^2 - 1) - (y^2 - 2^2)}{(y+2)(y+1)}$$

Step 3: Let's simplify.

$$= \frac{y^2 - 1 - y^2 + 4}{(y+2)(y+1)} = \frac{3}{(y+2)(y+1)}$$

$$\therefore \frac{y-1}{y+2} - \frac{y-2}{y+1} = \frac{3}{(y+2)(y+1)}$$

e) $3 - \dfrac{q}{p+q} - \dfrac{p}{p-q}$

Solution

We have three fractions here so let's do this differently.

Step 1: Determine the LCM.

Note that 3 is the same as $\dfrac{3}{1}$. Thus, the denominators are 1, $p + q$, and $p - q$ and their LCM is $(p + q)(p - q)$.

Step 2: Divide the LCM by the denominator of the first fraction $\dfrac{(p+q)(p-q)}{1} = (p + q)(p - q)$ and multiply the numerator and denominator of the first fraction by $(p + q)(p - q)$. Do the same for the second and third fractions as:

$$3 - \frac{q}{p+q} - \frac{p}{p-q} = \frac{3 \times (p+q)(p-q)}{1 \times (p+q)(p-q)} - \frac{q \times (p-q)}{(p+q) \times (p-q)} - \frac{p \times (p+q)}{(p-q) \times (p+q)}$$

$$= \frac{3(p+q)(p-q)}{(p+q)(p-q)} - \frac{q(p-q)}{(p+q)(p-q)} - \frac{p(p+q)}{(p-q)(p+q)}$$

Step 3: Let's combine the numerators since the denominators are now the same.

$$= \frac{3(p+q)(p-q) - q(p-q) - p(p+q)}{(p+q)(p-q)}$$

Step 4: Let's simplify.

$$= \frac{3(p^2 - q^2) - pq + q^2 - p^2 - pq}{(p+q)(p-q)}$$

$$= \frac{3p^2 - 3q^2 - pq + q^2 - p^2 - pq}{(p+q)(p-q)}$$

$$= \frac{2p^2 - 2q^2 - 2pq}{(p+q)(p-q)} = \frac{2(p^2 - q^2 - pq)}{(p+q)(p-q)}$$

We cannot factorise $p^2 - q^2 - pq$ so all done!

$$\therefore 3 - \frac{q}{p+q} - \frac{p}{p-q} = \frac{2(p^2 - q^2 - pq)}{(p+q)(p-q)}$$

---

f) $\dfrac{5}{x} + \dfrac{1}{x+5} - \dfrac{2}{x-1}$

Solution

Step 1: Determine the LCM.

The denominators are x, $x + 5$, and $x - 1$ and their LCM is $x(x + 5)(x - 1)$.

Step 2: Divide the LCM by the denominator of the first fraction $\dfrac{x(x+5)(x-1)}{x} = (x + 5)(x - 1)$ and multiply the numerator and denominator of the first fraction by $(x + 5)(x - 1)$. Do the same for the second and third fractions as:

$$\frac{5}{x} + \frac{1}{x+5} - \frac{2}{x-1} = \frac{5 \times (x+5)(x-1)}{x \times (x+5)(x-1)} + \frac{1 \times x(x-1)}{(x+5) \times x(x-1)} - \frac{2 \times x(x+5)}{(x-1) \times x(x+5)}$$

$$= \frac{5(x+5)(x-1)}{x(x+5)(x-1)} + \frac{x(x-1)}{x(x+5)(x-1)} - \frac{2x(x+5)}{x(x-1)(x+5)}$$

Step 3: Open the brackets in the numerator. We could leave this to the next step though.

$$= \frac{5(x^2 - x + 5x - 5)}{x(x+5)(x-1)} + \frac{x^2 - x}{x(x+5)(x-1)} - \frac{2x^2 + 10x}{x(x-1)(x+5)}$$

$$= \frac{5(x^2 + 4x - 5)}{x(x+5)(x-1)} + \frac{x^2 - x}{x(x+5)(x-1)} - \frac{2x^2 + 10x}{x(x-1)(x+5)}$$

Step 4: Let's combine (add/subtract) the numerators, since the denominators are now the same.

$$= \frac{5(x^2 + 4x - 5) + (x^2 - x) - (2x^2 + 10x)}{x(x+5)(x-1)}$$

Step 5: Let's simplify.

$$= \frac{5(x^2 + 4x - 5) + (x^2 - x) - (2x^2 + 10x)}{x(x+5)(x-1)}$$

$$= \frac{5x^2 + 20x - 25 + x^2 - x - 2x^2 - 10x}{x(x+5)(x-1)}$$

$$= \frac{4x^2 + 9x - 25}{x(x+5)(x-1)}$$

We will not be able to factorise $4x^2 + 9x - 25$ here so we will leave it as it is. All done!

$$\therefore \frac{5}{x} + \frac{1}{x+5} - \frac{2}{x-1} = \frac{4x^2 + 9x - 25}{x(x+5)(x-1)}$$

g) $\frac{1}{w+1} - \frac{2w}{(w+1)^2} + \frac{w^2-2}{(w+1)^3}$

Solution

Step 1: Determine the LCM.

The denominators are $w + 1$, $(w + 1)^2$, and $(w + 1)^3$ and their LCM is $(w + 1)^3$.

Step 2: Divide the LCM by the denominator of the first fraction $\frac{(w+1)^3}{w+1} = (w + 1)^2$ and multiply the numerator and denominator of the first fraction by $(w + 1)^2$. Do the same for the second and third fractions as:

$$\frac{1}{w+1} - \frac{2w}{(w+1)^2} + \frac{w^2-2}{(w+1)^3} = \frac{1 \times (w+1)^2}{(w+1) \times (w+1)^2} - \frac{2w \times (w+1)}{(w+1)^2 \times (w+1)} + \frac{w^2-2}{(w+1)^3}$$

$$= \frac{(w+1)^2}{(w+1)^3} - \frac{2w(w+1)}{(w+1)^3} + \frac{w^2-2}{(w+1)^3}$$

Step 3: Open the brackets and simplify.

$$= \frac{(w+1)^2 - 2w(w+1) + w^2 - 2}{(w+1)^3}$$

$$= \frac{w^2 + 2w + 1 - 2w^2 - 2w + w^2 - 2}{(w+1)^3}$$

$$= -\frac{1}{(w+1)^3}$$

$$\therefore \frac{1}{w+1} - \frac{2w}{(w+1)^2} + \frac{w^2 - 2}{(w+1)^3} = -\frac{1}{(w+1)^3}$$

3.6.2 MULTIPLICATION AND DIVISION

In this section we will focus on multiplication and division. Note that in numerical fractions, the division is carried out by changing it to multiplication and flipping the second fraction (i.e., the divisor). This is the same way we will treat algebraic fractions when it comes to division. Let's try some examples.

Example 22

Simplify each of the following algebraic expressions.

a) $\dfrac{6}{xy} \times \dfrac{x^2}{2y}$

b) $\dfrac{3pq}{5r} \div \dfrac{q^2}{10pr^2}$

c) $\dfrac{7}{v} \times \dfrac{u}{uv} \times \dfrac{v}{w}$

d) $\dfrac{b^2-1}{b-2} \div \dfrac{b-1}{2b-4}$

e) $\dfrac{3}{x^2-5x+4} \times \dfrac{x^3+x^2-2x}{x+2}$

f) $\dfrac{y}{y^2-25} \times \dfrac{y+5}{y+2} \div \dfrac{3y^2}{2y^2-8}$

What did you get? Find the solution below to double-check your answer.

Solution to Example 22

a) $\dfrac{6}{xy} \times \dfrac{x^2}{2y}$

Solution

$$\frac{6}{xy} \times \frac{x^2}{2y} = \frac{6 \times x^2}{xy \times 2y} = \frac{6 \times x \times x}{x \times y \times 2 \times y}$$

$$\therefore \frac{6}{xy} \times \frac{x^2}{2y} = \frac{3x}{y^2}$$

b) $\dfrac{3pq}{5r} \div \dfrac{q^2}{10pr^2}$

Solution

$$\frac{3pq}{5r} \div \frac{q^2}{10pr^2} = \frac{3pq}{5r} \times \frac{10pr^2}{q^2} = \frac{3pq \times 10pr^2}{5r \times q^2}$$

$$= \frac{3 \times p \times q \times 10 \times p \times r \times r}{5 \times r \times q \times q}$$

$$= \frac{3 \times p \times 2 \times p \times r}{q}$$

$$\therefore \; \frac{3pq}{5r} \div \frac{q^2}{10pr^2} = \frac{6p^2r}{q}$$

---

c) $\dfrac{7}{v} \times \dfrac{u}{uv} \times \dfrac{v}{w}$

Solution

$$\frac{7}{v} \times \frac{u}{uv} \times \frac{v}{w} = \frac{7 \times u \times v}{v \times uv \times w} = \frac{7}{v \times w}$$

$$\therefore \; \frac{7}{v} \times \frac{u}{uv} \times \frac{v}{w} = \frac{7}{vw}$$

---

d) $\dfrac{b^2-1}{b-2} \div \dfrac{b-1}{2b-4}$

Solution

This will require a bit of work, so let's go slowly.

Step 1: Factorise where necessary.

$$\frac{b^2-1}{b-2} \div \frac{b-1}{2b-4} = \frac{(b+1)(b-1)}{b-2} \div \frac{b-1}{2(b-2)}$$

Step 2: Change $\div$ to $\times$ and flip the fraction to the right of the $\div$ sign.

$$= \frac{(b+1)(b-1)}{b-2} \times \frac{2(b-2)}{b-1}$$

Step 3: Cancel the common factors. $(b-1)$ and $(b-2)$ appear on the numerator and denominator so 'delete' them.

$$= \frac{(b+1)}{1} \times \frac{2}{1}$$

Step 4: Simplify what is left by multiplying the numerators together. Do the same for the denominators.

$$= 2(b+1)$$

All done!

$$\therefore \; \frac{b^2-1}{b-2} \div \frac{b-1}{2b-4} = 2(b+1)$$

e) $\dfrac{3}{x^2-5x+4} \times \dfrac{x^3+x^2-2x}{x+2}$

Solution

Step 1: Factorise where necessary.

$x^2 - 5x + 4$ is a quadratic expression so we can factorise as:

$$x^2 - 5x + 4 = (x - 1)(x - 4)$$

and $x^3 + x^2 - 2x$ is a cubic expression, but x is common to all the terms, so let's take it out. Thus:

$$x^3 + x^2 - 2x = x(x^2 + x - 2)$$

So x is multiplied by a quadratic expression, let's simplify the latter. We have

$$x^3 + x^2 - 2x = x(x - 1)(x + 2)$$

Let's plug all these in as:

$$\frac{3}{x^2 - 5x + 4} \times \frac{x^3 + x^2 - 2x}{x + 2} = \frac{3}{(x - 1)(x - 4)} \times \frac{x(x - 1)(x + 2)}{x + 2}$$

Step 2: Cancel the common factors. $(x - 1)$ and $(x + 2)$ appear on the numerator and denominator so 'delete' them.

$$= \frac{3}{(x - 4)} \times \frac{x}{1}$$

Simplify what is left by multiplying the numerators together. Do the same for the denominators.

$$= \frac{3 \times x}{(x - 4) \times 1}$$

All done!

$$\therefore \frac{3}{x^2 - 5x + 4} \times \frac{x^3 + x^2 - 2x}{x + 2} = \frac{3x}{x - 4}$$

---

f) $\dfrac{y}{y^2-25} \times \dfrac{y+5}{y+2} \div \dfrac{3y^2}{2y^2-8}$

Solution

Step 1: Factorise where necessary.

$y^2 - 25$ is a difference of two squares so we can factorise as:

$$y^2 - 25 = y^2 - 5^2 = (y + 5)(y - 5)$$

and $2y^2 - 8$ is a quadratic expression but 2 is common to both terms, let's take it out:

$$2y^2 - 8 = 2(y^2 - 4)$$

2 is multiplied by a quadratic expression, let's simplify the latter. We have

$$2(y^2 - 4) = 2(y + 2)(y - 2)$$

Let's plug all these in.

$$\frac{y}{y^2 - 25} \times \frac{y + 5}{y + 2} \div \frac{3y^2}{2y^2 - 8} = \frac{y}{(y + 5)(y - 5)} \times \frac{y + 5}{y + 2} \div \frac{3y^2}{2(y + 2)(y - 2)}$$

Step 2: Change $\div$ to $\times$ and flip the fraction to the right of the $\div$

$$= \frac{y}{(y+5)(y-5)} \times \frac{y+5}{y+2} \times \frac{2(y+2)(y-2)}{3y^2}$$

Step 3: Cancel the common factors. $(y+2)$ and $(y+5)$ appear on the numerator and denominator so 'delete' them.

$$= \frac{y}{(y-5)} \times \frac{1}{1} \times \frac{2(y-2)}{3y^2}$$

Oh! We can cancel y too, so let's do it and it becomes:

$$= \frac{1}{(y-5)} \times \frac{1}{1} \times \frac{2(y-2)}{3y}$$

Step 4: Simplify what is left by multiplying the numerators together. Do the same for the denominators.

$$= \frac{1 \times 2(y-2)}{(y-5) \times 3y}$$

All done!

$$\therefore \frac{y}{y^2-25} \times \frac{y+5}{y+2} \div \frac{3y^2}{2y^2-8} = \frac{2(y-2)}{3y(y-5)}$$

3.7 FURTHER SIMPLIFICATION

As we've now introduced and covered fundamental rules and tips relating to algebraic expressions, we will look at further examples on simplification, but will be brief in our workings. Let's try some examples.

Example 23

Simplify the following algebraic expressions.

a) $\dfrac{4x^2+x}{x}$

b) $\dfrac{x-x^2-5x^3}{3x}$

c) $\dfrac{(7-2x)(x+2)}{(x+2)}$

d) $\dfrac{x^2+5x-24}{(x-3)}$

e) $\dfrac{3x^3-x^2-4x}{4-3x}$

f) $\dfrac{2x^3-4x^2-30x}{15+2x-x^2}$

g) $\dfrac{3x^3-\frac{1}{3}x}{3x^2-\frac{5}{2}x+\frac{1}{2}}$

What did you get? Find the solution below to double-check your answer.

Solution to Example 23

a) $\dfrac{4x^2+x}{x}$

Solution

$$\frac{4x^2 + x}{x} = \frac{x(4x + 1)}{x}$$

$$= 4x + 1$$

$$\therefore \frac{4x^2 + x}{x} = 4x + 1$$

ALTERNATIVE METHOD

$$\frac{4x^2 + x}{x} = \frac{4x^2}{x} + \frac{x}{x} = 4x + 1$$

NOTE

In this method, we divided each term of the numerator by the denominator and then simplified.

---

b) $\dfrac{x-x^2-5x^3}{3x}$

Solution

$$\frac{x - x^2 - 5x^3}{3x} = \frac{x\left(1 - x - 5x^2\right)}{3x}$$

$$= \frac{\left(1 - x - 5x^2\right)}{3}$$

$$\therefore \frac{x - x^2 - 5x^3}{3x} = \frac{1}{3}\left(1 - x - 5x^2\right)$$

---

c) $\dfrac{(7-2x)(x+2)}{(x+2)}$

Solution

$$\frac{(7 - 2x)(x + 2)}{(x + 2)} = 7 - 2x$$

$$\therefore \frac{(7 - 2x)(x + 2)}{(x + 2)} = 7 - 2x$$

---

d) $\dfrac{x^2+5x-24}{(x-3)}$

Solution

$$\frac{x^2 + 5x - 24}{(x - 3)} = \frac{(x + 8)(x - 3)}{(x - 3)}$$

$$= x + 8$$

$$\therefore \frac{x^2 + 5x - 24}{(x - 3)} = x + 8$$

---

e) $\dfrac{3x^3 - x^2 - 4x}{4 - 3x}$

Solution

$$\frac{3x^3 - x^2 - 4x}{4 - 3x} = \frac{x\left(3x^2 - x - 4\right)}{4 - 3x}$$

$$= \frac{x\left(3x - 4\right)\left(x + 1\right)}{-\left(3x - 4\right)}$$

$$= -x\left(x + 1\right)$$

$$\therefore \frac{3x^3 - x^2 - 4x}{4 - 3x} = -x\left(x + 1\right)$$

---

f) $\dfrac{2x^3 - 4x^2 - 30x}{15 + 2x - x^2}$

Solution

$$\frac{2x^3 - 4x^2 - 30x}{15 + 2x - x^2} = \frac{2x\left(x^2 - 2x - 15\right)}{-1\left(x^2 - 2x - 15\right)}$$

$$= \frac{2x}{-1} = -2x$$

$$\therefore \frac{2x^3 - 4x^2 - 30x}{15 + 2x - x^2} = -2x$$

---

g) $\dfrac{3x^3 - \frac{1}{3}x}{3x^2 - \frac{5}{2}x + \frac{1}{2}}$

Solution

$$\frac{3x^3 - \frac{1}{3}x}{3x^2 - \frac{5}{2}x + \frac{1}{2}} = \frac{\left[3x^3 - \frac{1}{3}x\right] \times 6}{\left[3x^2 - \frac{5}{2}x + \frac{1}{2}\right] \times 6}$$

$$= \frac{18x^3 - 2x}{18x^2 - 15x + 3} = \frac{2x\left(9x^2 - 1\right)}{3\left(6x^2 - 5x + 1\right)}$$

$$= \frac{2x\left(3x - 1\right)\left(3x + 1\right)}{3\left(3x - 1\right)\left(2x - 1\right)} = \frac{2x\left(3x + 1\right)}{3\left(2x - 1\right)}$$

$$\therefore \frac{3x^3 - \frac{1}{3}x}{3x^2 - \frac{5}{2}x + \frac{1}{2}} = \frac{2x\left(3x + 1\right)}{3\left(2x - 1\right)}$$

3.8 FORMULA, SUBSTITUTION, AND TRANSPOSITION

Formula is neither an expression nor an equation per se. It is like a recipe (or a general rule or a 'standardised' equation) which is a customised combination of things needed to obtain a particular result; it thus establishes a relationship between items or variables.

A formula usually takes the form of $y = f(x)$, where y is called the subject of the formula. Its value depends on the value of one or more variables on the right-hand side, given compactly by $f(x)$. For

TABLE 3.6

Formulas (Dependent and Independent Variables)

Equation or Formula	Dependent Variable	Independent Variable	Constant
$y = mx + c$	y	x	m, c
$F = kx$	F	x	k
$V = IR$	V	I	R
$T = 2\pi\sqrt{\dfrac{l}{g}}$	T	l	$2, \pi, g$
$T = 2\pi\sqrt{\dfrac{m}{k}}$	T	m	$2, \pi, k$
$f = \dfrac{1}{2\pi\sqrt{LC}}$	f	L, C	$2, \pi$

example, the formula for calculating the displacement by a body undergoing a uniform acceleration is given as:

$$s = ut + \frac{1}{2}at^2 \tag{3.4}$$

where s is the subject of the above formula, and u, t, and a are the independent variables. Technically speaking, u is a constant in this case, and a is also a constant if the motion is uniform.

The value(s) of these variables must be known and substituted into the RHS to obtain the corresponding value of s. When you replace the letter(s) and constant on the right-hand side of a formula with numerical values to obtain the particular value on the LHS, we say we '**substitute**' the values and the process is called **substitution**.

In general, a formula consists of a variable or variables (dependent and independent), relational sign, and constant. Table 3.6 shows some common formulas and their respective constituents.

The constant(s) shown for each formula is only valid for an ideal or specified condition. The relational sign in the above formulas is the equal sign ($=$), which makes a formula become a special equation. Whilst this is the common situation, a formula can however have other relationships connecting the subject on the LHS with the expression on the RHS. Other relational symbol includes 'approximately equal' ($\approx$), 'less than' ($<$), and 'far greater than' ($\gg$) among others.

Let's try some examples on substitution.

Example 24

Determine the value of the dependent variable (or subject of the formula) on the LHS at the given value of the independent variable(s) on the RHS, shown in square brackets in the following. Present answers correct to 3 s.f.

a) $A = \pi r^2$ $[r = 6]$ b) $y = mx + c$ $[m = 2,\ x = 3,\ c = 5]$

c) $S = 2\pi r^2 + 2\pi rh$ $[r = 8,\ h = 10]$ d) $s = ut + \frac{1}{2}at^2$ $[u = 1.5,\ t = 12,\ a = 4]$

e) $W = \frac{1}{2}CV^2$ $\left[C = 6 \times 10^{-6},\ V = 240\right]$

f) $F = BIL\sin\theta$ $[B = 0.35,\ I = 2,\ L = 0.5,\ \theta = 30]$

g) $R_1 = R_0(1 + \alpha\theta_1)$ $[R_0 = 350,\ \alpha = 0.00428,\ \theta_1 = 20]$

What did you get? Find the solution below to double-check your answer.

Solution to Example 24

a) $A = \pi r^2$ $[r = 6]$
Solution

$$A = \pi r^2$$
$$= \pi(6)^2 = 36\pi = 113.1$$
$$\therefore A = 113$$

---

b) $y = mx + c$ $[m = 2,\ x = 3,\ c = 5]$
Solution

$$y = mx + c$$
$$= 2 \times 3 + 5 = 6 + 5 = 11$$
$$\therefore y = 11.0$$

---

c) $S = 2\pi r^2 + 2\pi rh$ $[r = 8,\ h = 10]$
Solution

$$S = 2\pi r^2 + 2\pi rh$$
$$= 2\pi(8)^2 + 2\pi(8)(10)$$
$$= 2\pi(64) + 2\pi(80)$$
$$= 128\pi + 160\pi$$
$$= 288\pi = 904.8$$
$$\therefore S = 905$$

---

d) $s = ut + \frac{1}{2}at^2$ $[u = 1.5,\ t = 12,\ a = 4]$
Solution

$$s = ut + \frac{1}{2}at^2$$
$$= 1.5 \times 12 + \frac{1}{2} \times 4 \times 12^2$$
$$= 18 + 288 = 306$$
$$\therefore s = 306$$

e) $W = \frac{1}{2}CV^2$ $[C = 6 \times 10^{-6}, \ V = 240]$

Solution

$$W = \frac{1}{2}CV^2$$

$$= \frac{1}{2} \times 6 \times 10^{-6} \times 240^2$$

$$= 3 \times 10^{-6} \times 57\ 600$$

$$= 172\ 800 \times 10^{-6} = 0.1728$$

$$\therefore W = 0.173$$

f) $F = BIL \sin \theta$ $[B = 0.35, \ I = 2, \ L = 0.5, \ \theta = 30]$

Solution

$$F = BI \sin \theta$$

$$= 0.35 \times 2 \times 0.5 \times \sin 30$$

$$= 0.35 \times 2 \times 0.5 \times 0.5$$

$$\therefore F = 0.175$$

g) $R_1 = R_0 (1 + \alpha\theta_1)$ $[R_0 = 350, \ \alpha = 0.00428, \ \theta_1 = 20]$

Solution

$$R_1 = R_0 (1 + \alpha\theta_1)$$

$$= 350 (1 + 0.00428 \times 20)$$

$$= 350 (1 + 0.0856)$$

$$= 350 (1.0856) = 380$$

$$\therefore R_1 = 380$$

Sometimes, we may be interested in the value of one of the independent variables. We therefore need to re-arrange the formula such that the required variable stands alone, preferably on the left-hand side. This process of re-arranging the formula to ensure that the variable we are interested in finding stands alone on the LHS while others are collected or moved to the RHS is called **transposition** or '**changing the subject of the formula**'.

Based on this process, we can derive three different formulas each for u, t, and a from the single formula (Equation 3.4). You can imagine the number of equations we will have in science and engineering without this process; instead of being provided with a single equation for Ohm's law, we will have to be provided with three distinct equations to represent the three variables, i.e., $V = IR$, $I = \frac{V}{R}$, and $R = \frac{V}{I}$.

Changing the subject of the formula is a simple process, which becomes intuitive with practice. However, below are a couple of key tips to bear in mind and use when transposing.

Note 1 Keep both sides of the equation (or formula) '**balanced**' by ensuring that whatever is done to one side is done to the other side. For example, if a constant c is added to the LHS of the equation, the same must be added to the RHS. This also applies to subtraction, multiplication, division, and power among others.

Note 2 To eliminate an operation on one side, apply its inverse to both sides. For example, addition is cancelled with subtraction, while power is eliminated using its equivalent root.

It's time to practise transposition.

Example 25

Transpose the following formulas to make the variable (or letter) in the square brackets the subject of the formula.

a) $V = IR$ $\qquad$ $[R]$ $\qquad$ b) $A = \pi r^2$ $\qquad$ $[r]$ $\qquad$ c) $y = mx + c$ $\qquad$ $[m]$

d) $v = u + at$ $\qquad$ $[t]$ $\qquad$ e) $v^2 = u^2 + 2gs$ $\qquad$ $[g]$ $\qquad$ f) $a^2 = b^2 + c^2 - 2bc\cos A$ $\qquad$ $[\cos A]$

g) $pV = \frac{1}{3}mv^2$ $\qquad$ $[m]$ $\qquad$ h) $M = k\sqrt{L_1 L_2}$ $\qquad$ $[L_1]$ $\qquad$ i) $e = mgh + \frac{1}{2}mv^2$ $\qquad$ $[m]$

What did you get? Find the solution below to double-check your answer.

Solution to Example 25

---

a) $V = IR$ $\qquad$ $[R]$
Solution

$$V = IR$$

Step 1: Divide both sides by I

$$\frac{V}{I} = \frac{IR}{I}$$

Step 2: Simplify

$$\frac{V}{I} = R$$

$$\therefore R = \frac{V}{I}$$

---

b) $A = \pi r^2$ $\qquad$ $[r]$
Solution

$$A = \pi r^2$$

Step 1: Divide both sides by π and simplify

$$\frac{A}{\pi} = \frac{\pi r^2}{\pi}$$

which implies that

$$\frac{A}{\pi} = r^2$$

Step 2: Take the square root of both sides and simplify

$$\sqrt{\frac{A}{\pi}} = \sqrt{r^2}$$

which implies that

$$\sqrt{\frac{A}{\pi}} = r$$

NOTE

Here we took the square root because this is the inverse of square, which can be used to obtain r from r^2. It is very important to know operators and their inverse in transposition.

$$\therefore r = \sqrt{\frac{A}{\pi}}$$

---

c) $y = mx + c$ $[m]$

Solution

$$y = mx + c$$

Step 1: Subtract c from both sides and simplify

$$y - c = mx + c - c$$
$$y - c = mx$$

NOTE

Here we could have simply said '*move c across the equal sign to the LHS and change the sign*'. This will give us $y - c = mx$ in one step. Notice that the sign of c has changed from $+$ to $-$. We will use this approach in our subsequent worked examples here.

Step 2: Divide both sides by x and simplify

$$\frac{y - c}{x} = \frac{mx}{x}$$

which implies that

$$\frac{y - c}{x} = m$$
$$\therefore m = \frac{y - c}{x}$$

---

d) $v = u + at$ $[t]$

Solution

$$v = u + at$$

Step 1: Move u to the LHS

$$v - u = at$$

Step 2: Divide both sides by a and simplify

$$\frac{v - u}{a} = \frac{at}{a}$$

which implies that

$$\frac{v - u}{a} = t$$

$$\therefore t = \frac{v - u}{a}$$

---

e) $v^2 = u^2 + 2gs$ [g]
Solution

$$v^2 = u^2 + 2gs$$

Step 1: Move u^2 to the LHS

$$v^2 - u^2 = 2gs$$

Step 2: Divide both sides by $2s$ and simplify

$$\frac{v^2 - u^2}{2s} = \frac{2gs}{2s}$$

which implies that

$$\frac{v^2 - u^2}{2s} = g$$

$$\therefore g = \frac{v^2 - u^2}{2s}$$

---

f) $a^2 = b^2 + c^2 - 2bc \cos A$ $[\cos A]$
Solution

$$a^2 = b^2 + c^2 - 2bc \cos A$$

Step 1: Move $2bc \cos A$ to the LHS

$$a^2 + 2bc \cos A = b^2 + c^2$$

Step 2: Now move a^2 to the RHS

$$2bc \cos A = b^2 + c^2 - a^2$$

Step 3: Divide both sides by $2bc$ and simplify

$$\frac{2bc \cos A}{2bc} = \frac{b^2 + c^2 - a^2}{2bc}$$

which implies that

$$\cos A = \frac{b^2 + c^2 - a^2}{2bc}$$

$$\therefore \cos A = \frac{b^2 + c^2 - a^2}{2bc}$$

---

g) $pV = \frac{1}{3}mv^2$ $[m]$

Solution

$$pV = \frac{1}{3}mv^2$$

Step 1: Multiply both sides by 3 and simplify

$$pV \times 3 = \frac{1}{3}mv^2 \times 3$$
$$3pV = mv^2$$

Step 2: Divide both sides by v^2 and simplify

$$\frac{3pV}{v^2} = \frac{mv^2}{v^2}$$

which implies that

$$\frac{3pV}{v^2} = m$$

NOTE

Here we could not reduce the LHS to $\frac{3p}{v}$ because V and v are not the same variables. V represents the volume and v is the velocity. It is thus essential to pay attention to the case of a variable in algebra.

$$\therefore m = \frac{3pV}{v^2}$$

---

h) $M = k\sqrt{L_1 L_2}$ $[L_1]$

Solution

$$M = k\sqrt{L_1 L_2}$$

Step 1: Square both sides and simplify

$$M^2 = \left(k\sqrt{L_1 L_2}\right)^2$$
$$M^2 = k^2 L_1 L_2$$

Step 2: Divide both sides by $k^2 L_2$ and simplify

$$\frac{M^2}{k^2 L_2} = \frac{k^2 L_1 L_2}{k^2 L_2}$$

which implies that

$$\frac{M^2}{k^2 L_2} = L_1$$

$$\therefore L_1 = \frac{M^2}{k^2 L_2}$$

---

i) $e = mgh + \frac{1}{2}mv^2$ $[m]$

Solution

$$e = mgh + \frac{1}{2}mv^2$$

Step 1: Multiply each term on both sides by 2 and simplify

$$2 \times e = 2 \times mgh + 2 \times \frac{1}{2}mv^2$$

which implies that

$$2e = 2mgh + mv^2$$

Step 2: Factorise the RHS by taking m out

$$2e = m\left(2gh + v^2\right)$$

Step 3: Divide both sides by $2gh + v^2$ and simplify

$$\frac{2e}{2gh + v^2} = \frac{m\left(2gh + v^2\right)}{2gh + v^2}$$

which implies that

$$\frac{2e}{2gh + v^2} = m$$

$$\therefore m = \frac{2e}{2gh + v^2}$$

Let's try a final set of examples.

Example 26

Transpose the following formulas to make the variable (or letter) in the square brackets the subject of the formula.

a) $T = 2\pi\sqrt{\frac{m}{k}}$ $[k]$

b) $\frac{a}{\sin A} = \frac{b}{\sin B}$ $[B]$

c) $f = \frac{1}{2\pi\sqrt{LC}}$ $[C]$

d) $\frac{1}{f} = \frac{1}{u} + \frac{1}{v}$ $[v]$

e) $C = \frac{\epsilon A(n-1)}{d}$ $[n]$

f) $F = \frac{Q_1 Q_2}{4\pi\epsilon r^2}$ $[r]$

g) $\frac{1}{R} = \frac{1}{R_1} + \frac{1}{R_2} + \frac{1}{R_3}$ $[R]$

h) $\frac{mv^2}{r} = QvB$ $[r]$

What did you get? Find the solution below to double-check your answer.

Solution to Example 26

a) $T = 2\pi\sqrt{\dfrac{m}{k}}$ $[k]$

Solution

$$T = 2\pi\sqrt{\frac{m}{k}}$$

Step 1: Divide both sides by 2π and simplify

$$\frac{T}{2\pi} = \frac{2\pi\sqrt{\frac{m}{k}}}{2\pi}$$

which implies that

$$\frac{T}{2\pi} = \sqrt{\frac{m}{k}}$$

Step 2: Square both sides to remove the square root of $\dfrac{m}{k}$ and simplify

$$\left(\frac{T}{2\pi}\right)^2 = \left(\sqrt{\frac{m}{k}}\right)^2$$

which implies that

$$\frac{T^2}{(2\pi)^2} = \frac{m}{k}$$

$$\frac{T^2}{4\pi^2} = \frac{m}{k}$$

Step 3: Take the reciprocal of both sides and simplify

To 'take the reciprocal' is to 'flip' both sides of the equation, by making the denominators to become the numerators. Thus, we have:

$$\frac{4\pi^2}{T^2} = \frac{k}{m}$$

NOTE

The approach we've taken here is a bit easier than multiplying by k and then dividing by $\dfrac{T^2}{4\pi^2}$ either at one go or in steps. Any time we see that our target variable is the denominator of a single fraction on one side of a formula and the other side is also a single fraction, we can take the reciprocal of both sides.

Step 4: Multiply both sides by m and simplify

$$\frac{4\pi^2}{T^2} \times m = \frac{k}{m} \times m$$

which implies

$$\frac{4\pi^2}{T^2}m = k$$

$$\therefore k = \frac{4\pi^2 m}{T^2}$$

b) $\frac{a}{\sin A} = \frac{b}{\sin B}$ [B]

Solution

$$\frac{a}{\sin A} = \frac{b}{\sin B}$$

Step 1: Take the reciprocal of both sides

$$\frac{\sin A}{a} = \frac{\sin B}{b}$$

Step 2: Multiply both sides by b and simplify

$$\frac{\sin A}{a} \times b = \frac{\sin B}{b} \times b$$

which implies that

$$\frac{b \sin A}{a} = \sin B$$

Step 3: Take the inverse sine of both sides and simplify

$$\sin^{-1}\left(\frac{b \sin A}{a}\right) = \sin^{-1}(\sin B)$$

which implies that

$$\sin^{-1}\left(\frac{b \sin A}{a}\right) = B$$

$$\therefore B = \sin^{-1}\left(\frac{b \sin A}{a}\right)$$

c) $f = \frac{1}{2\pi\sqrt{LC}}$ [C]

Solution

$$f = \frac{1}{2\pi\sqrt{LC}}$$

Step 1: Multiply both sides by 2π and simplify

$$f \times 2\pi = \frac{1}{2\pi\sqrt{LC}} \times 2\pi$$

which implies that

$$2\pi f = \frac{1}{\sqrt{LC}}$$

Step 2: Take the reciprocal of both sides

$$\frac{1}{2\pi f} = \sqrt{LC}$$

Step 3: Square both sides to remove the square root of LC and simplify

$$\left(\frac{1}{2\pi f}\right)^2 = \left(\sqrt{LC}\right)^2$$

which implies that

$$\frac{1}{4\pi^2 f^2} = LC$$

Step 4: Divide both sides by L and simplify

$$\frac{1}{4\pi^2 f^2} \div L = \frac{LC}{L}$$

which implies that

$$\frac{1}{4\pi^2 f^2} \times \frac{1}{L} = C$$

$$\therefore C = \frac{1}{4\pi^2 f^2 L}$$

---

d) $\frac{1}{f} = \frac{1}{u} + \frac{1}{v}$ $[v]$

Solution

$$\frac{1}{f} = \frac{1}{u} + \frac{1}{v}$$

Step 1: Move $\frac{1}{u}$ to the LHS

$$\frac{1}{f} - \frac{1}{u} = \frac{1}{v}$$

Step 2: Simplify the LHS using cross-multiplication

$$\frac{u - f}{fu} = \frac{1}{v}$$

Step 3: Take the reciprocal of both sides

$$\frac{fu}{u - f} = v$$

$$\therefore v = \frac{fu}{u - f}$$

---

e) $C = \frac{\varepsilon A(n-1)}{d}$ $[n]$

Solution

$$C = \frac{\varepsilon A(n - 1)}{d}$$

Step 1: Multiply both sides by d and simplify

$$C \times d = \frac{\epsilon A (n-1)}{d} \times d$$

which implies that

$$Cd = \epsilon A(n-1)$$

Step 2: Divide both sides by ϵA

$$\frac{Cd}{\epsilon A} = n - 1$$

Step 3: Add 1 to both sides

$$\frac{Cd}{\epsilon A} + 1 = n$$

Optionally, we can simplify the LHS as

$$\frac{Cd}{\epsilon A} + \frac{\epsilon A}{\epsilon A} = n$$

$$\frac{Cd + \epsilon A}{\epsilon A} = n$$

$$\therefore n = \frac{Cd + \epsilon A}{\epsilon A}$$

---

f) $F = \frac{Q_1 Q_2}{4\pi \epsilon r^2}$ $[r]$

Solution

$$F = \frac{Q_1 Q_2}{4\pi \epsilon r^2}$$

Step 1: Multiply both sides by $4\pi\epsilon$ and simplify

$$F \times 4\pi\epsilon = \frac{Q_1 Q_2}{4\pi \epsilon r^2} \times 4\pi\epsilon$$

which implies that

$$4\pi\epsilon F = \frac{Q_1 Q_2}{r^2}$$

Step 2: Take the reciprocal of both sides

$$\frac{1}{4\pi\epsilon F} = \frac{r^2}{Q_1 Q_2}$$

Step 3: Multiply both sides by Q_1Q_2 and simplify

$$\frac{1}{4\pi\epsilon F} \times Q_1Q_2 = \frac{r^2}{Q_1Q_2} \times Q_1Q_2$$

which implies that

$$\frac{Q_1Q_2}{4\pi\epsilon F} = r^2$$

Step 4: Take the square root of both sides

$$\sqrt{\frac{Q_1Q_2}{4\pi\epsilon F}} = \sqrt{r^2}$$

$$\sqrt{\frac{Q_1Q_2}{4\pi\epsilon F}} = r$$

Let's apply the radical separately to the numerator and denominator, which implies

$$\frac{\sqrt{Q_1Q_2}}{\sqrt{4\pi\epsilon F}} = r$$

$$\frac{\sqrt{Q_1Q_2}}{2\sqrt{\pi\epsilon F}} = r$$

$$\frac{1}{2}\sqrt{\frac{Q_1Q_2}{\pi\epsilon F}} = r$$

$$\therefore r = \frac{1}{2}\sqrt{\frac{Q_1Q_2}{\pi\epsilon F}}$$

---

g) $\frac{1}{R} = \frac{1}{R_1} + \frac{1}{R_2} + \frac{1}{R_3}$ **[R]**

Solution

$$\frac{1}{R} = \frac{1}{R_1} + \frac{1}{R_2} + \frac{1}{R_3}$$

Step 1: The LCM of the RHS is $R_1R_2R_3$, so let's multiply each term of the above by the LCM.

$$\frac{1}{R} \times R_1R_2R_3 = \frac{1}{R_1} \times R_1R_2R_3 + \frac{1}{R_2} \times R_1R_2R_3 + \frac{1}{R_3} \times R_1R_2R_3$$

Step 2: Simplify the above

$$\frac{R_1R_2R_3}{R} = \frac{R_1R_2R_3}{R_1} + \frac{R_1R_2R_3}{R_2} + \frac{R_1R_2R_3}{R_3}$$

Thus, we have

$$\frac{R_1R_2R_3}{R} = R_2R_3 + R_1R_3 + R_1R_2$$

Step 3: Take the reciprocal of both sides.

$$\frac{R}{R_1R_2R_3} = \frac{1}{R_2R_3 + R_1R_3 + R_1R_2}$$

Step 4: Multiply both sides by $R_1R_2R_3$ and simplify

$$\frac{R}{R_1R_2R_3} \times R_1R_2R_3 = \frac{1}{R_2R_3 + R_1R_3 + R_1R_2} \times R_1R_2R_3$$

Thus, we have

$$R = \frac{R_1R_2R_3}{R_2R_3 + R_1R_3 + R_1R_2}$$

$$\therefore R = \frac{R_1R_2R_3}{R_2R_3 + R_1R_3 + R_1R_2}$$

NOTE

Often this type of transposition is achieved by simplifying the RHS using the LCM and then taking the reciprocal, but we decided to take a different approach.

---

h) $\frac{mv^2}{r} = QvB$ $[r]$

Solution

$$\frac{mv^2}{r} = QvB$$

Step 1: Take the reciprocal of both sides

$$\frac{r}{mv^2} = \frac{1}{QvB}$$

Step 2: Multiply both sides by mv^2

$$\frac{r}{mv^2} \times mv^2 = \frac{1}{QvB} \times mv^2$$

which implies that

$$r = \frac{mv^2}{QvB}$$

Step 3: Simplify the RHS

Given that

$$r = \frac{mv^2}{QvB}$$

Dividing both numerator and denominator by v, we have

$$r = \frac{mv}{QB}$$

$$\therefore r = \frac{mv}{QB}$$

3.9 CHAPTER SUMMARY

1) Algebra is a branch of mathematics involving the manipulation of variables and numbers.

2) The **variable** (also called an '**unknown**') is a quantity that we cannot assign a fixed numerical value to and generally represents quantities (time, amount, force, height, etc.) that change in values.

3) There are two types of variables: dependent and independent variables.

4) There are certain letters or symbols that have fixed numerical values and are referred to as **constants** as opposed to variables. These include *pi* (π), acceleration due to gravity (g), and Euler's number (e) among others.

5) Algebra can be used to represent a situation; this is called an **algebraic expression**.

6) A **binomial expression** is an algebraic expression with two terms.

7) The **coefficient** of a term is a number (or a numerical value) that appears in front of a variable.

8) Opening the bracket(s) is sometimes referred to as '**expanding the brackets**' or '**multiplying out**' or '**removing the brackets**'.

9) Factorisation is the inverse of opening the brackets.

10) Difference of two squares implies '**a squared term minus another squared term**'.

11) A quadratic expression in one variable is an expression of the form $ax^2 + bx + c$ where a, b, and c are any rational numbers. The value of b or c can be zero, however a must not be zero.

12) A formula usually takes the form of $y = f(x)$, where y is called the subject of the formula.

****

3.10 FURTHER PRACTICE

To access complementary contents, including additional exercises, please go to www.dszak.com.

4 Coordinate Geometry I

Learning Outcomes

Once you have studied the content of this chapter, you should be able to:

- Represent a point in a coordinate form
- Determine the mid-point of a line
- Calculate the distance between two points
- Determine the mid-point of two points along a straight line
- Calculate the gradient of a straight line
- Understand the relationship between a line and its gradient and intercept
- Write the equation of a straight line

4.1 INTRODUCTION

A coordinate system is a way of specifying the position of an object on a line (or a one-dimension), on a surface (or a two-dimension), and in a space (or a three-dimension). Coordinate systems are all around us, we use them, and we even live them. For example, you may want to describe the position of a passenger or customer in a straight-line queue or that of a student in a lecture theatre with chairs arranged in rows and columns or that of a book on a library shelf or a product in a big supermarket. All these are coordinate systems. We need to be able to specify a position precisely, especially in computing where the memory address should be accurately referenced. The way to do this is what we are about to learn in this and succeeding chapter.

4.2 CARTESIAN COORDINATE SYSTEM

It consists of three axes labelled x, y, and z. When we are interested in the position of a two-dimensional item, we call this a plane surface or simply a plane. If we identify it by specifying its x and y coordinates, we call this the Cartesian coordinate system. By convention, we are required to specify the x coordinate followed by the y coordinate, both included in a bracket. In other words, a position on a plane is specified by a pair of coordinates as:

$$(x, y)$$

where x is the length of a perpendicular line from the y-axis to the position and y is the length of a perpendicular line from the x-axis to the position.

For example, coordinates given as $(3, 2)$ show that the point is 3 units from the y-axis and 2 units from the x-axis. Alternatively, we can say that the item is a point where $x = 3$ and $y = 2$, i.e., third column

DOI:10.1201/9781003027928-4

and second row. While columns and rows can only be positive in practice, however, in Mathematics, both positive and negative coordinates are possible, as shown in Figure 4.1.

Note that the x-axis is the horizontal line and the y-axis is the vertical line. The point at which the two lines or axes cross is called the **Origin** and has the coordinates **(0, 0)**. The y-axis is a vertical line of symmetry such that the numbers to the right are mirrored to the left of the line. The numbers to the right of the y-axis are positive whilst those to its left are considered negative. Similarly, the x-axis is a horizontal line of symmetry; the numbers above the line are considered positive and the ones below are negative.

We can see that the **x−y** plane, as it is technically referred to (because it is two-dimensional consisting of the x-axis and y-axis), is divided into four. Each part is called a **quadrant** and numbered counter clockwise from the top right. Although the x- and y-axes are used, but the principle can be applied to any two related variables. For example, voltage (V) and current (I) become a **voltage–current** plane (or graph) or simply V–I plane (or graph), likewise stress and strain are called a **stress–strain** plane (or graph), and velocity and time are known as a **velocity–time** plane (or graph).

We would like to finish this section by explaining how to write the coordinates of a given point. Follow these steps to write the coordinates of point P given in Figure 4.2:

Step 1: Draw a vertical line parallel to the y-axis until you reach the horizontal line or x-axis.
Step 2: Read the number on the x-axis, taking into consideration the sign. This is your x-coordinate.
Step 3: Draw a horizontal line parallel to the x-axis until it reaches the vertical line or y-axis.
Step 4: Read the number on the y-axis, taking into consideration the sign. This is your y-coordinate.
Step 5: Write this in the standard form as (x_1, y_1). This is the coordinates of point P expressed in the Cartesian format.

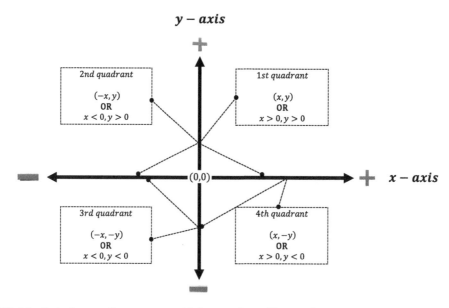

FIGURE 4.1 Cartesian coordinate system with four quadrants illustrated.

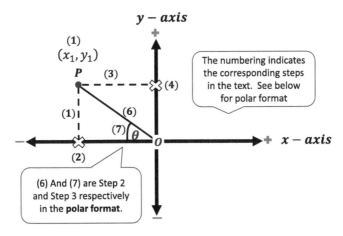

FIGURE 4.2 Locating the coordinates of a given point.

There is another format called the **polar format**, which can be written by following these steps:

Step 1: Find the length from the origin $(0,0)$ to point P using any suitable method of finding the distance between two points (as explained in Section 4.4).
Step 2: Find the angle between the line OP and the x-axis.
Step 3: Write the position P in polar form as (OP, θ).

Great! Let's try some examples.

Example 1

Straight line equations were used to draw the two lines shown on the graph below.

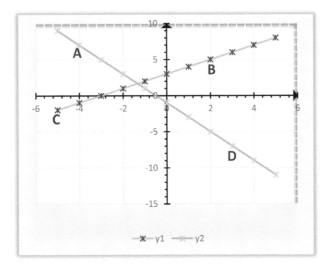

FIGURE 4.3 Example 1.

Using Figure 4.3, determine the coordinates of the following points:

a) *A*   **b)** *B*   **c)** *C*   **d)** *D*

What did you get? Find the solution below to double-check your answer.

Solution to Example 1

HINT

In this example, we need to follow the five steps described to obtain the coordinates of the stated point.

a) Point *A*
Solution

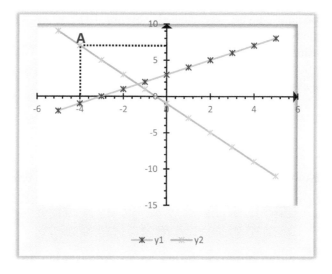

FIGURE 4.4 Solution to Example 1(a).

The coordinates of point *A* (see Figure 4.4) are:

$$A\ (-4, 7)$$

b) Point B
Solution

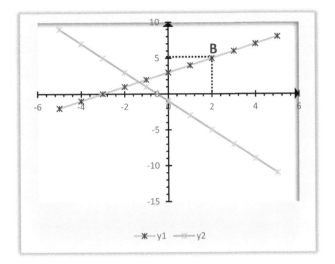

FIGURE 4.5   Solution to Example 1(b).

The coordinates of point B (see Figure 4.5) are:

$$B \ (2, 5)$$

c) Point C
Solution

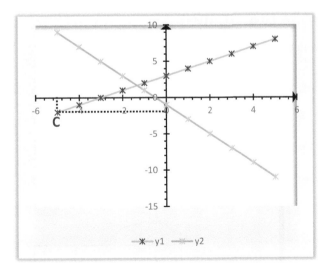

FIGURE 4.6   Solution to Example 1(c).

The coordinates of point C (see Figure 4.6) are:

$$C \ (-5, -2)$$

d) Point D
Solution

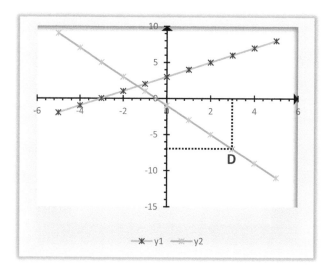

FIGURE 4.7   Solution to Example 1(d).

The coordinates of point D (see Figure 4.7) are:

$$D\ (3, -7)$$

4.3 MID-POINT OF A LINE SEGMENT

Consider a line AB with the coordinates of its end points as shown in Figure 4.8.

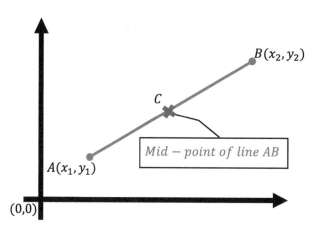

FIGURE 4.8   Mid-point of a line segment illustrated.

Imagine we are interested in the coordinates of the mid-way between A and B. Let's call this point C and its coordinates represented by (x_m, y_m). The coordinates of point C are related to those of points A and B as:

$$(x_m, y_m) = \left(\frac{x_1 + x_2}{2}, \frac{y_1 + y_2}{2} \right) \qquad (4.1)$$

In other words, the coordinates of the mid-point of line AB are the average of the corresponding coordinates of the line segments obtained as follows:

Step 1: Add the two x-coordinates.
Step 2: Divide the result of step 1 by two.
Step 3: Repeat the same for the two y-coordinates.

Good! Let's try some examples.

Example 2

Determine the coordinates of the mid-points joining a line AB in the following cases:

a) $A(6, \ 3)$ **and** $B(-4, \ 7)$

b) $A(\sqrt{3}, -\sqrt{6})$ **and** $B(\sqrt{3}, 5\sqrt{6})$

c) $A(3a, \ -15b)$ **and** $B(7a, \ 11b)$

What did you get? Find the solution below to double-check your answer.

Solution to Example 2

a) $A(6,3)$ **and** $B(-4,7)$
Solution
Let's label the points as:

$$A(6,3) = (x_1, y_1)$$

and

$$B(-4,7) = (x_2, y_2)$$

Therefore, the mid-point is

$$\begin{aligned}(x_m, y_m) &= \left(\frac{x_1 + x_2}{2}, \frac{y_1 + y_2}{2} \right) \\ &= \left(\frac{6-4}{2}, \frac{3+7}{2} \right) \\ &= \left(\frac{2}{2}, \frac{10}{2} \right) \\ &= (\mathbf{1,5})\end{aligned}$$

b) $A(\sqrt{3}, -\sqrt{6})$ **and** $B(\sqrt{3}, 5\sqrt{6})$

Solution

Let's label the points as:

$$A\left(\sqrt{3}, -\sqrt{6}\right) = (x_1, y_1)$$

and

$$B\left(\sqrt{3}, 5\sqrt{6}\right) = (x_2, y_2)$$

Therefore, the mid-point is

$$
\begin{aligned}
(x_m, y_m) &= \left(\frac{x_1 + x_2}{2}, \frac{y_1 + y_2}{2}\right) \\
&= \left(\frac{\sqrt{3} + \sqrt{3}}{2}, \frac{-\sqrt{6} + 5\sqrt{6}}{2}\right) \\
&= \left(\frac{2\sqrt{3}}{2}, \frac{4\sqrt{6}}{2}\right) \\
&= \left(\sqrt{3}, 2\sqrt{6}\right)
\end{aligned}
$$

---

c) $A(3a, \ -15b)$ **and** $B(7a, \ 11b)$

Solution

Let's label the points as:

$$A\left(3a, \ -15b\right) = (x_1, y_1)$$

and

$$B\left(7a, \ 11b\right) = (x_2, y_2)$$

Therefore, the mid-point is

$$
\begin{aligned}
\left(\frac{x_1 + x_2}{2}, \frac{y_1 + y_2}{2}\right) &= \left(\frac{3a + 7a}{2}, \frac{-15b + 11b}{2}\right) \\
&= \left(\frac{10a}{2}, \frac{-4b}{2}\right) \\
&= (5a, -2b)
\end{aligned}
$$

Good stuff there! Before we move on, let's try another example.

Example 3

P and Q are two points on the circumference of a circle and line PQ is the diameter of the circle with centre coordinates of $C(0, \ -2)$. Determine the coordinates of point Q, given that the coordinates of P are $(5, \ 7)$.

What did you get? Find the solution below to double-check your answer.

Solution to Example 3

Let's represent the coordinates of Q as:

$$Q(x_q, y_q)$$

Let's label the points as:

$$P(5, 7) = (x_1, y_1)$$

and

$$Q(x_q, y_q) = (x_2, y_2)$$

Therefore, the mid-point is

$$\left(\frac{x_1 + x_2}{2}, \frac{y_1 + y_2}{2}\right) = \left(\frac{5 + x_q}{2}, \frac{7 + y_q}{2}\right)$$

Thus

$$\left(\frac{5 + x_q}{2}, \frac{7 + y_q}{2}\right) = (0, -2)$$

We then have that

$$\frac{5 + x_q}{2} = 0$$
$$5 + x_q = 0$$
$$x_q = -5$$

Similarly,

$$\frac{7 + y_q}{2} = -2$$
$$7 + y_q = -4$$
$$y_q = -4 - 7$$
$$y_q = -11$$

Therefore, the coordinates of Q are:

$$(-5, -11)$$

4.4 DISTANCE BETWEEN TWO POINTS

Suppose we are given a line AB with the coordinates of its end points as shown in Figure 4.9.

The distance between points A and B (or the length of line AB) given as d can be found using:

$$\boxed{d = \sqrt{(x_2 - x_1)^2 + (y_2 - y_1)^2}}$$ (4.2)

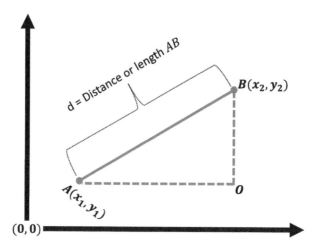

FIGURE 4.9 Distance between two points illustrated.

Note the following about the above equation:

Note 1 The equation is very similar to Pythagoras' theorem. In fact, it is the theorem in action, since line AB is the hypotenuse of the triangle AOB.

Note 2 As the square of any number is always positive, it is less relevant which of the x- and y-coordinates is subtracted from the other. In other words, using $d = \sqrt{(x_1 - x_2)^2 + (y_2 - y_1)^2}$ will be equally fine.

Let's try some examples.

Example 4

Calculate the distance between these pairs of points joining a line AB. Present your answer correct to 1 d.p. where applicable.

    a) $A(-9, \ 11)$ **and** $B(-13, \ -1)$

    b) $A(a, -ab)$ **and** $B(3a, 4ab)$

    c) $A(-\sqrt{2}, \ -2\sqrt{6})$ **and** $B(\sqrt{2}, \ \sqrt{6})$

What did you get? Find the solution below to double-check your answer.

Solution to Example 4

a) $A(-9, \ 11)$ **and** $B(-13, \ -1)$
Solution
Let's label the points as:

$$A\,(-9, \ 11) = (x_1, y_1)$$

and

$$B(-13, \ -1) = (x_2, y_2)$$

Therefore, the length of AB is:

$$d = \sqrt{(x_2 - x_1)^2 + (y_2 - y_1)^2}$$

$$= \sqrt{(-13 - (-9))^2 + (-1 - 11)^2}$$

$$= \sqrt{(-13 + 9)^2 + (-12)^2} = \sqrt{(-4)^2 + (-12)^2}$$

$$= \sqrt{16 + 144} = \sqrt{160}$$

$$\therefore d = 4\sqrt{10} = 12.6$$

---

b) $A(a, -ab)$ **and** $B(3a, 4ab)$
Solution
Let's label the points as:

$$A(a, -ab) = (x_1, y_1)$$

and

$$B(3a, 4ab) = (x_2, y_2)$$

Therefore, the length of AB is:

$$d = \sqrt{(x_2 - x_1)^2 + (y_2 - y_1)^2}$$

$$= \sqrt{(3a - a)^2 + (4ab - (-ab))^2}$$

$$= \sqrt{(2a)^2 + (4ab + ab)^2}$$

$$= \sqrt{(2a)^2 + (5ab)^2} = \sqrt{4a^2 + 25a^2b^2}$$

$$= \sqrt{a^2(4 + 25b^2)}$$

$$\therefore d = a\sqrt{(4 + 25b^2)}$$

---

c) $A(-\sqrt{2}, \ -2\sqrt{6})$ **and** $B(\sqrt{2}, \ \sqrt{6})$
Solution
Let's label the points as:

$$A\left(-\sqrt{2}, \ -2\sqrt{6}\right) = (x_1, y_1)$$

and

$$B\left(\sqrt{2}, \ \sqrt{6}\right) = (x_2, y_2)$$

Remember we said that the order of the coordinates can be reversed. This is very useful for manipulation. Therefore, the length of *AB* is

$$d = \sqrt{(x_2 - x_1)^2 + (y_2 - y_1)^2}$$
$$= \sqrt{\left(\sqrt{2} + \sqrt{2}\right)^2 + \left(\sqrt{6} + 2\sqrt{6}\right)^2}$$
$$= \sqrt{\left(2\sqrt{2}\right)^2 + \left(3\sqrt{6}\right)^2}$$
$$= \sqrt{4\,(2) + 9\,(6)} = \sqrt{8 + 54}$$
$$= \sqrt{62}$$
$$\therefore d = \sqrt{62} = 7.9$$

Let's try another example.

Example 5

Line *QS* is the diameter of a circle *C*, where the coordinates of points *Q* and *S* are $(-5, -b)$ and $(-1, b)$ respectively. Calculate the radius of the circle, giving the answer in terms of *b*.

What did you get? Find the solution below to double-check your answer.

Solution to Example 5

HINT

A sketch will be very helpful to visualise this problem and the next one and they have been provided accordingly.

Let's label the points (Figure 4.10) as:

$$Q(-5, -b) = (x_1, y_1)$$

and

$$S(-1, b) = (x_2, y_2)$$

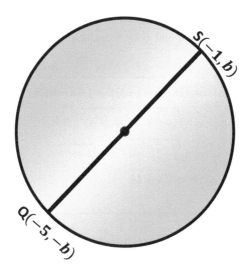

FIGURE 4.10 Solution to Example 5.

Therefore, the length of QS is

$$d = \sqrt{(x_2 - x_1)^2 + (y_2 - y_1)^2}$$

$$= \sqrt{(-5 - (-1))^2 + (b - (-b))^2}$$

$$= \sqrt{(-5 + 1)^2 + (b + b)^2}$$

$$= \sqrt{(-4)^2 + (2b)^2} = \sqrt{16 + 4b^2}$$

$$= \sqrt{4(4 + b^2)} = 2\sqrt{(4 + b^2)}$$

$$\therefore d = 2\sqrt{4 + b^2}$$

Radius is half of the diameter hence,

$$\therefore r = \frac{1}{2}d = \sqrt{4 + b^2}$$

ALTERNATIVE METHOD

We can determine the mid-point of the diameter QS as

$$\left(\frac{x_1 + x_2}{2}, \frac{y_1 + y_2}{2}\right) = \left(\frac{-5 - 1}{2}, \frac{-b + b}{2}\right)$$

$$= \left(\frac{-6}{2}, \frac{0}{2}\right) = (-3, 0)$$

Now use the mid-point with either point Q or S to calculate the radius. Using the mid-point and point Q, we have:

$$r = \sqrt{(-5 - (-3))^2 + (-b - 0)^2}$$
$$= \sqrt{(-5 + 3)^2 + (-b)^2}$$
$$= \sqrt{(-2)^2 + b^2} = \sqrt{4 + b^2}$$
$$\therefore r = \sqrt{4 + b^2}$$

One more example to try before we bid farewell to distance between two points.

Example 6

$C(2, -2)$ is the centre coordinates of a circle with radius 12.5 units. Show that $D(-10, 1.5)$ is a point on the circumference of the circle.

What did you get? Find the solution below to double-check your answer.

Solution to Example 6

If D is a point on the circumference of the circle, then length CD is equal to the radius of the circle (Figure 4.11).

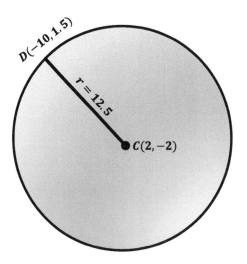

FIGURE 4.11   Solution to Example 6.

Let's label the points as:

$$C(2, -2) = (x_1, y_1)$$

and

$$D(-10, 1.5) = (x_2, y_2)$$

Therefore, the length of CD is

$$\begin{aligned}
d &= \sqrt{(x_2 - x_1)^2 + (y_2 - y_1)^2} \\
&= \sqrt{(-10 - 2)^2 + (1.5 - (-2))^2} \\
&= \sqrt{(-12)^2 + (3.5)^2} \\
&= \sqrt{144 + 12.25} = \sqrt{156.25} \\
&= \frac{25}{2} \\
\therefore r &= 12.5
\end{aligned}$$

Thus, point D is on the circumference of the circle.

4.5 SLOPE OF A LINE

Slope, denoted by m, is also called gradient. When used on its own, it generally refers to the slope of a straight line. We can think of slope as a measure of the steepness of a line, with a numerical value. The larger the value of m, the greater the steepness or tendency to fall. In a more loosed usage, it is the tendency to fall when standing on a straight line or road. A slope of m implies that for every one unit of translation in the horizontal direction, there will be m units of translation in the vertical direction.

Slope can be positive or negative. In Figure 4.12, l_1 and l_3 have positive slopes, (i.e., $m = +ve$) whilst l_2 and l_4 have negative slopes (i.e., $m = -ve$). As a principle, a line that is pointing from the bottom left to the top right (or vice versa) will have a positive slope, since an increase in the x-value results in a corresponding increase in y-value. On the other hand, a line will have a negative slope if it points from the bottom right to the top left (or vice versa), such that an increase in x-value will mean a decrease in y-value and vice versa.

A horizontal line has a slope of zero as there is no 'steepness' associated with it. On the other hand, a vertical line has an infinite slope, because the tendency to fall is forbiddingly large (Figure 4.13). As the horizontal line moves towards the vertical line, the slope increases accordingly, as further illustrated in Figure 4.13.

The slope of a straight line is given by:

$$m = \frac{\text{change in vertical}}{\text{change in horizontal}} = \frac{\text{difference in } y \text{ coordinates}}{\text{difference in } x \text{ coordinates}}$$

If the vertical line is labelled as the y-axis and the horizontal line the x-axis, then

$$m = \frac{\textbf{change in } y}{\textbf{change in } x}$$

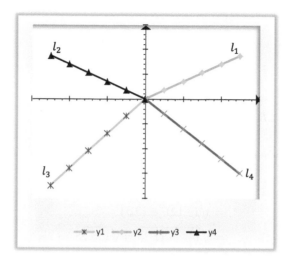

FIGURE 4.12 Positive and negative slopes of a line illustrated.

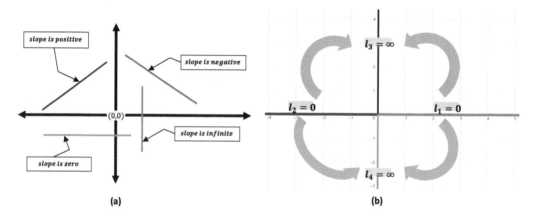

FIGURE 4.13 Slope of a line: (a) four situations showing – a slope of zero (horizontal line), a slope of infinity (a vertical line), a positive slope and a negative slope, (b) shows how the slope of a line changes from zero to infinity (as it is turned from being parallel to the x-axis to being parallel to the y-axis in each of the four quadrants).

This is commonly referred to as '**rise over run**' and can be written compactly as:

$$m = \frac{\Delta y}{\Delta x} = \frac{y_2 - y_1}{x_2 - x_1} \qquad (4.3)$$

We also have that

$$m = \tan\theta \qquad (4.4)$$

where θ is the angle the line makes with the positive x-axis as shown in Figure 4.14.

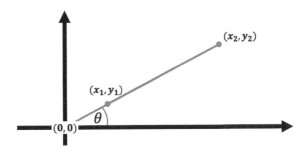

FIGURE 4.14 Coordinates of two points used to calculate the slope illustrated.

The following should be noted about the slope of a line:

Note 1 Any quantity (force, stress, etc.) can represent *y*, and the same can be said about *x*.
Note 2 Two points on a line are required to determine a gradient.
Note 3 The coordinates of the points can be represented as (x_1, y_1) and (x_2, y_2). Any of the two points can be taken as 1 or 2.
Note 4 When calculating the slope, ensure that exact corresponding values are used. In other words, x_2 and y_2 do not necessarily stand for the bigger number in each case. This will become clearer in the worked examples.

Let's try a couple of examples.

Example 7

Calculate the gradient of a line joining points $A(1, -2)$ and $B(-7,\ 8)$.

What did you get? Find the solution below to double-check your answer.

Solution to Example 7

Let's label the points as:

$$A(1,\ -2) = (x_1, y_1)$$

and

$$B(-7,\ 8) = (x_2, y_2)$$

Now let's find the slope

$$m = \frac{y_2 - y_1}{x_2 - x_1}$$

$$= \frac{8 - (-2)}{-7 - 1} = \frac{8 + 2}{-8}$$

$$m = \frac{10}{-8}$$

$$\therefore m = -\frac{5}{4}$$

Example 8

The gradient of a line which passes through points $A(b, -1)$ and $B(3b, 5)$ is 2. Determine the value of b.

What did you get? Find the solution below to double-check your answer.

Solution to Example 8

Let's label the points as:

$$A(b, -1) = (x_1, y_1)$$

and

$$B(3b, 5) = (x_2, y_2)$$

Now let's determine the slope

$$m = \frac{y_2 - y_1}{x_2 - x_1}$$
$$= \frac{5 - (-1)}{3b - b} = \frac{5 + 1}{2b}$$

Therefore,

$$2 = \frac{6}{2b}$$
$$4b = 6$$
$$b = \frac{6}{4} = \frac{3}{2}$$
$$\therefore b = 1.5$$

4.6 SLOPE OF A CURVE

We've just looked at gradient with a focus on straight lines. A pertinent question here is what about the slope of curves, such as the ones shown in Figure 4.15. Well, as earlier noted, slope is associated with steepness, which is the same at any point along a straight line. However, for a curve, it varies from one point to another. As a result, it is only valid to determine the slope of a curve at a particular point on the curve.

It therefore becomes necessary to be able to determine the slope of a curve. You will be pleased to learn that there are ways to deal with this including calculus, which is covered in *Advanced Mathematics for Engineers and Scientists with Worked Examples* by the same author.

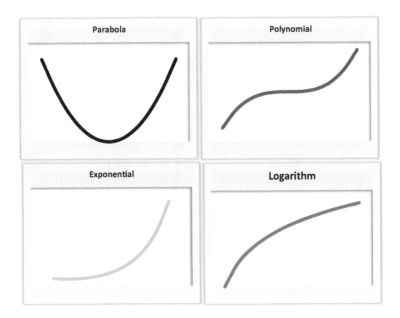

FIGURE 4.15 Slope of curves.

The focus here is an extension to the approach used to determine the slope of straight lines, but including these additional steps:

Step 1: Identify the point along the curve whose slope is required.

Step 2: Locate and clearly mark this point.

Step 3: Draw a tangent to the curve at the point. A tangent to a curve at a point is a straight line perpendicular to the curve at this point.

Step 4: Use any suitable method to find the gradient of the tangent line drawn. The commonly used approach is to find two coordinates and then use the formula $\frac{y_2 - y_1}{x_2 - x_1}$.

Step 5: The value in step 4 is the slope of the curve at the stated point.

Note that the answer that will be obtained by following the above steps is only an approximation. It largely depends on how best the tangent is drawn and the scale of the chart. but it is covered in *Advanced Mathematics for Engineers and Scientists with Worked Examples* by the same author.

Let's try an example.

Example 9

The graph in Figure 4.16 is the curve of a quadratic function given by $y = x^2 + 3x - 5$. Determine the slope of the curve at a point when $x = 2$.

What did you get? Find the solution below to double-check your answer.

Solution to Example 9

We've drawn the tangent at point $x = 2$ and identified two points $A\,(2, 5)$ and $B\,(4, 19)$ as shown in Figure 4.17.

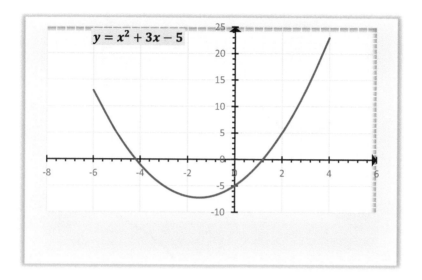

FIGURE 4.16 Example 9.

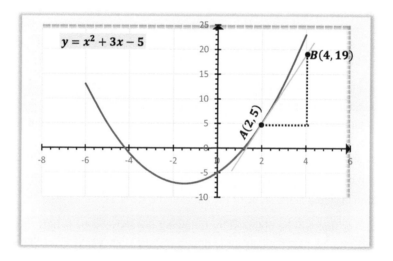

FIGURE 4.17 Solution to Example 9.

Using the coordinates of the points $A(2,5)$ and $B(4,19)$, the slope of the tangent is

$$m = \frac{y_2 - y_1}{x_2 - x_1}$$

$$= \frac{19 - 5}{4 - 2} = \frac{14}{2}$$

$$\therefore m = 7$$

NOTE

The slope can be calculated analytically using calculus, which is beyond the scope of this book as previously noted.

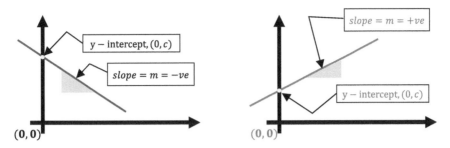

FIGURE 4.18 Equation of a line using the slope and y-intercept illustrated.

4.7 EQUATION OF A LINE

The equation of a straight line is characterised by three terms and represented as:

$$y = mx + c \qquad (4.5)$$

where

- m is the slope of the line (it is the number in front of x or any similar independent variable);
- c is the y-intercept (where the line crosses the vertical line or y-axis, as shown in Figure 4.18). This is the number that is not attached to x or any similar independent variable.

Alternatively

$$ax + by + d = 0 \qquad (4.6)$$

where a, b, c, d, and m are constant or rational numbers.

We have used variable d in the second equation above although c is generally used; this is to ensure that a variable is not repeated for clarity. Although the first equation is very popular, the second can be re-written in the format of the first one as:

$$y = -\frac{a}{b}x - \frac{d}{b} \qquad (4.7)$$

Comparing both equations, it can be seen that

$$m = -\frac{a}{b}$$

and

$$c = -\frac{d}{b}$$

The following should be noted about these equations:

Note 1 m and c are constants including zero.

Note 2 The power of x must be 1 or 0 to have a straight line. Ideally, a linear equation is characterised by the power of x being 1.

Note 3 If the term c is zero, the line will pass through the origin, $(0,0)$, such as in $y = 3x$ and $y = -5x$. If in addition, $m = \pm 1$, the line will bisect the quadrants as shown in Figure 4.19.

Note 4 If $mx = 0$ or $m = 0$, then the line is horizontal and passes through c or through coordinates $(0, c)$ such as in $y = 3$ and $y = -2$ (Figure 4.20).

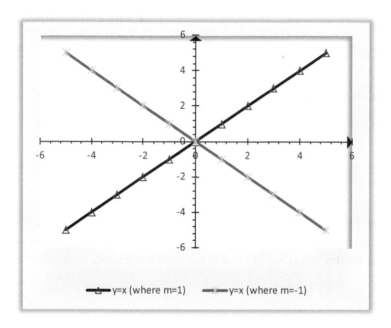

FIGURE 4.19 Illustrating when $m = \pm 1$ and $c = 0$.

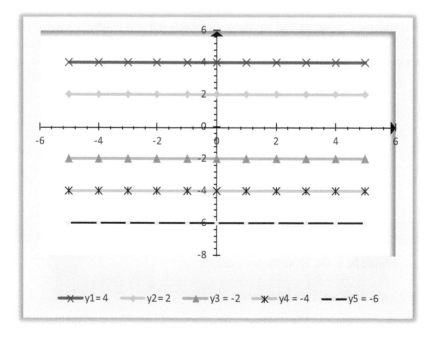

FIGURE 4.20 Illustrating when $m = 0$ and c is a real number.

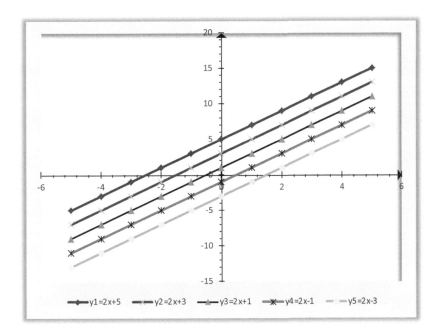

FIGURE 4.21 Illustrating when **m** is the same for two or more lines with different values of **c**.

Note 5 If the **y** term (or **y**) is zero, then the line is vertical, going through $-\frac{c}{m}$ or through the coordinates $(-\frac{c}{m}, 0)$. The slope is said to be infinite in this case, i.e., $m = \infty$.

Note 6 When two or more equations have the same value of **m** but different value of **c**, their lines will be parallel and cross the vertical line at different points. In Figure 4.21, the slope is the same ($m = 2$) while c varies.

Note 7 When two or more equations have the same value of **c** but different value of **m**, their lines will not be parallel but cross the vertical line at the same point. In Figure 4.22, the y-intercept is the same ($c = -6$) while **m** varies.

Note 8 For a pair of lines where the slope of one is the negative of the other, the graph will mirror each other. In Figure 4.23, each pair have the same y-intercept but the positive slope of one is the negative of the other.

Note 9 Finally, if a point is on the line, then it must satisfy the equation of the line. What this means is that if (x_n, y_n) is a point on a line defined by $y = mx + c$ then if we substitute x_n in the line equation, we will obtain y_n and vice versa.

Let's try some examples.

Example 10

Determine the gradient and y-intercept of each of the following lines:

a) $3y - 2x + 5 = 0$ **b)** $x - 6y - 6 = 7x - 3$ **c)** $11y + x - 5 = 13y - 3 - 2x$
d) $V = IR$ **e)** $F = kx$ **f)** $v = u + at$

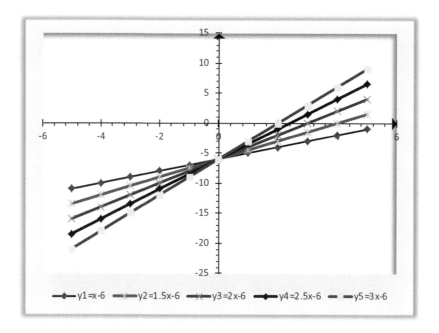

FIGURE 4.22 Illustrating when c is the same for two or more lines with different values of m.

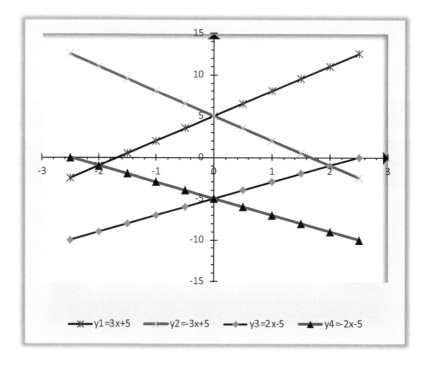

FIGURE 4.23 Illustrating when a pair of lines have slopes that differ only in sign and the same values of c.

What did you get? Find the solution below to double-check your answer.

Solution to Example 10

HINT

In this example, we need to write the equation in the form $y = mx + c$ and then compare with the given equation.

a) $3y - 2x + 5 = 0$
Solution

$$3y - 2x + 5 = 0$$
$$3y = 2x - 5$$
$$y = \frac{2}{3}x - \frac{5}{3}$$

Hence,

$$m = \frac{2}{3} \qquad c = -\frac{5}{3}$$

b) $x - 6y - 6 = 7x - 3$
Solution

$$x - 6y - 6 = 7x - 3$$
$$-6y = 7x - x - 3 + 6$$
$$-6y = 6x + 3$$
$$6y = -6x - 3$$
$$y = -x - \frac{1}{2}$$

Hence,

$$m = -1 \qquad c = -\frac{1}{2}$$

c) $11y + x - 5 = 13y - 3 - 2x$
Solution

$$11y + x - 5 = 13y - 3 - 2x$$
$$11y - 13y = -2x - x - 3 + 5$$
$$-2y = -3x + 2$$
$$2y = 3x - 2$$
$$y = \frac{3}{2}x - 1$$

Hence,

$$m = \frac{3}{2} \qquad c = -1$$

d) $V = IR$
Solution

$$V = IR$$
$$V = IR + 0$$

Hence,

$$\boldsymbol{m = R} \qquad \boldsymbol{c = 0}$$

NOTE
In this case, current (I) is the independent variable and resistance (R) is the constant. As such, R is our slope. This is called **Ohm's law**.

---

e) $F = kx$
Solution

$$F = kx$$
$$F = kx + 0$$

Hence,

$$\boldsymbol{m = k} \qquad \boldsymbol{c = 0}$$

NOTE
In this case, extension (x or sometimes given as e) is the independent variable and stiffness or elastic constant (k) is the constant. As such, k is our slope. This is called **Hooke's law**.

---

f) $v = u + at$
Solution
Write the equation in the form $y = mx + c$

$$v = u + at$$
$$v = at + u$$

Hence,

$$\boldsymbol{m = a} \qquad \boldsymbol{c = u}$$

NOTE
In this case, time (t) is the independent variable and a is the constant. As such, a is our slope and the initial velocity (u) is the y-intercept. This is one of the equations of linear motion used in mechanics.

Let's attempt another example.

Example 11

Show that the points $A(3,\ 2f)$, $B(-2,\ -3f)$, and $C(1,\ 0)$ are collinear, given that f represents a real number.

What did you get? Find the solution below to double-check your answer.

Solution to Example 11

If points A, B, and C are collinear (i.e., lie on the same straight line), then the gradient of the line using any pair of points must be the same. Let's label the points as:

$$A\,(3,\ 2f) = (x_1, y_1)$$

and

$$B\,(-2,\ -3f) = (x_2, y_2)$$

and

$$C\,(1,\ 0) = (x_3, y_3)$$

In other words, we should be able to show that the slope of the line using points A and B is the same as using A and C and B and C. Let's do this now.

$$m_{AB} = \frac{y_2 - y_1}{x_2 - x_1}$$
$$= \frac{-3f - 2f}{-2 - 3} = \frac{-5f}{-5}$$
$$m_{AB} = f$$

Also

$$m_{AC} = \frac{y_3 - y_1}{x_3 - x_1}$$
$$= \frac{0 - 2f}{1 - 3} = \frac{-2f}{-2}$$
$$m_{AC} = f$$

Also

$$m_{BC} = \frac{y_3 - y_2}{x_3 - x_2}$$
$$= \frac{0 + 3f}{1 + 2} = \frac{3f}{3}$$
$$m_{BC} = f$$

Therefore, points A, B, and C are on the same straight line or collinear, since the gradient is the same when we used any pair of points.

ALTERNATIVE METHOD

We can determine the equation of the straight line using points A and B, after finding $m_{AB} = f$ and check if point C satisfies this equation.

Using point $A\,(3, 2f)$, we have $x_1 = 3$ and $y_1 = 2f$. Substitute these in $(y - y_1) = m\,(x - x_1)$ as:

$$y - 2f = f(x - 3)$$
$$y - 2f = fx - 3f$$
$$y = fx - f$$

Now let's use point $C\,(1, 0)$ and substitute for x in the equation above as:

$$y = f(1) - f$$
$$y = 0$$

The result is $y = 0$, implying that point C satisfies the line equation $y = fx - f$ or point C is on the line $y = fx - f$.

Example 12

The coordinates of the diameter AB of a circle are points $A(-2,\ 8)$ and $B(4,\ -3)$. Show that the line $6x - 2y - 1 = 0$ passes through the centre of the circle.

What did you get? Find the solution below to double-check your answer.

Solution to Example 12

Let's label the points as:

$$A\,(-2,\ 8) = (x_1, y_1)$$

and

$$B\,(4,\ -3) = (x_2, y_2)$$

Therefore, the centre of the circle is the mid-point of the diameter AB and is

$$\left(\frac{x_1 + x_2}{2}, \frac{y_1 + y_2}{2}\right) = \left(\frac{-2+4}{2}, \frac{8-3}{2}\right) = \left(\frac{2}{2}, \frac{5}{2}\right)$$
$$= \left(1, 2\frac{1}{2}\right)$$

If the line $6x - 2y - 1 = 0$ passes through the centre, then the centre coordinates above will satisfy the equation.

When $x = 1$, we have:

$$6x - 2y - 1 = 0$$
$$6(1) - 2y - 1 = 0$$
$$6 - 2y - 1 = 0$$
$$5 - 2y = 0$$
$$5 = 2y$$
$$\therefore y = \frac{5}{2} = 2\frac{1}{2}$$

Hence, we can conclude that the line $6x - 2y - 1 = 0$ passes through the centre of the circle.

Let's try another example.

Example 13

The line $\frac{1}{2}y - 3x + 5 = 0$ crosses the x-axis at point B. Another line l, with a gradient of -1, is drawn such that it also passes through point B. Determine the equation of l and write it in the form $ax + by + c = 0$ such that a, b, and c are integers.

What did you get? Find the solution below to double-check your answer.

Solution to Example 13

At point B, $\frac{1}{2}y - 3x + 5 = 0$ has $y = 0$, therefore

$$\frac{1}{2}y - 3x + 5 = 0$$
$$\frac{1}{2}(0) - 3x + 5 = 0$$
$$0 - 3x + 5 = 0$$
$$3x = 5$$
$$x = \frac{5}{3}$$

Thus, the coordinates of point B are

$$\left(\frac{5}{3}, 0\right)$$

We can now find the equation of l using $m = -1$ and the coordinates of point B. Using point $B\left(\frac{5}{3}, 0\right)$, we have $x_1 = \frac{5}{3}$ and $y_1 = 0$. Substitute these in $(y - y_1) = m(x - x_1)$ as:

$$y - 0 = -1\left(x - \frac{5}{3}\right)$$
$$y = -x + \frac{5}{3}$$
$$3y = -3x + 5$$
$$\therefore 3x + 3y - 5 = 0$$

Example 14

Lines $l_1 : 2x - y = 0$, and $l_2 : y - 7x = 8$ intersects at point M.

a) Determine the coordinates of point M.

b) Show that the equation of another line l_3, which has a gradient -3 and passes through point M, is represented by the equation $3x + y + 8 = 0$.

What did you get? Find the solution below to double-check your answer.

Solution to Example 14

a) Coordinates of M
Solution
At point M, the equation of l_1 is equal to that of l_2. In other words, we need to solve these equations simultaneously. First, let's write the equations as follows:

- Line l_1

$$2x - y = 0$$
$$y = 2x$$

- Line l_2

$$y - 7x = 8$$
$$y = 7x + 8$$

Now equate both as:

$$2x = 7x + 8$$
$$2x - 7x = 8$$
$$-5x = 8$$
$$x = -\frac{8}{5}$$

- This is the x-coordinate of point M. Substitute this in the equation of line l_1 above to find the y-axis of point M as:

$$y = 2x$$

$$y = 2\left(-\frac{8}{5}\right) = -\frac{16}{5}$$

Hence, the coordinates of point M is $(-\frac{8}{5}, -\frac{16}{5})$.

---

b) Show that l_3 is $3x + y + 8 = 0$.

Solution

We know that line l_3 has $m = -3$ and passes through point M. Using point $M\left(-\frac{8}{5}, -\frac{16}{5}\right)$, we have $x_1 = -\frac{8}{5}$ and $y_1 = -\frac{16}{5}$. Substitute these in $(y - y_1) = m(x - x_1)$ as:

$$y - \left(-\frac{16}{5}\right) = -3\left(x - \left(-\frac{8}{5}\right)\right)$$

$$y + \frac{16}{5} = -3\left(x + \frac{8}{5}\right)$$

$$y + \frac{16}{5} = -3x - \frac{24}{5}$$

$$3x + y + \frac{16}{5} + \frac{24}{5} = 0$$

$$\therefore 3x + y + 8 = 0$$

Hence, the equation of line l_3 with $m = -3$, which also passes through M, is $3x + y + 8 = 0$.

Let's consider another example.

Example 15

Line $x - 7y - 9 = 0$ meets x-axis at point P and line $\frac{1}{3}y - 5x - 1 = 0$ meets the y-axis at Q. Determine the equation of the straight line PQ, joining the two points.

What did you get? Find the solution below to double-check your answer.

Solution to Example 15

At point P, line $x - 7y - 9 = 0$ has $y = 0$, therefore

$$x - 7(0) - 9 = 0$$

$$x - 9 = 0$$

$$x = 9$$

Thus, the coordinates of point P are:

$$(9, 0)$$

At point Q, line $\frac{1}{3}y - 5x - 1 = 0$ has $x = 0$, therefore

$$\frac{1}{3}y - 5(0) - 1 = 0$$

$$\frac{1}{3}y - 0 - 1 = 0$$

$$\frac{1}{3}y = 1$$

$$y = 3$$

Thus, the coordinates of point Q are:

$$(0, 3)$$

Let's find the slope of line PQ

$$m_{PQ} = \frac{y_2 - y_1}{x_2 - x_1}$$

$$= \frac{3 - 0}{0 - 9} = \frac{3}{-9}$$

$$m_{PQ} = -\frac{1}{3}$$

The equation of PQ can be found using

$$y = mx + c$$

where $m = -\frac{1}{3}$ and $c = 3$. Thus

$$y = -\frac{1}{3}x + 3$$

ALTERNATIVE METHOD

We can determine the equation of line PQ using $m = -\frac{1}{3}$ and the coordinates of point P. Using point P, $(9, 0)$ we have $x_1 = 9$ and $y_1 = 0$. Substitute these in $(y - y_1) = m(x - x_1)$ as:

$$y - 0 = -\frac{1}{3}(x - 9)$$

$$\therefore y = -\frac{1}{3}x + 3$$

4.8 CHAPTER SUMMARY

1) A plane surface (or simply a plane) is identified by specifying its x and y coordinates and are given as (x, y).

2) The point at which the x- and y-axes cross is called the **Origin** and has the coordinates **(0, 0)**.

3) The $x - y$ plane can be divided into four parts and each part is called a **quadrant**.

4) The coordinates of the mid-way between points $A(x_1, y_1)$ and $B(x_2, y_2)$, denoted as $C(x_m, y_m)$, are given as:

$$(x_m, y_m) = \left(\frac{x_1 + x_2}{2}, \frac{y_1 + y_2}{2} \right)$$

5) The distance between points $A(x_1, y_1)$ and $B(x_2, y_2)$, or the length of the line AB, denoted as d, is given by:

$$d = \sqrt{(x_2 - x_1)^2 + (y_2 - y_1)^2}$$

6) Slope (also called gradient), denoted by m, is a measure of the steepness of a line. The larger the value of m, the greater the steepness or tendency to fall. Slope can be positive or negative.

7) Slope is commonly referred to as '**rise over run**'.

8) Two points on a line are required to determine a gradient. If these points are $A(x_1, y_1)$ and $B(x_2, y_2)$, then the slope is given by:

$$m = \frac{\Delta y}{\Delta x} = \frac{y_2 - y_1}{x_2 - x_1}$$

9) Slope can also be given as

$$m = \tan\theta$$

where θ is the angle the line makes with the positive x-axis

10) The slope of a curve varies from one point to another. As a result, it is only valid to determine the slope of a curve at a particular point on the curve.

11) The equation of a straight line is characterised by three terms and represented as:

$$y = mx + c$$

where m is the slope of the line and c is the y-intercept.

****

4.9 FURTHER PRACTICE

To access complementary contents, including additional exercises, please go to www.dszak.com.

5 Coordinate Geometry II

Once you have studied the content of this chapter, you should be able to:

- Evaluate the equation of a straight line
- Determine the equation of a line parallel to a given line
- Determine the equation of a line perpendicular to a given line
- Determine the point of intersection of a pair of straight lines

5.1 INTRODUCTION

This chapter will, in continuation of our discussion on geometry, look at methods of determining the equation of a straight line, parallel and perpendicular lines, and the intersection of straight lines.

5.2 DETERMINING THE EQUATION OF A LINE

The method used to determine the equation of a straight line depends on the information available. We will go through some of these now.

5.2.1 GRADIENT AND y-INTERCEPT

If the gradient m and the y-intercept c of a line are given (Figure 5.1), follow these steps to determine the equation of the line:

Step 1: Use the general equation $y = mx + c$
Step 2: Substitute m and c, and simplify to have the equation.

Easy, right? This is the simplest situation.

Let's try some examples.

Example 1

Write the equation of the straight line for the given slope and y-intercept.

a) $m = -1$, $c = 4$ **b)** $m = 3$, $c = -11$ **c)** $m = \frac{3}{5}$, $c = -\frac{1}{5}$ **d)** $m = -1.5$, $c = 3.2$

DOI:10.1201/9781003027928-5

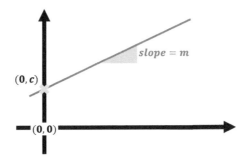

FIGURE 5.1   Determining the equation of a line using the gradient and *y*-intercept illustrated.

What did you get? Find the solution below to double-check your answer.

Solution to Example 1

HINT

In this example, use $y = mx + c$ and substitute for m and c. Simplify if necessary.

a) $m = -1$, $c = 4$
Solution

$$y = mx + c$$
$$\therefore y = -x + 4$$

b) $m = 3$, $c = -11$
Solution

$$y = mx + c$$
$$\therefore y = 3x - 11$$

c) $m = \frac{3}{5}$, $c = -\frac{1}{5}$
Solution

$$y = mx + c$$
$$\therefore y = \frac{3}{5}x - \frac{1}{5}$$

d) $m = -1.5$, $c = 3.2$
Solution

$$y = mx + c$$
$$\therefore y = -1.5x + 3.2$$

Another similar set of examples to try.

Example 2

Determine the equation of the straight line for the given slope and y-intercept. Write the equation in the form $ax + by + c = 0$, such that a, b, and c are integers.

a) $m = -2$, $c = 7$ **b)** $m = \frac{2}{3}$, $c = -1$ **c)** $m = 0.5$, $c = -1\frac{1}{2}$ **d)** $m = -4$, $c = 1.6$

What did you get? Find the solution below to double-check your answer.

Solution to Example 2

HINT

In this example, use $y = mx + c$ and substitute for m and c. Simplify and write the final answer in the required form $ax + by + c = 0$ and ensure that a, b, and c are integers. This is very important.

a) $m = -2$, $c = 7$
Solution

$$y = mx + c$$
$$y = -2x + 7$$
$$\therefore 2x + y - 7 = 0$$

b) $m = \frac{2}{3}$, $c = -1$
Solution

$$y = mx + c$$
$$y = \frac{2}{3}x - 1$$
$$\frac{2}{3}x - y - 1 = 0$$
$$\therefore 2x - 3y - 3 = 0$$

c) $m = 0.5$, $c = -1\frac{1}{2}$
Solution

$$y = mx + c$$
$$y = 0.5x - 1\frac{1}{2}$$

But $-1\frac{1}{2}$ is a mixed fraction, so we need to change it to an improper fraction. Thus we have:

$$y = 0.5x - \frac{3}{2}$$

$$0.5x - y - \frac{3}{2} = 0$$

$$\frac{1}{2}x - y - \frac{3}{2} = 0$$

$$\therefore x - 2y - 3 = 0$$

d) $m = -4$, $c = 1.6$

Solution

$$y = mx + c$$

$$y = -4x + 1.6$$

$$4x + y - 1.6 = 0$$

$$4x + y - \frac{8}{5} = 0$$

$$\therefore 20x + 5y - 8 = 0$$

Another set of examples to try.

Example 3

Determine the gradient and the y-intercept for each of the following straight-line equations.

a) $y - x - 4 = 0$ **b)** $3x - 4y + 1 = 0$
c) $2y + 3 - 6x = 1$ **d)** $7y - 1 - x = 2y - 16 - 4x$

What did you get? Find the solution below to double-check your answer.

Solution to Example 3

HINT

In this example, write the equation in the form $y = mx + c$ and then compare with the given equation to determine m and c.

a) $y - x - 4 = 0$

Solution

$$y - x - 4 = 0$$

$$y = x + 4$$

$$\therefore m = 1, \ c = 4$$

b) $3x - 4y + 1 = 0$

Solution

$$3x - 4y + 1 = 0$$
$$-4y = -3x - 1$$
$$4y = 3x + 1$$
$$y = \frac{3}{4}x + \frac{1}{4}$$
$$\therefore m = \frac{3}{4}, \ c = \frac{1}{4}$$

---

c) $2y + 3 - 6x = 1$

Solution

$$2y + 3 - 6x = 1$$
$$2y = 6x + 1 - 3$$
$$2y = 6x - 2$$
$$y = \frac{6}{2}x - \frac{2}{2}$$
$$y = 3x - 1$$
$$\therefore m = 3, \ c = -1$$

---

d) $7y - 1 - x = 2y - 16 - 4x$

Solution

$$7y - 1 - x = 2y - 16 - 4x$$
$$7y - 2y = x - 4x + 1 - 16$$
$$5y = -3x - 15$$
$$y = -\frac{3}{5}x - \frac{15}{5}$$
$$y = -\frac{3}{5}x - 3$$
$$\therefore m = -\frac{3}{5}, \ c = -3$$

5.2.2 One Point and Gradient

If a point (x_1, y_1) and the gradient m of a line are given (Figure 5.2), follow these steps to determine the equation of the line:

Step 1: Use the general equation $y = mx + c$.

Step 2: Substitute point (x_1, y_1) in the line equation such that $x = x_1$ and $y = y_1$. Also, substitute for m.

Step 3: Simplify to determine the y-intercept c.

Step 4: Remember to write the equation in the required format.

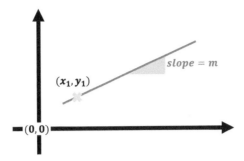

FIGURE 5.2 Determining the equation of a line using the gradient and a point on the line illustrated.

Alternatively, we can simply use the formula below:

$$\boxed{(y - y_1) = m(x - x_1)}$$ (5.1)

Let's try an example.

Example 4

Determine the equation of the line with a gradient of 2 which passes through a point $(5, -3)$.

What did you get? Find the solution below to double-check your answer.

Solution to Example 4

Given that

$$y = mx + c$$

we know that $m = 2$. Using point $(5, -3)$, we have $x = 5$ and $y = -3$. Substitute these in the above equation to determine the intercept as:

$$-3 = 2(5) + c$$
$$-3 = 10 + c$$
$$c = -13$$

Thus, the equation of the line is:

$$y = 2x - 13$$

ALTERNATIVE METHOD

We know that

$$(y - y_1) = m(x - x_1)$$

Using point $(5, -3)$, we have $x_1 = 5$ and $y_1 = -3$. Substitute these in the above equation to determine the equation as before.

$$y - (-3) = 2(x - 5)$$
$$y + 3 = 2x - 10$$
$$y = 2x - 13$$

5.2.3 ONE POINT AND y-INTERCEPT

If a point (x_1, y_1) and the y-intercept c of a line are given (Figure 5.3), follow these steps to determine the equation of the line:

Step 1: Use the general equation $y = mx + c$.
Step 2: Substitute point (x_1, y_1) in the line equation such that $x = x_1$ and $y = y_1$. Also, substitute for c.
Step 3: Determine the gradient, m (see Chapter 4).
Step 4: Write the equation.

Let's try an example.

Example 5

Determine the equation of the line passing through a point $(-4, -1)$ and which crosses the y-axis at $(0, 5)$.

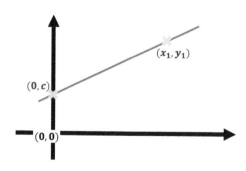

FIGURE 5.3 Determining the equation of a line using a point on the line and y-intercept illustrated.

What did you get? Find the solution below to double-check your answer.

Solution to Example 5

$$y = mx + c$$

Using point $(-4, -1)$, we have $x = -4$ and $y = -1$. Point $(0,\ 5)$ implies that $c = 5$.
Substitute these in the above equation to find the gradient as:

$$-1 = m(-4) + 5$$

$$-1 - 5 = -4m$$

$$-6 = -4m$$

$$m = \frac{-6}{-4} = \frac{3}{2}$$

Thus, the equation of the line is

$$y = \frac{3}{2}x + 5$$

5.2.4 Two Points

If two points (x_1, y_1) and (x_2, y_2) on a line are given (Figure 5.4), follow these steps to determine the equation of the line:

Step 1: Label the points. The choice is entirely yours.
Step 2: Determine the gradient, m (see Chapter 4).
Step 3: Now follow the steps for when a point and the gradient are given (since you have two points, you can use either). Alternatively, use either of the point and the gradient to find c.
Step 4: Write the equation.

Before we conclude, there is a special formula for a two-point situation, and it is given as:

$$\boxed{\frac{y - y_1}{x - x_1} = \frac{y_2 - y_1}{x_2 - x_1}} \quad \text{OR} \quad \boxed{\frac{y - y_1}{y_2 - y_1} = \frac{x - x_1}{x_2 - x_1}} \tag{5.2}$$

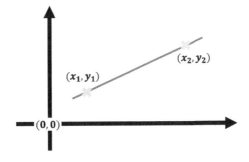

FIGURE 5.4 Determining the equation of a line using two points on the line illustrated.

The above is very handy to use. You will however realise that it is a manipulation of the one-point formula. Let's start with $(y - y_1) = m(x - x_1)$, divide both sides by $x - x_1$, we have:

$$\frac{y - y_1}{x - x_1} = m$$

Substituting for m, we have:

$$\frac{y - y_1}{x - x_1} = \frac{y_2 - y_1}{x_2 - x_1}$$

Re-arrange to have the second version of the 2-point equation as:

$$\frac{y - y_1}{y_2 - y_1} = \frac{x - x_1}{x_2 - x_1}$$

Both formulas are useful, though some may prefer using $y - y_1 = m(x - x_1)$ and then simplify.

Let's try some examples.

Example 6

Two points $(3, 5)$ and $(-1, 7)$ are located on a line l. Determine the equation of the line in the form $ax + by + c = 0$.

What did you get? Find the solution below to double-check your answer.

Solution to Example 6

Let's label the points as:

$$(3, 5) = (x_1, y_1)$$

and

$$(-1, 7) = (x_2, y_2)$$

Now let's find the slope

$$m = \frac{y_2 - y_1}{x_2 - x_1}$$
$$= \frac{7 - 5}{-1 - 3} = \frac{2}{-4}$$
$$m = -\frac{1}{2}$$

It is time to determine the equation using:

$$(y - y_1) = m(x - x_1)$$

With point $(3, 5)$, we have $x_1 = 3$ and $y_1 = 5$. Therefore

$$y - 5 = -\frac{1}{2}(x - 3)$$
$$2y - 10 = -(x - 3)$$
$$2y - 10 = -x + 3$$
$$2y + x - 10 - 3 = 0$$
$$x + 2y - 13 = 0$$

ALTERNATIVE METHOD

Given that

$$y = mx + c$$

we know that $m = -\frac{1}{2}$. With point $(-1, 7)$, we have $x = -1$ and $y = 7$. Substitute these in the above equation to find the intercept as:

$$7 = -\frac{1}{2}(-1) + c$$
$$7 = \frac{1}{2} + c$$
$$c = 7 - \frac{1}{2} = \frac{13}{2}$$

Thus, the equation of the line is

$$y = -\frac{1}{2}x + \frac{13}{2}$$
$$y + \frac{1}{2}x - \frac{13}{2} = 0$$

Multiply through by 2, we have

$$x + 2y - 13 = 0$$

Another example to try.

Example 7

Use the 2-point formula $\frac{y - y_1}{x - x_1} = \frac{y_2 - y_1}{x_2 - x_1}$ to determine the equation of a line joining points $A(-2, 9)$ and $B(7, -3)$. Write the equation in the form $y = mx + c$.

What did you get? Find the solution below to double-check your answer.

Solution to Example 7

Let's label the points as:

$$(-2, 9) = (x_1, y_1)$$

and

$$(7, -3) = (x_2, y_2)$$

Now let's find the equation using the 2-point equation as:

$$\frac{y-9}{x-(-2)} = \frac{-3-9}{7-(-2)}$$
$$\frac{y-9}{x+2} = \frac{-12}{7+2} = \frac{-12}{9}$$
$$\frac{y-9}{x+2} = \frac{-4}{3}$$
$$3(y-9) = -4(x+2)$$
$$3y - 27 = -4x - 8$$
$$3y = -4x - 8 + 27$$
$$3y = -4x + 19$$
$$y = -\frac{4}{3}x + \frac{19}{3}$$

Let's try another example.

Example 8

Line p passes through points $(-1, 2)$ and $(3, -1)$. Line q passes through points $(4, -5)$ and $(9, 10)$. If the two lines p and q intersect at point A, determine the coordinates of A.

What did you get? Find the solution below to double-check your answer.

Solution to Example 8

- **To determine the equation of line p**

Let's label the points as:

$$(-1, 2) = (x_1, y_1)$$

and

$$(3, -1) = (x_2, y_2)$$

Now let's find the slope

$$m = \frac{y_2 - y_1}{x_2 - x_1}$$

$$= \frac{-1 - 2}{3 + 1} = \frac{-3}{4}$$

$$m = -\frac{3}{4}$$

It is time to determine the equation using:

$$(y - y_1) = m(x - x_1)$$

With point $(-1, 2)$, we have $x_1 = -1$ and $y_1 = 2$

$$y - 2 = -\frac{3}{4}(x + 1)$$

$$y - 2 = -\frac{3}{4}x - \frac{3}{4}$$

$$y = -\frac{3}{4}x - \frac{3}{4} + 2$$

$$y = -\frac{3}{4}x + \frac{5}{4}$$

- **To determine the equation of line q**

We will use the 2-point equation $\frac{y-y_1}{x-x_1} = \frac{y_2-y_1}{x_2-x_1}$ for this, just to vary our approach. Let's label the points as:

$$(4, -5) = (x_1, y_1)$$

and

$$(9, 10) = (x_2, y_2)$$

$$\frac{y - (-5)}{x - 4} = \frac{10 - (-5)}{9 - 4}$$

$$\frac{y + 5}{x - 4} = \frac{10 + 5}{5} = \frac{15}{5}$$

$$\frac{y + 5}{x - 4} = 3$$

$$y + 5 = 3(x - 4)$$

$$y + 5 = 3x - 12$$

$$y = 3x - 12 - 5$$

$$y = 3x - 17$$

- **To determine the point of intersection**

When the two lines p and q meet, their equations are equal, then we have:

$$-\frac{3}{4}x + \frac{5}{4} = 3x - 17$$

Let's solve for x as follows:

$$-\frac{3}{4}x - 3x = -\frac{5}{4} - 17$$

$$-\frac{15}{4}x = -\frac{73}{4}$$

$$\frac{15}{4}x = \frac{73}{4}$$

$$15x = 73$$

$$x = \frac{73}{15}$$

This is the x-coordinate of point A. Substitute this in either equation for p or q to determine the y-coordinate of A. We will use the equation of p, thus

$$y = 3x - 17$$

$$= 3\left(\frac{73}{15}\right) - 17$$

$$= \frac{73}{5} - 17$$

$$y = -\frac{12}{5}$$

Therefore, the coordinates of point A are:

$$\left(\frac{73}{15}, \; -\frac{12}{5}\right)$$

Let's try one last example.

Example 9

The vertices of a triangle ABC are $A\,(-4, \; -3)$, $B\,(3, 2)$, and $C(-3, 5)$.

   a) Determine the equations of the three sides, AB, BC, and AC.

   b) Show that triangle ABC is scalene (i.e., the 3 sides are different in length).

What did you get? Find the solution below to double-check your answer.

Solution to Example 9

a) Equations of lines *AB*, *BC*, and *AC*
Solution

- **To determine the equation of the side *AB*.**

Let's label the points as:

$$A(-4, \ -3) = (x_1, y_1)$$

and

$$B(3, \ 2) = (x_2, y_2)$$

Now let's find the slope

$$m = \frac{y_2 - y_1}{x_2 - x_1}$$
$$= \frac{2 + 3}{3 + 4} = \frac{5}{7}$$
$$m = \frac{5}{7}$$

It is time to determine the equation of side *AB* using:

$$(y - y_1) = m(x - x_1)$$

With the point (3, 2), we have, in reference to the line equation above, $x_1 = 3$ and $y_1 = 2$. However, these labels do not align with our original 'tags', as the labeling is solely for reference.

$$y - 2 = \frac{5}{7}(x - 3)$$
$$y - 2 = \frac{5}{7}x - \frac{15}{7}$$
$$y = \frac{5}{7}x - \frac{15}{7} + 2$$
$$\therefore y = \frac{5}{7}x - \frac{1}{7}$$

- **To determine the equation of side *BC***

We will use the 2-point equation $\frac{y - y_1}{x - x_1} = \frac{y_2 - y_1}{x_2 - x_1}$ to vary our approach. Let's label the points as:

$$B(3, \ 2) = (x_1, y_1)$$

and

$$C(-3, 5) = (x_2, y_2)$$

This implies that

$$\frac{y-2}{x-3} = \frac{5-2}{-3-3}$$

$$\frac{y-2}{x-3} = \frac{3}{-6} = -\frac{1}{2}$$

$$\frac{y-2}{x-3} = -\frac{1}{2}$$

$$y-2 = -\frac{1}{2}(x-3)$$

$$y-2 = -\frac{1}{2}x + \frac{3}{2}$$

$$y = -\frac{1}{2}x + \frac{3}{2} + 2$$

$$\therefore y = -\frac{1}{2}x + \frac{7}{2}$$

- **To determine the equation of side AC**

Again, we will use the 2-point equation $\frac{y-y_1}{x-x_1} = \frac{y_2-y_1}{x_2-x_1}$ to get used to it as a viable approach. Let's label the points as:

$$A(-4,\ -3) = (x_1, y_1)$$

and

$$C(-3, 5) = (x_2, y_2)$$

This implies that

$$\frac{y+3}{x+4} = \frac{5+3}{-3+4}$$

$$\frac{y+3}{x+4} = \frac{8}{1}$$

$$\frac{y+3}{x+4} = 8$$

$$y+3 = 8(x+4)$$

$$y+3 = 8x + 32$$

$$y = 8x + 32 - 3$$

$$\therefore y = 8x + 29$$

---

b) Sides AB, BC, and AC

Solution

The length between two points can be found using:

$$\sqrt{(y_2 - y_1)^2 + (x_2 - x_1)^2}$$

- **Length AB**

Let's label the points as:

$$A(-4, \ -3) = (x_1, y_1)$$

and

$$B(3, \ 2) = (x_2, y_2)$$

Thus

$$|AB| = \sqrt{(y_2 - y_1)^2 + (x_2 - x_1)^2}$$
$$= \sqrt{(2 + 3)^2 + (3 + 4)^2}$$
$$= \sqrt{(5)^2 + (7)^2} = \sqrt{25 + 49}$$
$$\therefore |AB| = \sqrt{74}$$

- **Length BC**

Let's label the points as:

$$B(3, \ 2) = (x_1, y_1)$$

and

$$C(-3, \ 5) = (x_2, y_2)$$

Thus

$$|BC| = \sqrt{(y_2 - y_1)^2 + (x_2 - x_1)^2}$$
$$= \sqrt{(5 - 2)^2 + (-3 - 3)^2}$$
$$= \sqrt{(3)^2 + (-6)^2} = \sqrt{9 + 36}$$
$$\therefore |BC| = 3\sqrt{5}$$

- **Length AC**

Let's label the points as:

$$A(-4, \ -3) = (x_1, y_1)$$

and

$$C(-3, \ 5) = (x_2, y_2)$$

Thus

$$|AC| = \sqrt{(y_2 - y_1)^2 + (x_2 - x_1)^2}$$
$$= \sqrt{(5 + 3)^2 + (-3 + 4)^2}$$
$$= \sqrt{(8)^2 + (1)^2} = \sqrt{64 + 1}$$
$$\therefore |AC| = \sqrt{65}$$

Since

$$|AB| \neq |BC| \neq |AC|$$

Therefore, the triangle ABC is a scalene triangle.

5.2.5 FINDING AN EQUATION FROM A GRAPH

If we are provided with the equation of a straight line, we can use it to sketch its graph. The reverse is also possible; that is, given the graph of a straight line, we can determine its equation. We will now demonstrate three ways of doing so here.

Follow these steps to determine the equation of a line from a graph:

Option 1

Step 1: If the line crosses the y-axis, read the value at which the line crosses the y-axis. This is the y-intercept or c.

Step 2: Choose another suitable point on the line as your (x_1, y_1).

Step 3: Determine m using (x_1, y_1) by substituting in $y = mx + c$. Alternatively, use $m = \frac{y_2 - y_1}{x_2 - x_1}$ where $(x_2, y_2) = (0, c)$.

Step 4: Substitute m and c in $y = mx + c$ to write the equation.

Option 2

Step 1: If the line does not cross the y-axis, extend it to cross it and find c.

Step 2: Follow option 1 above to determine the equation of the line.

Option 3

Step 1: If the line does not cross y-axis and it is impossible to do so, choose two convenient points on the line and read their coordinates. You should now have (x_1, y_1) and (x_2, y_2).

Step 2: Determine gradient m from the graph.

Step 3: Use $(y - y_1) = m(x - x_1)$ or otherwise write the equation using the 2-point formula as previously discussed.

The above seems easy. Let's try a few examples, first on sketching graphs from an equation and then working our way back by deriving an equation from a graph.

Example 10

Sketch the graph of the following equations showing their vertical and horizontal intercepts.

a) $y = 2x - 1$ b) $y = -5x + 3$ c) $2x - 3y = 0$

What did you get? Find the solution below to double-check your answer.

Solution to Example 10

HINT

A line can be drawn if two coordinates are known. The best way in this case is to determine and use the **y**-intercept and **x**-intercept as the two points.

a) $y = 2x - 1$
Solution
For the y-intercept, put $x = 0$

$$y = 2x - 1$$
$$y = 2(0) - 1$$
$$y = -1$$
$$\therefore (\mathbf{0}, \mathbf{-1})$$

For the x-intercept, put $y = 0$

$$y = 2x - 1$$
$$0 = 2x - 1$$
$$2x = 1$$
$$x = \frac{1}{2}$$
$$\therefore \left(\frac{\mathbf{1}}{\mathbf{2}}, \mathbf{0}\right)$$

Locate these coordinates (or points) and draw a straight line through them (Figure 5.5).

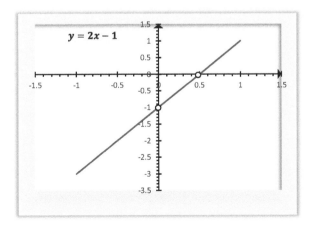

FIGURE 5.5 Solution to Example 10(a).

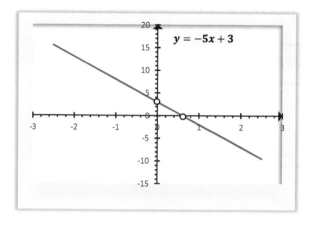

FIGURE 5.6   Solution to Example 10(b).

b) $y = -5x + 3$

Solution

For the y-intercept, put $x = 0$

$$y = -5x + 3$$
$$y = -5\,(0) + 3$$
$$y = 3$$
$$\therefore (0, 3)$$

For the x-intercept, put $y = 0$

$$y = -5x + 3$$
$$0 = -5x + 3$$
$$5x = 3$$
$$x = \frac{3}{5}$$
$$\therefore \left(\frac{3}{5},\, 0\right)$$

Locate these coordinates (or points) and draw a straight line through them (Figure 5.6).

---

c) $2x - 3y = 0$

Solution

For the y-intercept, put $x = 0$

$$2x - 3y = 0$$
$$2\,(0) - 3y = 0$$
$$0 - 3y = 0$$
$$3y = 0$$
$$y = 0$$
$$\therefore (0, 0)$$

For the *x*-intercept, put $y = 0$

$$2x - 3y = 0$$
$$2x - 3(0) = 0$$
$$2x - 0 = 0$$
$$2x = 0$$
$$x = 0$$
$$\therefore (0,\ 0)$$

We've only managed to produce a single point with this approach for this equation, so we need to use another value, let's say $y = 1$. Put $y = 1$, we have:

$$2x - 3y = 0$$
$$2x - 3(1) = 0$$
$$2x - 3 = 0$$
$$2x = 3$$
$$x = \frac{3}{2} = 1.5$$
$$\therefore (1.5,\ 1)$$

Locate these coordinates (or points) and draw a straight line through them (Figure 5.7).

NOTE
Let's write the equation in the form $y = mx + c$ as $2x = 3y$ or $3y = 2x$. This implies that $y = \frac{2}{3}x$ or $y = \frac{2}{3}x + 0$. It is therefore obvious that $c = 0$. This explains why the graph passes the origin.

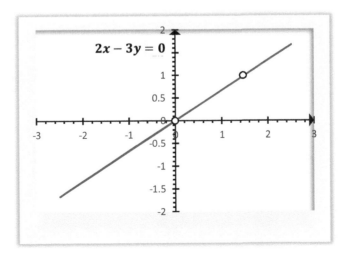

FIGURE 5.7 Solution to Example 10(c).

Let's try an example on determining an equation from a graph.

Example 11

Using Figure 5.8, determine the equation of the line l in the form $ax + by + c = 0$.

What did you get? Find the solution below to double-check your answer.

Solution to Example 11

From the graph, $c = 3$. Also, the coordinates of x-intercept are

$$(x_1, y_1) = (-2, 0)$$

Substitute these in $y = mx + c$ to determine the gradient as:

$$0 = m(-2) + 3$$
$$0 = -2m + 3$$
$$2m = 3$$
$$m = \frac{3}{2}$$

Thus, the equation of the line is:

$$y = \frac{3}{2}x + 3$$
$$2y = 3x + 6$$
$$\mathbf{-3x + 2y - 6 = 0}$$

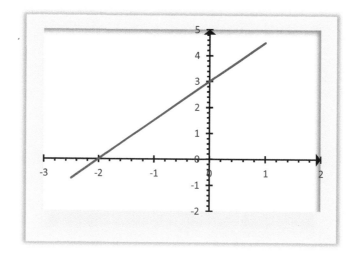

FIGURE 5.8 Example 11.

ALTERNATIVE METHOD

We have a point:

$$(x_1, y_1) = (-2, 0)$$

Let's find a second point on the graph as:

$$(x_2, y_2) = (0, c) = (0, 3)$$

Now let's find the slope using:

$$m = \frac{y_2 - y_1}{x_2 - x_1}$$
$$= \frac{3 - 0}{0 - (-2)}$$
$$m = \frac{3}{2}$$

Using point $(0, 3)$ we have:

$$(y - y_1) = m(x - x_1)$$
$$y - 3 = \frac{3}{2}(x - 0)$$
$$y - 3 = \frac{3}{2}x$$
$$y - \frac{3}{2}x - 3 = 0$$
$$\therefore -3x + 2y - 6 = 0$$

Another example to try.

Example 12

$3y + x - 9 = 0$ is the equation of line l which crosses the x-axis at A and the y-axis at point B as shown in Figure 5.9. Determine the coordinates of points A and B.

What did you get? Find the solution below to double-check your answer.

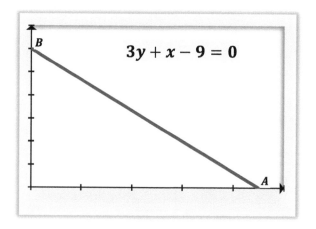

FIGURE 5.9 Example 12.

Solution to Example 12

At point A, $y = 0$. Thus

$$3y + x - 9 = 0$$
$$3(0) + x - 9 = 0$$
$$0 + x - 9 = 0$$
$$x = 9$$

The x-intercept is at $x = 9$, hence the coordinates of A are $(\mathbf{9, 0})$.

At point B, $x = 0$, thus

$$3y + x - 9 = 0$$
$$3y + 0 - 9 = 0$$
$$3y - 9 = 0$$
$$3y = 9$$
$$y = 3$$

The y-intercept is at $y = 3$, hence the coordinates of B are $(\mathbf{0, 3})$.

ALTERNATIVE METHOD

Let's write l in the form $y = mx + c$

$$3y + x - 9 = 0$$
$$3y = -x + 9$$
$$y = -\frac{1}{3}x + 3$$

The y-intercept is $c = 3$, hence the coordinates of B are $(\mathbf{0, 3})$.

One last example to try.

Example 13

$6x - 4y - 12 = 0$ is a straight-line equation. Determine the slope and the y-intercept of the line.

What did you get? Find the solution below to double-check your answer.

Solution to Example 13

The equation should be written in the form $y = mx + c$ as:

$$6x - 4y - 12 = 0$$
$$-4y = -6x + 12$$

Divide both sides by -4

$$\frac{-4}{-4}y = -\frac{6}{-4}x + \frac{12}{-4}$$
$$y = 1.5x - 3$$

Therefore,

$$c = -3, \; m = 1.5$$

ALTERNATIVE METHOD

$$6x - 4y - 12 = 0$$

Therefore,

$$a = 6, \; b = -4, \; d = -12$$

Thus

$$m = -\frac{a}{b} = -\frac{6}{-4}$$
$$m = 1.5$$

then

$$c = -\frac{d}{b} = -\frac{-12}{-4}$$
$$c = -3$$

5.3 PARALLEL AND PERPENDICULAR LINES

When lines are parallel, their slope is the same. On the other hand, when two lines l_1 and l_2 are perpendicular to each other, the slope of one is the negative reciprocal of the other or the product of their slopes equals -1. Parallel and perpendicular lines are shown in Figure 5.10.

Let's make this clearer. Given that the slope of l_1 and l_2 are m_1 and m_2 respectively, then if the two lines are:

a) Parallel

$$\boxed{m_1 = m_2} \tag{5.3}$$

b) Perpendicular

$$\boxed{m_1 = -\frac{1}{m_2}} \text{ OR } \boxed{m_1 m_2 = -1} \tag{5.4}$$

Note that all we need to do when two lines are perpendicular is to:

1) change the sign (from positive to negative and vice versa), and

2) flip the slope, i.e., change the numerator to become the denominator and vice versa. For example, $\frac{2}{3}$ becomes $\frac{3}{2}$, $\frac{1}{7}$ becomes 7, and 2 becomes $\frac{1}{2}$. Note that 2 is $\frac{2}{1}$ and 7 is $\frac{7}{1}$ too.

Let's try some examples on this.

Example 14

Two lines l_1 and l_2 are perpendicular to each other with slopes m_1 and m_2 respectively. Determine m_2 for each of the following cases given that:

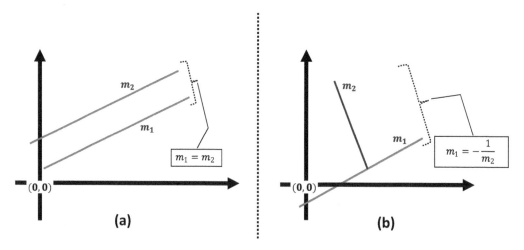

(a)   **(b)**

FIGURE 5.10 (a) Parallel lines and (b) perpendicular lines.

a) $m_1 = 2$ **b)** $m_1 = -\dfrac{2}{3}$ **c)** $m_1 = 2.5$ **d)** $m_1 = -1.6$

What did you get? Find the solution below to double-check your answer.

Solution to Example 14

a) $m_2 = -\dfrac{1}{2}$ **b)** $m_2 = \dfrac{3}{2}$ **c)** $m_1 = 2.5 = \dfrac{25}{10} = \dfrac{5}{2}$ **d)** $m_1 = -1.6 = -\dfrac{16}{10} = -\dfrac{8}{5}$

$$\therefore m_2 = -\dfrac{2}{5}$$ $$\therefore m_2 = \dfrac{5}{8}$$

5.3.1 EQUATION OF PARALLEL AND PERPENDICULAR LINES

We can write the equation of a line parallel or perpendicular (also known as '**normal**') to a given line by following the procedure given. By default, however, we will not be told the gradient of the parallel or perpendicular line in the question, as this can be obtained from the equation of the given line. This is everything! Let's try some examples.

Example 15

Line l, with an intercept of $(0, 2)$, is drawn perpendicular to $7y - 3x + 4 = 0$. Write the equation of l in the form $ax + by + d = 0$.

What did you get? Find the solution below to double-check your answer.

Solution to Example 15

The equation of the straight line $7y - 3x + 4 = 0$ should be written first in the form $y = mx + c$ and then compared with the standard format.

$$7y = 3x - 4$$
$$y = \frac{3}{7}x - \frac{4}{7}$$

From the above, we have:

$$m_1 = \frac{3}{7}$$

Therefore, the slope of the perpendicular line l is:

$$m_2 = -\frac{7}{3}$$

The intercept $(0, 2)$ implies that $c = 2$. Thus, the equation of l is:

$$y = -\frac{7}{3}x + 2$$

Let's re-write this in the specified format.

$$3y = -7x + 6$$
$$7x + 3y - 6 = 0$$
$$\therefore 7x + 3y - 6 = 0$$

Another example to try.

Example 16

l_1 and l_2 are parallel to each other. Line l_1 is defined by $5 - 6x - 2y = 0$. Determine the value of g if l_2 passes through points $(-2, 1)$ and $(g, 2g)$.

What did you get? Find the solution below to double-check your answer.

Solution to Example 16

Write the equation l_1 $5 - 6x - 2y = 0$ in the form $y = mx + c$ and then compare it with the standard format.

$$5 - 6x - 2y = 0$$
$$-2y = 6x - 5$$
$$y = -3x + \frac{5}{2}$$

Thus

$$m_1 = m_{l_1} = -3$$

Since l_1 and l_2 are parallel, we have:

$$m_2 = m_{l_2} = -3$$

Now let's find the slope of l_2 using the two points given

$$m = \frac{y_2 - y_1}{x_2 - x_1}$$

It follows that

$$m_2 = \frac{2g - 1}{g + 2}$$

Since $m_2 = -3$, we have:

$$-3 = \frac{2g - 1}{g + 2}$$

Simplify this, we have:

$$-3(g + 2) = 2g - 1$$
$$-3g - 6 = 2g - 1$$
$$-5g = 5$$
$$\therefore g = -1$$

Example 17

l_1 is a line that passes through the centre C of a circle and meets the circumference at points $P(7, 3)$ and $Q(2, -4)$. Determine the equation of line l_2 which also passes through C and is perpendicular to l_1.

What did you get? Find the solution below to double-check your answer.

Solution to Example 17

To find the equation of l_2 in this case, we need point C and the gradient of the line.

- **Point C**

Let's label the points as:

$$P(7, 3) = (x_1, y_1)$$

and

$$Q(2, -4) = (x_2, y_2)$$

Therefore, the mid-point (or point C) is

$$\left(\frac{x_1 + x_2}{2}, \frac{y_1 + y_2}{2}\right) = \left(\frac{7 + 2}{2}, \frac{3 + (-4)}{2}\right) = \left(\frac{9}{2}, \frac{-1}{2}\right)$$
$$= \left(\frac{9}{2}, -\frac{1}{2}\right)$$

- **Slope of line l_1**

Since l_1 and l_2 are perpendicular, we have:

$$m = \frac{y_2 - y_1}{x_2 - x_1}$$
$$= \frac{-4 - 3}{2 - 7} = \frac{-7}{-5}$$
$$m_{l_1} = \frac{7}{5}$$

- **Slope of line l_2**

Now we can find the slope of l_2

$$m_{l_2} = -\frac{5}{7}$$

- **Equation of l_2**

Finally, it is time to write the equation of l_2 using:

$$(y - y_1) = m(x - x_1)$$

With point $C\left(\frac{9}{2}, -\frac{1}{2}\right)$, we have $x_1 = \frac{9}{2}$ and $y_1 = -\frac{1}{2}$

$$y + \frac{1}{2} = -\frac{5}{7}\left(x - \frac{9}{2}\right)$$
$$y + \frac{1}{2} = -\frac{5}{7}x + \frac{45}{14}$$
$$y + \frac{5}{7}x + \frac{1}{2} - \frac{45}{14} = 0$$
$$y + \frac{5}{7}x - \frac{19}{7} = 0$$
$$\therefore 7y + 5x - 19 = 0$$

Example 18

Line LM is a chord of a circle which passes through points $L(-1, 3)$ and $M(5, 7)$ and line PQ is a perpendicular bisector of LM. Determine the equation of the line PQ in the form $ax + by + c = 0$, where a, b, and c are integers.

What did you get? Find the solution below to double-check your answer.

Solution to Example 18

Solution

- **Slope of line LM**

Let's label the points as:

$$L(-1, 3) = (x_1, y_1)$$

and

$$M(5, 7) = (x_2, y_2)$$

Thus

$$m = \frac{y_2 - y_1}{x_2 - x_1}$$

$$m_{LM} = \frac{7 - 3}{5 - (-1)}$$

$$= \frac{4}{5 + 1} = \frac{4}{6}$$

$$\therefore m_{LM} = \frac{2}{3}$$

- **Slope of line PQ**

Since LM and PQ are perpendicular, we have

$$\therefore m_{PQ} = -\frac{3}{2}$$

Since PQ is a perpendicular bisector of LM, it follows that it will pass through the mid-point of LM. Therefore, the mid-point is

$$\left(\frac{x_1 + x_2}{2}, \frac{y_1 + y_2}{2} \right) = \left(\frac{-1 + 5}{2}, \frac{3 + 7}{2} \right)$$

$$= \left(\frac{4}{2}, \frac{10}{2} \right)$$

$$= (2,\ 5)$$

Let's now determine the equation of line PQ using the slope and the mid-point found above as:

$$(y - y_1) = m(x - x_1)$$

With point $(2,\ 5)$, we have $x_1 = 2$ and $y_1 = 5$. Therefore

$$y - 5 = -\frac{3}{2}(x - 2)$$

$$2y - 10 = -3(x - 2)$$

$$2y - 10 = -3x + 6$$

$$3x + 2y - 10 - 6 = 0$$

$$\therefore 3x + 2y - 16 = 0$$

Another example to try.

Example 19

$3x - y - 5 = 0$ is the equation of a straight line l_1 that is perpendicular to another line l_2. The two lines meet at a point where $y = -2$. Determine the equation of the perpendicular line l_2.

What did you get? Find the solution below to double-check your answer.

Solution to Example 19

The equation of l_1 $3x - y - 5 = 0$ should first be written in the form $y = mx + c$ and then compared with the standard format.

$$3x - 5 = y \text{ or } y = 3x - 5$$

Thus

$$m_1 = 3$$

Therefore, the slope of the perpendicular line l_2 is

$$\boldsymbol{m_2 = -\frac{1}{3}}$$

We need to find the coordinates of intersection. Since $y = -2$ at the point of intersection, this is our y-coordinate. We can substitute this in the equation of l_1 to obtain the x-coordinate as:

$$3x - y - 5 = 0$$
$$3x - (-2) - 5 = 0$$
$$3x + 2 - 5 = 0$$
$$3x = 3$$
$$x = 1$$

Hence, the coordinates of the point of intersection are $(1, -2)$ and it is a point on the perpendicular line. Since we have a point and the slope, we can find the equation of the perpendicular line l_2 using:

$$(y - y_1) = m(x - x_1)$$

With point $(1, -2)$, we have $x_1 = 1$ and $y_1 = -2$

$$y - (-2) = -\frac{1}{3}(x - 1)$$
$$y + 2 = -\frac{1}{3}(x - 1)$$
$$3y + 6 = -(x - 1)$$
$$3y + 6 = -x + 1$$
$$3y + x + 6 - 1 = 0$$
$$\boldsymbol{\therefore x + 3y + 5 = 0}$$

One final example to try on this.

Example 20

Lines l_1 and l_2 are perpendicular to each other. l_1 passes through point $A(1, 2)$ and $B(3, -2)$ while line l_2 passes through point $C(7, 5)$.

a) Determine the value of c if l_2 crosses the y-axis at $(0, c)$.

b) Write the equation of l_2 in the form $ax + by + d = 0$.

What did you get? Find the solution below to double-check your answer.

Solution to Example 20

a) Value of c
Solution
Let the gradient of l_1 be m_1. Using $A(1, 2)$ and $B(3, -2)$, we have:

$$m_1 = \frac{y_2 - y_1}{x_2 - x_1}$$

$$= \frac{-2 - 2}{3 - 1} = \frac{-4}{2}$$

$$m_1 = -2$$

Since l_1 and l_2 are perpendicular to each other, if m_2 is the slope of l_2 then:

$$m_2 = -\frac{1}{m_1}$$

$$= -\frac{1}{-2} = \frac{1}{2}$$

Now we need to use $C(7, 5)$, m_2, and $y = mx + c$. Thus we have:

$$y = mx + c$$

$$5 = \frac{1}{2}(7) + c$$

$$5 - \frac{7}{2} = c$$

$$\therefore c = \frac{3}{2}$$

b) Equation of l_2
Solution
Using c, m_2 and $y = mx + c$, we have:

$$y = mx + c$$

$$y = \frac{1}{2}x + \frac{3}{2}$$

Let's re-write this in the required format as:

$$2y = x + 3$$

$$\therefore x - 2y + 3 = 0$$

5.3.2 CONDITIONS FOR PARALLELISM AND PERPENDICULARITY

Two or more lines are parallel if we can establish that their slopes are the same, such that:

$$\boxed{m_1 = m_2 = m_3 = m_4 \dots m_n}$$

(5.5)

To do this analytically, re-arrange their equations in the standard form and compare the slopes. If m (or the coefficients of x term or the independent variable) is the same, then the lines are parallel. We can also check this graphically, by confirming that the angle between the lines is either zero or 180 degrees.

Alternatively, the lines are parallel if they cannot meet no matter how much they are extended. This test can be assumed, otherwise it will become an exhaustive process depending on the separation distance between the lines and how much space is available to extend them.

Similarly, if two lines are perpendicular to each other, we must be able to show that:

1) The angle between them is 90 degrees (or they are at a right angle to each other).

2) They meet only once no matter how far they are extended.

3) Their slope is such that $m_1 = -\dfrac{1}{m_2}$. Alternatively, when their slopes are multiplied together, the answer is minus 1, i.e., $m_1 \times m_2 = -1$.

Let's try an example.

Example 21

State which of the following lines is parallel or perpendicular to $y = 2x - 13$.

a) $2y - 4x - 1 = 0$ b) $4x + 8y - 16 = 0$ c) $x - y + 3 = x - 1$

What did you get? Find the solution below to double-check your answer.

Solution to Example 21

HINT

The given equation is $y = 2x - 13$, thus $m = 2$. We are not interested in the y-intercept but want to establish whether the two lines are parallel or perpendicular.

a) $2y - 4x - 1 = 0$
Solution
Write the equation in the form $y = mx + c$

$$2y - 4x - 1 = 0$$
$$2y = 4x + 1$$
$$y = 2x + \frac{1}{2}$$

Hence,

$$m = 2$$
$$\therefore \text{it's PARALLEL}$$

---

b) $4x + 8y - 16 = 0$
Solution
Write the equation in the form $y = mx + c$

$$4x + 8y - 16 = 0$$
$$8y = -4x + 16$$
$$y = -\frac{4}{8}x + \frac{16}{8}$$
$$y = -\frac{1}{2}x + 2$$

Hence,

$$m = -\frac{1}{2}$$
$$\therefore \text{it's PERPENDICULAR because } -\frac{1}{2} \times 2 = -1$$

---

c) $x - y + 3 = x - 1$
Solution
Write the equation in the form $y = mx + c$

$$x - y + 3 = x - 1$$
$$-y = x - x - 1 - 3$$
$$-y = 0x - 4$$
$$y = 0x + 4$$

Hence,

$$m = 0$$

$\therefore$ it's neither **PARALLEL** nor **PERPENDICULAR** because **0 $\neq$ 2 (not parallel)** and **0 $\times$ 2 $\neq$ –1 (not perpendicular)**

Another example to try.

Example 22

Line l_1 passes through points $(0, \ -4)$ and $\left(2, \ -2\frac{1}{2}\right)$ and line l_2 passes through points $(3, \ -3)$ and $\left(\frac{3}{4}, \ 0\right)$. Show that the lines are perpendicular to each other.

What did you get? Find the solution below to double-check your answer.

Solution to Example 22

- **Slope of line l_1**

Let's label the points as:

$$(0, \ -4) = (x_1, y_1)$$

and

$$\left(2, -2\frac{1}{2}\right) = (x_2, y_2)$$

Thus, we have:

$$m = \frac{y_2 - y_1}{x_2 - x_1}$$

$$m_{l_1} = \frac{-2\frac{1}{2} - (-4)}{2 - 0}$$

$$= \frac{-\frac{5}{2} + 4}{2} = \frac{\left(\frac{3}{2}\right)}{2}$$

$$\therefore m_{l_1} = \frac{3}{4}$$

- **Slope of line l_2**

Let's label the points as:

$$\left(\frac{3}{4}, \ 0\right) = (x_1, y_1)$$

and

$$(3, -3) = (x_2, y_2)$$

Thus, we have:

$$m = \frac{y_2 - y_1}{x_2 - x_1}$$

$$m_{l_2} = \frac{-3 - 0}{3 - \frac{3}{4}}$$

$$= \frac{-3}{\left(\frac{9}{4}\right)} = -3 \times \frac{4}{9}$$

$$\therefore m_{l_2} = -\frac{4}{3}$$

If lines l_1 and l_2 are perpendicular, then the product of their slopes must be equal to -1. In other words:

$$m_{l_1} \times m_{l_2} = -1$$

That is

$$\frac{3}{4} \times -\frac{4}{3} = -1$$

5.4 INTERSECTION OF TWO LINES

When lines are not parallel, they are expected to meet at a point. If they are in a plane, they can only meet once and only once. We may be interested in knowing the coordinates of the point of intersection (x_i, y_i) shown in Figure 5.11.

To find the coordinates of the point of intersection of two lines, follow these steps:

Step 1: If the equations are in the form $ax + by + d = 0$ or another non-standard form, re-arrange and present them in the form $y = mx + c$.

Step 2: Equate the right-hand side of both equations. In other words, if after applying step 1, the two equations are: $y = m_1x + c_1$ and $y = m_2x + c_2$ then we say

$$m_1x + c_1 = m_2x + c_2$$

Step 3: Solve for x in step 2. This is the x-coordinate or x_i.

Step 4: Substitute the value of x in step 3 in either $y = m_1x + c_1$ or $y = m_2x + c_2$ to find y. This is the y-coordinate or y_i.

The above is a simultaneous linear equation solved in a slightly different way. You will meet this again in Chapter 7.

Let's try an example.

Example 23

The equations of lines l_1 and l_2 are $y - x + 2 = 0$ and $y + 2x - 1 = 0$ respectively.

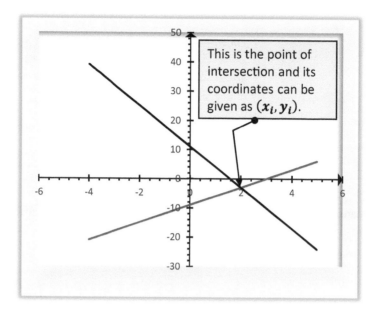

FIGURE 5.11   Coordinates of the point of intersection of two lines illustrated.

    a) Using the interval $-3 \leq x \leq 3$, plot the two equations on the same graph.

    b) Determine the coordinates of the point of intersection.

    c) Use an alternative (or analytical) method to find the point of intersection of the lines.

What did you get? Find the solution below to double-check your answer.

Solution to Example 23

a) Plot of $y - x + 2 = 0$ and $y + 2x - 1 = 0$
Solution
First, we need to re-arrange the equations and present them in the form $y = mx + c$ as:

$$y - x + 2 = 0$$
$$y = x - 2$$

and

$$y + 2x - 1 = 0$$
$$y = -2x + 1$$

We now need to prepare the table for each line equation in the stated interval as presented in Table 5.1. Let's illustrate how to fill the first row for each equation.

- $y = x - 2$

When $x = -3$, we have $y = -3 - 2 = -5$, hence the point A_1 is $(-3, -5)$.

- $y = -2x + 1$

When $x = -3$, we have $y = -2 \times -3 + 1 = 6 + 1 = 7$, hence the point A_2 is $(-3, 7)$.

To plot the graph, locate points A_1, B_1, ..., G_1 for $y = x - 2$ and join them together. Similarly, locate points A_2, B_2, ..., G_2 for $y = -2x + 1$ and join them together. These must be done on the same scale as shown in Figure 5.12.

---

b) The coordinates of the intersection point
Solution
Draw a vertical line to cross the x-axis and a horizontal line to cross the y-axis to obtain x-coordinate and y-coordinate respectively, as shown in Figure 5.13.

Hence, the coordinates are

$$(1, -1)$$

---

c) Analytical method
Solution
In this case, we have these two equations

$$y = x - 2 \ ----- (i)$$
$$y = -2x + 1 \ ----- (ii)$$

TABLE 5.1
Solution to Example 23(a)

	$y = x - 2$			$y = -2x + 1$	
x	y	(x_i, y_i)	x	y	(x_i, y_i)
-3	-5	$A_1(-3, -5)$	-3	7	$A_2(-3, 7)$
-2	-4	$B_1(-2, -4)$	-2	5	$B_2(-2, 5)$
-1	-3	$C_1(-1, -3)$	-1	3	$C_2(-1, 3)$
0	-2	$D_1(0, -2)$	0	1	$D_2(0, 1)$
1	-1	$E_1(1, -1)$	1	-1	$E_2(1, -1)$
2	0	$F_1(2, 0)$	2	-3	$F_2(2, -3)$
3	1	$G_1(3, 1)$	3	-5	$G_2(3, -5)$

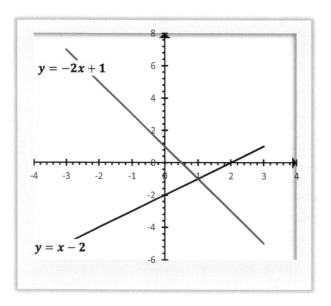

FIGURE 5.12   Solution to Example 23(a).

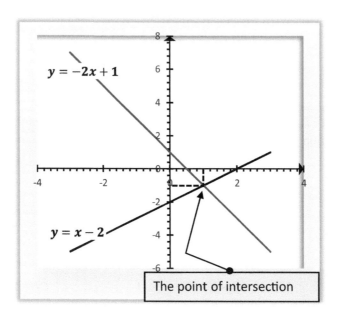

FIGURE 5.13   Solution to Example 23(b).

Equate the right-hand sides of both equations, we have:

$$x - 2 = -2x + 1$$

Simplify the above as:

$$x + 2x = 1 + 2$$
$$3x = 3$$
$$\therefore x = 1$$

The x-coordinate is 1. Now substitute $x = 1$ in either of the equations to obtain the value of y. From equation (i)

$$y = x - 2$$
$$= 1 - 2$$
$$\therefore y = -1$$

Or from equation (ii)

$$y = -2x + 1$$
$$= -2(1) + 1$$
$$= -2 + 1$$
$$\therefore y = -1$$

Hence, the coordinates of the point of intersection are

$$(1, -1)$$

5.5 CHAPTER SUMMARY

1) The method used to determine the equation of a straight line depends on the information available.

- If the gradient m and the y-intercept c are known, we use

$$\boxed{y = mx + c}$$

- If a point (x_1, y_1) and the gradient m are known, we use

$$\boxed{(y - y_1) = m(x - x_1)}$$

- If a point (x_1, y_1) and the y-intercept c are known, we use $y = mx + c$ to determine m and then follow the procedure for when we have the gradient m and the y-intercept c.

- If two points (x_1, y_1) and (x_2, y_2) on a line are known, we use

$$\boxed{\frac{y - y_1}{x - x_1} = \frac{y_2 - y_1}{x_2 - x_1}} \quad \text{OR} \quad \boxed{\frac{y - y_1}{y_2 - y_1} = \frac{x - x_1}{x_2 - x_1}}$$

2) We can determine the equation of a line from a given graph using one of the following options:

Option 1

- If the line crosses the y-axis, read the value at which the line crosses the y-axis. This is the y-intercept or c.
- Choose another suitable point on the line as your (x_1, y_1).
- Determine m using (x_1, y_1) by substituting in $y = mx + c$. Alternatively, use $m = \frac{y_2 - y_1}{x_2 - x_1}$ where $(x_2, y_2) = (0, c)$.
- Substitute m and c in $y = mx + c$ to write the equation.

Option 2

- If the line does not cross the y-axis, extend it to cross it and find c.
- Follow option 1 above to determine the equation of the line.

Option 3

- If the line does not cross y-axis and it is impossible to do so, choose two convenient points on the line and read their coordinates. You should now have (x_1, y_1) and (x_2, y_2).
- Determine gradient m from the graph.
- Use $(y - y_1) = m(x - x_1)$ or otherwise to write the equation as previously discussed.

3) When lines are parallel, their slopes are the same. This is given as:

$$\boxed{m_1 = m_2} \ \text{ OR } \ \boxed{m_1 = m_2 = m_3 = m_4 \ldots m_n}$$

4) When two lines l_1 and l_2 are perpendicular to each other, the slope of one is the negative reciprocal of the other or the product of their slopes equals -1.

$$\boxed{m_1 = -\frac{1}{m_2}} \ \text{ OR } \ \boxed{m_1 m_2 = -1}$$

5) When lines are not parallel, they are expected to meet at a point. If they are in a plane, they can only meet once and only once. The coordinates of the point of intersection (x_i, y_i) are obtained by solving the two equations of the lines simultaneously.

****

5.6 FURTHER PRACTICE

To access complementary contents, including additional exercises, please go to www.dszak.com.

6 Algebraic Equations I

Learning Outcomes

Once you have studied the content of this chapter, you should be able to:

- Differentiate between an expression and an equation
- Solve linear equations
- Solve quadratic equations by factorisation

6.1 INTRODUCTION

Earlier, we introduced the use of letters and symbols to represent unknown values, and this forms the basis of algebraic expressions. When these expressions are equated with another expression or a number, it becomes an algebraic equation. Consequently, the unknown letter or variable can only assume specific value(s) if the statement (of the equality) is to be true. We are often interested in determining these values, which is the focus of this chapter. We will cover, among other things, the difference between an expression and an equation, the method of solving linear equations, and solving quadratic equations by factorisation.

6.2 EXPRESSION AND EQUATION

An algebraic expression (or simply an expression) is a function of an independent variable or variables with one or more terms. The independent variable, an unknown quantity, can assume a range of values as set by the experimenter.

An algebraic term (or simply a term) is the simplest algebraic unit (like cells in biology, atoms in chemistry, and bits in computing) and may consist of:

a) a constant (number) or constants

b) a variable or variables

c) a combination of constant(s) and variable(s)

The above three parts must be connected using suitable mathematical operator, such as multiplication, division, power, and root, among others. Examples of algebraic terms are given in Table 6.1.

On the other hand, an algebraic equation is an expression that is equated to a constant k (k can be another number or another expression) such that the equality is valid only for a limited number of cases or answers. Table 6.2 provides a few examples with comments.

The key difference between an **expression** and an **equation** is the presence of the equality sign ($=$) in the latter. There is always an unknown quantity (or variable) in both cases.

DOI:10.1201/9781003027928-6

If an equation is valid for all values of the variable, this is technically called an **identity** and is represented with a three-lined symbol $\equiv$. Examples of identities are given in Table 6.3.

Of course, you will often find that the equal sign is used where the identity sign should have been used, this is common and acceptable. It will however not be valid to use the identity sign when the equal sign is intended. We can therefore say that all identities are equations but not vice versa.

TABLE 6.1

Constituents of a Term in an Algebraic Expression Described

Term	Component		Remarks
	Constant	**Variable**	
$2x^2$	2	x	Connected by multiplication as: $2x^2 = 2 \times x \times x$
$2\pi\sqrt{\dfrac{l}{g}}$	$2, \pi, g$	l	Connected by multiplication, division, and root as: $2\pi\sqrt{\dfrac{l}{g}} = 2 \times \pi \times \dfrac{\sqrt{l}}{\sqrt{g}}$
$\pi r l$	π	r, l	Connected by multiplication as: $\pi r l = \pi \times r \times l$

TABLE 6.2

Algebraic Equation

Equation	Remark	Validity
1) $v + 1 = 0$	It is an equation that is equal to 0.	It is valid only if $v = -1$.
2) $2x^2 - 5 = 3$	It is an equation that is equal to 3.	It is valid only if $x = 2$ or $x = -2$.
3) $x^3 + 8 = 0$	It is an equation that is equal to 0.	It is valid only if $x = -2$. There are two other values, but we are less concerned with them at this stage, as they are complex numbers, which is beyond the scope of this book.

TABLE 6.3

Remarks on Identities

Identity	Remark
1) $x^2 - 1 \equiv (x + 1)(x - 1)$	It is an identity because the value of the left-hand side (LHS) is equal to the right-hand side (RHS) for all values of x.
2) $\sin^2\theta + \cos^2\theta \equiv 1$	This is an identity because the LHS is always equal to 1 for all values of θ (measured in degrees, radians, or other units).

6.3 SOLVING LINEAR EQUATIONS

Solving an equation is a process of finding the value or values of the variable for which the expression on the LHS equals that on the RHS. These values are also called **solutions** or **roots** of the equation.

You can always verify if these values are correct by substituting the variables with these values and see if the RHS equals the LHS. Great idea, right?

6.3.1 LINEAR EXPRESSIONS

A linear expression in one variable is an algebraic expression that has the form:

$$\boxed{ax + b} \tag{6.1}$$

where a and b are real numbers. a is also called the coefficient of x.

The following should be noted about the above expression:

Note 1 The highest power of x in the expression must be 1. For this reason, $x^2 + 2x - 7$ is not a linear expression because the highest power is 2.

Note 2 The power of x should be positive, as such $x^{-1} + 3$ is not a linear expression.

Note 3 The constant b can be zero but a cannot. For example:

    a) $3x + 11$ is a linear expression where $a = 3$ and $b = 11$.

    b) πx is a linear expression where $a = \pi$ and $b = 0$.

6.3.2 LINEAR EQUATIONS

A linear equation in one variable (or sometimes called a simple equation) will generally take the form:

$$\boxed{ax + b = 0} \tag{6.2}$$

where x is the unknown quantity to be found.

Solving this equation implies finding the value of x that satisfies the above equality condition. It is a matter of making the unknown variable standalone as per the process of transposition covered in Chapter 3.

In another technical jargon, a linear equation is referred to as a **polynomial of degree 1**. This is covered in detail under polynomials in Chapter 9. For this reason, every linear equation must have only one answer (root or solution).

The graph of a linear equation is always a straight line and nothing else. The straight line can be plotted by joining two points (or coordinates) that satisfy the equation. We've shown this in Chapter 4, where this was represented by $y = mx + c$.

That's all for now. Let's take a break and try some examples.

Example 1

With reason(s), state whether each of the following is a linear equation or not.

a) $x + 3 = 0$ **b)** $x^{-1} + 4 = 0$ **c)** $x + 5$

d) $\frac{1}{x} - 7 = 0$ **e)** $13\sqrt{x} + 2 = 0$ **f)** $6x - y + 2 = 0$

What did you get? Find the solution below to double-check your answer.

Solution to Example 1

a) $x + 3 = 0$
Solution
This is a linear equation because:

- It can be written in the standard form $ax + b = 0$ and the highest power of x is 1.

- There is an $=$ sign connecting the left and right sides.

b) $x^{-1} + 4 = 0$
Solution
This is NOT a linear equation because:

- Although it appears in the 'standard form-like', the power of x is however negative.

- This will not give a straight-line graph when drawn.

c) $x + 5$
Solution
This is NOT an equation because:

- There is no $=$ sign. It is however a linear expression.

d) $\frac{1}{x} - 7 = 0$
Solution
This is NOT a linear equation because:

- $\frac{1}{x} - 7 = 0$ is the same as $x^{-1} - 7 = 0$. It becomes obvious in the transformed format that the power of x is negative.

- This will not give a straight-line graph when drawn.

NOTE
Let's modify the equation a bit as:

$$\frac{1}{x} - 7 = 0$$
$$1 - 7x = 0$$
$$7x - 1 = 0$$

After modification, it appears that the equation is linear. This is true but the original equation remains non-linear.

e) $13\sqrt{x} + 2 = 0$
Solution
This is NOT a linear equation because:

- $13\sqrt{x} + 2 = 0$ is the same as $13x^{\frac{1}{2}} + 2 = 0$. It becomes obvious in the transformed format that the power of x is not 1.

- This will not give a straight-line graph when drawn.

f) $6x - y + 2 = 0$
Solution
This is a linear equation because:

- $6x - y + 2 = 0$ can be re-written as $6x + 2 = y$. In this case, the LHS is clearly a linear expression and we can write $y = f(x)$.

- $6x - y + 2 = 0$ can also be re-written as $6x - y = -2$. In this case, it is the same as $ax + by = c$, which is a standard form for a linear equation in two variables.

- There is an $=$ sign connecting the left and right sides.

NOTE

- Although our focus here is a linear equation in one variable, it is helpful to know that a linear equation (or expression) can have more than a variable, provided that the power of each of the variables is 1.

- $3x + 4y = 5$ and $x - y - z = 10$ are examples of linear equations in two and three variables respectively. The variables here are x, y, and z.

Let's try another set of examples.

Example 2

State the coefficient of the unknown variable in each of the following linear equations.

a) $x + 3 = 0$ **b)** $2x - 3 = 17 + 8x$ **c)** $2(x - 3) = 3$ **d)** $9(1 - x) = -6x$

What did you get? Find the solution below to double-check your answer.

Solution to Example 2

HINT

We need to compare the given equation with the standard form $y = ax + b$.

a) $x + 3 = 0$
Solution
Comparing this with the standard form:

$$x + 3 = 0$$

which implies that

$$y = x + 3$$

Thus, $a = 1$.

∴ The coefficient of x is 1.

---

b) $2x - 3 = 17 + 8x$

Solution

Let us first simplify and write the equation in the standard form as:

$$2x - 3 = 17 + 8x$$
$$2x - 8x - 3 - 17 = 0$$
$$-6x - 20 = 0$$

which implies that

$$y = -6x - 20$$

Thus, $a = -6$.

∴ The coefficient of x is -6.

NOTE

We could have approached this as

$$2x - 3 = 17 + 8x$$
$$0 = 17 + 8x - 2x + 3$$
$$0 = 6x + 20$$

which implies that

$$y = 6x + 20$$

Thus, $a = 6$.

∴ The coefficient of x is 6.

---

c) $2(x - 3) = 3$

Solution

Let us open the brackets, simplify, and write the equation in the standard form as:

$$2(x - 3) = 3$$
$$2x - 6 = 3$$
$$2x - 6 - 3 = 0$$
$$2x - 9 = 0$$

which implies that

$$y = 2x - 9$$

Thus, $a = 2$.

$$\therefore \text{ The coefficient of } x \text{ is } 2.$$

---

d) $9(1 - x) = -6x$

Solution

Let us open the brackets, simplify, and write the equation in the standard form as:

$$9(1 - x) = -6x$$
$$9 - 9x = -6x$$
$$-9x + 6x + 9 = 0$$
$$-3x + 9 = 0$$

which implies that

$$y = -3x + 9$$

Thus, $a = -3$.

$$\therefore \text{ The coefficient of } x \text{ is } -3.$$

NOTE

For the same reason stated in (b), $y = 3x - 9$.

$$\therefore \text{The coefficient of } x \text{ is } 3.$$

One more set of examples to try.

Example 3

Solve the following linear equations.

a) $3s - 4 = 0$ **b)** $7t - 5 = 1 - t$ **c)** $\dfrac{u+8}{5} = 5u$

d) $\dfrac{1}{3}x - \dfrac{2}{5}x = \dfrac{1}{2}$ **e)** $5v + 7 - 4(8 - v) = 2$ **f)** $\dfrac{1}{5}w - 2 = \dfrac{2w}{3}$

g) $\dfrac{4y-1}{3} + \dfrac{y-5}{2} = \dfrac{2y}{4} - \dfrac{5}{6}$ **h)** $3(4 - x) - 7[8x - 5(x - 1)] = 1$ **i)** $\dfrac{5}{x-2} - \dfrac{7}{2x-4} = \dfrac{10}{3x}$

What did you get? Find the solution below to double-check your answer.

Solution to Example 3

a) $3s - 4 = 0$

Solution

$$3s - 4 = 0$$
$$3s = 4$$
$$\therefore s = \frac{4}{3}$$

b) $7t - 5 = 1 - t$
Solution

$$7t - 5 = 1 - t$$
$$7t + t = 1 + 5$$
$$8t = 6$$
$$t = \frac{6}{8}$$
$$\therefore t = \frac{3}{4}$$

c) $\frac{u+8}{5} = 5u$
Solution

$$\frac{u + 8}{5} = 5u$$
$$u + 8 = 25u$$
$$u - 25u = -8$$
$$-24u = -8$$
$$24u = 8$$
$$u = \frac{8}{24}$$
$$\therefore u = \frac{1}{3}$$

d) $\frac{1}{3}x - \frac{2}{5}x = \frac{1}{2}$
Solution

$$\frac{1}{3}x - \frac{2}{5}x = \frac{1}{2}$$

LCM of 3, 5, and 2 is 30, so we need to multiply each term by the LCM as:

$$\frac{1}{3}x \times 30 - \frac{2}{5}x \times 30 = \frac{1}{2} \times 30$$

Now simplify

$$x \times 10 - 2x \times 6 = 15$$
$$10x - 12x = 15$$
$$-2x = 15$$
$$x = -\frac{15}{2}$$
$$\therefore x = -7.5$$

e) $5v + 7 - 4(8 - v) = 2$

Solution

$$5v + 7 - 4(8 - v) = 2$$
$$5v + 7 - 32 + 4v = 2$$
$$5v + 4v + 7 - 32 - 2 = 0$$
$$9v - 27 = 0$$
$$9v = 27$$
$$v = \frac{27}{9}$$
$$\therefore v = 3$$

---

f) $\frac{1}{5}w - 2 = \frac{2w}{3}$

Solution

$$\frac{1}{5}w - 2 = \frac{2w}{3}$$
$$3w - 30 = 10w$$
$$3w - 10w = 30$$
$$-7w = 30$$
$$w = -\frac{30}{7}$$
$$\therefore w = -4\frac{2}{7}$$

---

g) $\frac{4y-1}{3} + \frac{y-5}{2} = \frac{2y}{4} - \frac{5}{6}$

Solution

$$\frac{4y - 1}{3} + \frac{y - 5}{2} = \frac{2y}{4} - \frac{5}{6}$$

LCM of 3, 2, 4, and 6 is 12, so we need to multiply each term by the LCM as:

$$12 \times \frac{4y - 1}{3} + 12 \times \frac{y - 5}{2} = 12 \times \frac{2y}{4} - 12 \times \frac{5}{6}$$
$$4 \times (4y - 1) + 6 \times (y - 5) = 3 \times 2y - 2 \times 5$$
$$4(4y - 1) + 6(y - 5) = 6y - 10$$
$$16y - 4 + 6y - 30 = 6y - 10$$
$$16y + 6y - 6y = 4 + 30 - 10$$
$$16y = 24$$
$$y = \frac{24}{16} = \frac{3}{2}$$
$$\therefore x = 1.5$$

h) $3(4 - x) - 7[8x - 5(x - 1)] = 1$
Solution

$$3(4 - x) - 7[8x - 5(x - 1)] = 1$$

Let's open the brackets as:

$$3(4 - x) - 7[8x - 5(x - 1)] = 1$$
$$12 - 3x - 7[8x - 5x + 5] = 1$$
$$12 - 3x - 7[3x + 5] = 1$$
$$12 - 3x - 21x - 35 = 1$$
$$-24x = 1 - 12 + 35$$
$$-24x = 24$$
$$x = -\frac{24}{24} = -1$$
$$\therefore x = -1$$

---

i) $\dfrac{5}{x-2} - \dfrac{7}{2x-4} = \dfrac{10}{3x}$
Solution
This question will require that we go gently, so we'll go step by step.

Step 1: Factorise the second fraction on the LHS

$$\frac{5}{x - 2} - \frac{7}{2(x - 2)} = \frac{10}{3x}$$

Step 2: Simplify the LHS by using the LCM $2(x - 2)$

$$\frac{10 - 7}{2(x - 2)} = \frac{10}{3x}$$
$$\frac{3}{2(x - 2)} = \frac{10}{3x}$$

Step 3: Take the reciprocal of both sides

$$\frac{2(x - 2)}{3} = \frac{3x}{10}$$

Step 4: Multiply both sides by 30 and simplify

$$\frac{2(x - 2)}{3} \times 30 = \frac{3x}{10} \times 30$$
$$2(x - 2) \times 10 = 3x \times 3$$
$$20(x - 2) = 9x$$

Step 5: Open the brackets

$$20x - 40 = 9x$$
$$20x - 9x = 40$$
$$11x = 40$$
$$\therefore x = \frac{40}{11}$$

Let's try another example.

Example 4

Show that $f(x) = (x-7)^2 - (x+3)^2$ can be written as $f(x) = ax + b$ where a and b are integers. Hence, determine the root of $f(x)$.

What did you get? Find the solution below to double-check your answer.

Solution to Example 4

HINT

- This does not look like a linear equation, but let's work through it.
- We will consider using the difference of two squares:

$$a^2 - b^2 = (a+b)(a-b)$$

Solution

$$f(x) = (x-7)^2 - (x+3)^2 = (x-7)^2 - (x+3)^2$$
$$= [(x-7) + (x+3)][(x-7) - (x+3)]$$
$$= (x - 7 + x + 3)(x - 7 - x - 3)$$
$$= (2x - 4)(-10)$$
$$= -10(2x - 4)$$
$$= -20x + 40$$

The root is when $f(x) = 0$, thus we have:

$$-20x + 40 = 0$$
$$-20x = -40$$
$$20x = 40$$
$$x = \frac{40}{20} = 2$$
$$\therefore x = 2$$

ALTERNATIVE METHOD

We will open the brackets in this case. Note that $(a \pm b)^2 = a^2 \pm 2ab + b^2$

$$\begin{aligned} f(x) &= (x-7)^2 - (x+3)^2 \\ &= (x^2 - 14x + 49) - (x^2 + 6x + 9) \\ &= x^2 - 14x + 49 - x^2 - 6x - 9 \\ &= -20x + 40 \end{aligned}$$

6.4 SOLVING QUADRATIC EQUATIONS

Solving quadratic equations will require some additional steps and notes, which we will be covering in this section.

6.4.1 QUADRATIC EXPRESSIONS

A quadratic expression in one variable is an algebraic expression that takes the form:

$$\boxed{ax^2 + bx + c} \text{ for } a \neq 0 \tag{6.3}$$

where a, b, and c are real numbers.

a and b are also called the coefficients of x^2 and x terms respectively, and c the constant term. The variable x is the unknown quantity and can represent any physical quantity. In other words, x can be force, time, speed, diffusion rate, etc.

Let's make the following notes about the above quadratic expression:

Note 1 The highest power of x in the expression must be 2. Consequently, $3x^3 + 2x - 7$ is not a quadratic expression because the highest power is 3.

Note 2 The power of x should be positive, as such $x^2 - x^{-1} + 3$ is not a quadratic expression.

Note 3 The constant b and c can be zero, but a cannot. For examples:

    a) $5x^2 - 4$ is a quadratic expression where $a = 5$, $b = 0$, and $c = -4$.

    b) $3x^2 + 5x$ is a quadratic expression where $a = 3$, $b = 5$, and $c = 0$.

    c) x^2 is a quadratic expression where $a = 1$, $b = 0$, and $c = 0$.

    d) $x + 12$ is NOT a quadratic expression because $a = 0$ though $b = 1$ and $c = 12$. In other words, $x + 12$ can be written as $0x^2 + x + 12$.

Let's illustrate the above with examples.

Example 5

With reason(s), state whether each of the following is a quadratic expression or not. Where applicable, also state the values of a, b, and c.

a) $x^2 + 4x - 3$ b) $3\phi^2 - 4\phi - \frac{1}{2}$ c) $\omega^2 - 5\omega^{-2} - 1$ d) $\mu^2 - \frac{3}{\sqrt{5}}$ e) $0.3t^2 + 0.25t - 1.1$

f) $3 - \theta^2 - \theta^3$ g) $1 + 2\beta - \sqrt{6}\beta^2$ h) $5\gamma - \frac{1}{\gamma^2}$ i) $\delta - \sqrt{3} - \delta^2$ j) $0.3 - 2\rho^2$

What did you get? Find the solution below to double-check your answer.

Solution to Example 5

a) $x^2 + 4x - 3$
Solution
This is a quadratic expression because:

- It can be written in the standard form $ax^2 + bx + c$.
- The highest power of x is 2, and all powers are positive integers.
- The coefficient a is not zero.

The coefficients are: $a = 1$, $b = 4$, $c = -3$.

b) $3\phi^2 - 4\phi - \frac{1}{2}$
Solution
This is a quadratic expression because:

- It can be written in the standard form $a\phi^2 + b\phi + c$.
- The highest power of ϕ is 2 and all powers are positive integers.
- The coefficient a is not zero.

The coefficients are: $a = 3$, $b = -4$, $c = -\frac{1}{2}$.

c) $\omega^2 - 5\omega^{-2} - 1$
Solution
This is NOT a quadratic expression because:

- It is not in the standard form $a\omega^2 + b\omega + c$, though the highest power of ω is 2.
- The power of the variable in the second term is negative.

d) $\mu^2 - \dfrac{3}{\sqrt{5}}$

Solution

This is a quadratic expression because:

- It can be written in the standard form $a\mu^2 + b\mu + c$.
- The highest power of μ is 2 and all powers are positive integers.
- The coefficient a is not zero.

The coefficients are: $a = 1$, $b = 0$, $c = -\dfrac{3}{\sqrt{5}}$.

---

e) $0.3t^2 + 0.25t - 1.1$

Solution

This is a quadratic expression because:

- It can be written in the standard form $at^2 + bt + c$.
- The highest power of x is 2 and all powers are positive integers.
- The coefficient a is not zero.

The coefficients are: $a = 0.3$, $b = 0.25$, $c = -1.1$.

---

f) $3 - \theta^2 - \theta^3$

Solution

This is NOT a quadratic expression because:

- The highest power of θ is 3.

---

g) $1 + 2\beta - \sqrt{6}\beta^2$

Solution

This is a quadratic expression because:

- It can be written in the standard form $a\beta^2 + b\beta + c$.
- The highest power of β is 2 and all powers are positive integers.
- The coefficient a is not zero.

The coefficients are: $a = -\sqrt{6}$, $b = 2$, $c = 1$.

---

h) $5\gamma - \dfrac{1}{\gamma^2}$

Solution

This is NOT a quadratic expression because:

- If we re-write $5\gamma - \dfrac{1}{\gamma^2}$ we will get $5\gamma - \gamma^{-2}$, which is clearly not in the standard form $a\gamma^2 + b\gamma + c$.

---

i) $\delta - \sqrt{3} - \delta^2$

Solution

This is a quadratic expression because:

- It can be written in the standard form $a\delta^2 + b\delta + c$.
- The highest power of δ is 2 and all powers are positive integers.
- The coefficient a is not zero.

The coefficients are: $a = -1$, $b = 1$, $c = -\sqrt{3}$.

---

j) $0.3 - 2\rho^2$

Solution

This is a quadratic expression because:

- It can be written in the standard form $a\rho^2 + b\rho + c$.
- The highest power of ρ is 2 and all powers are positive integers.
- The coefficient a is not zero.

The coefficients are: $a = -2$, $b = 0$, $c = 0.3$.

6.4.2 SUBSTITUTION

Before we move further, it will be helpful to briefly cover the idea of substitution. We mentioned earlier that to establish that the zeros obtained for an equation are correct, we only need to replace the unknown variable with them and check that it satisfies the equation, i.e., LHS = RHS.

This idea can be generalised to include expressions, where one may be interested in what would be the numerical value of an expression, if a numerical value is assigned to the independent variable in the expression. For example, the expression $x^2 + x - 3$ has a numerical value of -3 when $x = 0$, because $0^2 + 4(0) - 3 = -3$. Similarly, if we assume x to be 3, the expression will have a value of 9.

There is a shorthand format to express this called **function**. A function is written as $f(x)$ and it is read as 'f **function of** x'. What it implies is that there is an expression that varies as the unknown x changes. We can also state that the function depends on x. Therefore, we can neatly write our quadratic expression as:

$$f(x) = x^2 + x - 3$$

When we desire to evaluate the above function at a particular value of x, say $x = 3$, we'll change the x in the brackets () for the desired value. Here we will write $f(x = 3)$ or simply $f(3)$ to imply that we want to evaluate the function at this value.

Let's show the full working now.

$$f(x) = x^2 + x - 3$$
$$f(3) = (3)^2 + (3) - 3$$
$$= 9 + 3 - 3$$
$$\therefore f(3) = 9$$

That's cool, right? So, the statement 'the value of $x^2 + x - 3$ when $x = -2$ is -1' can be written as:

$$f(-2) = -1$$

The f in $f(x)$, by custom, stands for function but we can equally choose a different letter. So $g(x)$ and $h(x)$ are also functions of x. It becomes imperative to choose a letter different from f when we have two or more functions to denote.

One more characteristic is that if the variable is a different letter or symbol, just use it instead of x. Thus, we have $f(t)$ and $g(\theta)$ as functions that vary with respect to t and θ respectively. It is important to mention that this is not limited or exclusive to quadratic or linear functions, rather it is for any function.

That's all for now, let's try some examples.

Example 6

Evaluate the following functions at the stated value(s) of the variable.

a) $f(x) = 3x - 5$ $[x = -1, \ x = 5]$ b) $v(t) = 2t^2 - 5t + 3$ $[t = 1, \ t = 2]$

c) $g(\alpha) = \ln(3\alpha + 7)$ $[\alpha = -2, \ \alpha = 1]$ d) $h(\omega) = 4e^{2\omega} - \omega$ $[\omega = 0, \ \omega = 0.5]$

e) $i(\theta) = \sin(2\theta) - \cos(2\theta)$ $\left[\theta = 0, \ \theta = \frac{\pi}{6}\right]$ f) $p(x) = 3x.\tan x$ $\left[x = \frac{\pi}{4}, \ x = \pi\right]$

What did you get? Find the solution below to double-check your answer.

Solution to Example 6

HINT

For Examples (e) and (f), ensure that you work in radians. You can either change the mode of your calculator to radians or convert π to $180°$. See Section 14.2 for further details regarding conversion between angles in degrees and radians.

a) $f(x) = 3x - 5$ $[x = -1, \ x = 5]$
Solution
At $x = -1$, we have:

$$f(x) = 3x - 5$$
$$f(-1) = 3(-1) - 5$$
$$= -3 - 5 = -8$$
$$\therefore f(-1) = -8$$

At $x = 5$, we have:

$$f(x) = 3x - 5$$
$$f(5) = 3(5) - 5$$
$$= 15 - 5 = 10$$
$$\therefore f(5) = 10$$

---

b) $v(t) = 2t^2 - 5t + 3$ $\qquad [t = 1, \; t = 2]$
Solution
At $t = 1$, we have:

$$v(t) = 2t^2 - 5t + 3$$
$$v(1) = 2(1)^2 - 5(1) + 3$$
$$= 2 - 5 + 3 = 0$$
$$\therefore v(1) = 0$$

At $t = 2$, we have:

$$v(x) = 2t^2 - 5t + 3$$
$$v(2) = 2(2)^2 - 5(2) + 3$$
$$= 2(4) - 10 + 3 = 1$$
$$\therefore v(2) = 1$$

---

c) $g(\alpha) = \ln(3\alpha + 7)$ $\qquad [\alpha = -2, \; \alpha = 1]$
Solution
At $\alpha = -2$, we have:

$$g(\alpha) = \ln(3\alpha + 7)$$
$$g(-2) = \ln(3 \times -2 + 7)$$
$$= \ln(-6 + 7) = \ln(1) = 0$$
$$\therefore g(-2) = 0$$

At $\alpha = 1$, we have:

$$g(\alpha) = \ln(3\alpha + 7)$$
$$g(1) = \ln(3 \times 1 + 7)$$
$$= \ln(3 + 7) = \ln(10)$$
$$\therefore g(1) = \ln(10)$$

---

d) $h(\omega) = 4e^{2\omega} - \omega$ $\qquad [\omega = 0, \; \omega = 0.5]$
Solution
At $\omega = 0$, we have:

$$h(\omega) = 4e^{2\omega} - \omega$$
$$h(0) = 4e^{2\omega} - \omega$$
$$= 4e^{2 \times 0} - 0 = 4e^0 - 0$$
$$= 4 \times 1 - 0 = 4$$
$$\therefore h(0) = 4$$

At $\omega = 0.5$, we have:

$$h(\omega) = 4e^{2\omega} - \omega$$
$$h(0.5) = 4e^{2\omega} - \omega$$
$$= 4e^{2 \times 0.5} - 0.5$$
$$= 4e^1 - 0.5 = 4e - 0.5$$
$$\therefore h(0.5) = 4e - \frac{1}{2}$$

---

e) $i(\theta) = \sin(2\theta) - \cos(2\theta)$ $\left[\theta = 0, \ \theta = \frac{\pi}{6}\right]$

Solution

At $\theta = 0$, we have:

$$i(\theta) = \sin(2\theta) - \cos(2\theta)$$
$$i(0) = \sin(2 \times 0) - \cos(2 \times 0)$$
$$= \sin(0) - \cos(0)$$
$$= 0 - 1 = -1$$
$$\therefore i(0) = -1$$

At $\theta = \frac{\pi}{6}$, we have:

$$i(\theta) = \sin(2\theta) - \cos(2\theta)$$
$$i\left(\frac{\pi}{6}\right) = \sin\left(2 \times \frac{\pi}{6}\right) - \cos\left(2 \times \frac{\pi}{6}\right)$$
$$= \sin\left(\frac{\pi}{3}\right) - \cos\left(\frac{\pi}{3}\right)$$
$$= \frac{\sqrt{3}}{2} - \frac{1}{2} = \frac{-1 + \sqrt{3}}{2}$$
$$\therefore i\left(\frac{\pi}{6}\right) = \frac{-1 + \sqrt{3}}{2}$$

---

f) $p(x) = 3x.\tan x$ $\left[x = \frac{\pi}{4}, \ x = \pi\right]$

Solution

At $x = \frac{\pi}{4}$, we have:

$$p(x) = 3x.\tan x$$
$$p\left(\frac{\pi}{4}\right) = 3\left(\frac{\pi}{4}\right).\tan\left(\frac{\pi}{4}\right)$$
$$= \frac{3\pi}{4}.1$$
$$= \frac{3\pi}{4}$$
$$\therefore p\left(\frac{\pi}{4}\right) = \frac{3\pi}{4}$$

At $x = \pi$, we have:

$$p(x) = 3x.\tan x$$
$$p(\pi) = 3(\pi).\tan(\pi)$$
$$= 3\pi.0 = 0$$
$$\therefore p(\pi) = 0$$

Let's try another example.

Example 7

Show that $t = 1$ and $t = -\dfrac{3}{2}$ satisfy the equation $2t^2 + t - 3 = 0$.

What did you get? Find the solution below to double-check your answer.

Solution to Example 7

HINT

In this example, all we need to demonstrate is that when the value of t is substituted in the LHS we will obtain zero (or whatever the RHS is).

$$f(t) = 2t^2 + t - 3$$

for $t = 1$, we need to substitute $t = 1$.

$$f(1) = 2(1)^2 + 1 - 3$$
$$= 2 + 1 - 3$$
$$\therefore f(1) = 0$$

for $t = -\dfrac{3}{2}$, we need to substitute $t = -\dfrac{3}{2}$:

$$f\left(-\frac{3}{2}\right) = 2\left(-\frac{3}{2}\right)^2 + \left(-\frac{3}{2}\right) - 3$$
$$= 2\left(\frac{9}{4}\right) - \frac{3}{2} - 3$$
$$= \frac{9}{2} - \frac{3}{2} - 3 = \frac{6}{2} - 3$$
$$\therefore f\left(-\frac{3}{2}\right) = 0$$

Because $f(t) = 0$ in both cases, these values satisfy the equation $f(t)$.

6.4.3 QUADRATIC EQUATIONS

A quadratic equation is simply a quadratic expression that is equated to zero, a constant, or an expression. In another technical term, a quadratic equation is referred to as a **polynomial of degree 2**.

Every quadratic equation must have two answers (roots or solutions), both real values or both complex values. In this section, we will only consider the real roots. Therefore, when we say no roots for a particular equation, it implies no real roots.

Although we say that a quadratic equation produces two solutions, it does not necessarily mean that the two are valid for the problem being solved. For example, solving for t (time) in a linear motion problem using $s = ut + \frac{1}{2}at^2$ may produce two positive values where one is not valid for the context. It can also yield two roots, with one being positive and the other negative. Obviously, time cannot be negative and as such the negative value is an invalid answer.

Unlike a linear equation, there are three (or four) methods to solve a quadratic equation, namely:

- Factorisation method
- Completing the square method
- Quadratic formula method
- Graphical method

We will discuss factorisation method in this chapter while the remaining methods will be covered in Chapter 7. Let's go.

6.4.4 FACTORISATION METHOD

Factorisation is one of the three common methods used to solve a quadratic equation analytically, and it is usually used if all that is needed is the roots of the equation. By factorisation, it implies that the quadratic equation is in (or should be placed into) two brackets, with each part called a **factor**, thus the name for the method (factorisation). Technically speaking, it is the quadratic expression in a quadratic equation that we are factorising.

Factorising a quadratic equation is illustrated below:

$$ax^2 + bx + c = 0$$
$$(\alpha x + \lambda)(\beta x + \mu) = 0$$

where α, β, λ, and μ are real numbers.

The two factors, $(\alpha x + \lambda)$ and $(\beta x + \mu)$, are multiplied together to give zero and we use the above fact to find the solutions of the equation.

Note that when two numbers are multiplied to give zero, it is either one or both numbers are zero. Therefore, the solution of the quadratic $ax^2 + bx + c = 0$ is

$$\alpha x + \lambda = 0 \quad \textbf{OR} \quad \beta x + \mu = 0$$
$$\alpha x = -\lambda \qquad \beta x = -\mu$$
$$x = -\frac{\lambda}{\alpha} \qquad x = -\frac{\mu}{\beta}$$

For the purpose of using this method, quadratic equations can be grouped into two, and each can be factorised slightly differently. Note that we cover this approach in Section 3.5.3, but we will re-visit it here.

6.4.4.1 Case 1: $a = 1$

For this case, the general expression $ax^2 + bx + c = 0$ becomes $x^2 + bx + c = 0$. An example of this case is $x^2 - 3x + 4 = 0$. To factorise a quadratic equation with $a = 1$, follow these steps:

Step 1: If not already expressed, write the equation in the standard form:

$$x^2 + bx + c = 0$$

Step 2: Write the two brackets and put x in each as:

$$(x \quad)(x \quad) = 0$$

Note that the coefficient of x in both brackets is 1, because when you open the brackets, only this combination will give x^2.

Step 3: Think about two numbers whose product is equal to c and whose sum is b. Let's say the numbers are m and n. Thus, $mn = c$ and $m + n = b$.

Step 4: Put each number along with their sign into each bracket as:

$$(x + m)(x + n) = 0$$

Step 5: Equate each factor to zero and solve for x as previously shown for linear equations. In other words, $(x + m)(x + n) = 0$ implies that

$$x + m = 0 \ \rightarrow x = -m \ \textbf{ OR } \ x + n = 0 \ \rightarrow \ x = -n.$$

Now let's illustrate with examples.

Example 8

Solve the following quadratic equations by factorisation method.

a) $x^2 - 4x + 3 = 0$ 　　　　b) $t^2 - 2t + 1 = 0$ 　　　　c) $w^2 + 7w = -12$

d) $144 - 7m = m^2$ 　　　　e) $36 - k^2 = 0$ 　　　　f) $p^2 = 13p$

What did you get? Find the solution below to double-check your answer.

Solution to Example 8

a) $x^2 - 4x + 3 = 0$
Solution

$$x^2 - 4x + 3 = 0$$
$$(x \quad)(x \quad) = 0$$

For this case, $b = -4$ and $c = 3$. Next, we need two numbers whose sum is -4 and product is 3. The answer is -3 and -1. Hence,

$$x^2 - 4x + 3 = 0$$
$$(x \quad)(x \quad) = 0$$
$$(x - 3)(x - 1) = 0$$

Now that we have completed factorisation, let's proceed to solve the equation as:

$$(x - 3) = 0 \quad \textbf{OR} \quad (x - 1) = 0$$
$$x - 3 = 0 \qquad\qquad x - 1 = 0$$
$$\therefore x = 3 \qquad\qquad \therefore x = 1$$

b) $t^2 - 2t + 1 = 0$
Solution

$$t^2 - 2t + 1 = 0$$
$$(t \quad)(t \quad) = 0$$

For this case, $b = -2$ and $c = 1$. Next, we need two numbers whose sum is -2 and product is 1. The answer is -1 and -1. Hence,

$$t^2 - 2t + 1 = 0$$
$$(t \quad)(t \quad) = 0$$
$$(t - 1)(t - 1) = 0$$

We have now completed factorisation, let's proceed to solve the equation as:

$$(t - 1) = 0 \quad \textbf{OR} \quad (t - 1) = 0$$
$$t - 1 = 0 \qquad\qquad t - 1 = 0$$
$$\therefore t = 1 \qquad\qquad \therefore t = 1$$

NOTE

- For this case, we say $t = 1$ (*twice*) or say this is a repeated root.
- As a result, we still consider that we have two roots.

c) $w^2 + 7w = -12$

Solution

Let us first write the equation in the standard form as:

$$w^2 + 7w = -12$$
$$w^2 + 7w + 12 = 0$$
$$(w \quad)(w \quad) = 0$$

For this case, $b = 7$ and $c = 12$. Next, we need two numbers whose sum is 7 and product is 12. The answer is 3 and 4. Hence,

$$w^2 + 7w + 12 = 0$$
$$(w \quad)(w \quad) = 0$$
$$(w + 3)(w + 4) = 0$$

We have completed factorisation, let's proceed to solve the equation as:

$$(w + 3) = 0 \qquad \textbf{OR} \qquad (w + 4) = 0$$
$$w + 3 = 0 \qquad\qquad\qquad w + 4 = 0$$
$$\therefore w = -3 \qquad\qquad\qquad \therefore w = -4$$

---

d) $144 - 7m = m^2$

Solution

Let us first write the equation in the standard form as:

$$144 - 7m = m^2$$
$$m^2 + 7m - 144 = 0$$
$$(m \quad)(m \quad) = 0$$

For this case, $b = 7$ and $c = -144$. Next, we need two numbers whose sum is 7 and product is -144. The answer is -9 and 16. Hence,

$$m^2 + 7m - 144 = 0$$
$$(m \quad)(m \quad) = 0$$
$$(m - 9)(m + 16) = 0$$

We have now completed factorisation, let's proceed to solve the equation as:

$$(m - 9) = 0 \qquad \textbf{OR} \qquad (m + 16) = 0$$
$$m - 9 = 0 \qquad\qquad\qquad m + 16 = 0$$
$$\therefore m = 9 \qquad\qquad\qquad \therefore m = -16$$

---

e) $36 - k^2 = 0$

Solution

Let us first write the equation in the standard form as:

$$36 - k^2 = 0$$

Multiply both sides by -1, we have:

$$-36 + k^2 = 0$$
$$k^2 - 36 = 0$$
$$k^2 + 0k - 36 = 0$$
$$(k \quad)(k \quad) = 0$$

For this case, $b = 0$ and $c = -36$. Next, we need two numbers whose sum is 0 and product is -36. The answer is -6 and 6. Hence,

$$k^2 - 36 = 0$$
$$(k \quad)(k \quad) = 0$$
$$(k - 6)(k + 6) = 0$$

We have now completed factorisation, let's proceed to solve the equation as:

$$(k - 6) = 0 \quad \textbf{OR} \quad (k + 6) = 0$$
$$k - 6 = 0 \qquad\qquad k + 6 = 0$$
$$\therefore k = 6 \qquad\qquad \therefore k = -6$$

ALTERNATIVE METHOD

This is called the difference of two squares. We've explained this in Section 3.5.2.

$$36 - k^2 = 0$$

becomes

$$6^2 - k^2 = 0$$

We can see that there are two squares with a minus between them.
To solve the difference of two squares, add the term in one factor and minus the subtrahend (the second term) from the first term in the second factor. Thus,

$$6^2 - k^2 = 0$$
$$(6 - k)(6 + k) = 0$$

We have now completed factorisation, let's proceed to solve the equation as:

$$(6 - k) = 0 \quad \textbf{OR} \quad (6 + k) = 0$$
$$6 = k \qquad\qquad 6 + k = 0$$
$$\therefore k = 6 \qquad\qquad \therefore k = -6$$

f) $p^2 = 13p$

Solution

Let's first write the equation in the standard form as:

$$p^2 = 13p$$
$$p^2 - 13p = 0$$
$$p^2 - 13p + 0 = 0$$
$$(p \quad)(p \quad) = 0$$

For this case, $b = -13$ and $c = 0$. Next, we need two numbers whose sum is -13 and product is 0. The answer is 0 and -13. Hence,

$$p^2 - 13p + 0 = 0$$
$$(p \quad)(p \quad) = 0$$
$$(p + 0)(p - 13) = 0$$

We've now completed factorisation; we are required to solve the equation as:

$$(p + 0) = 0 \quad \textbf{OR} \quad (p - 13) = 0$$
$$p + 0 = 0 \qquad\qquad p - 13 = 0$$
$$\therefore p = 0 \qquad\qquad \therefore p = 13$$

Let's state the following observations:

Note 1 It is common to solve this as follows:

$$p^2 - 13p = 0$$
$$p(p - 13) = 0$$

$$p = 0 \quad \textbf{OR} \quad (p - 13) = 0$$
$$\therefore p = 0 \qquad\qquad p - 13 = 0$$
$$\qquad\qquad\qquad \therefore p = 13$$

This is a straightforward and easy approach for this type of equation.

Note 2 It is a common mistake to approach the question as:

$$p^2 = 13p$$

Divide both sides by p

$$p = 13$$

Whilst the answer above is correct, we have however lost another answer in the process. One may argue that it is zero. Yes, it is. However, stating zero as a solution should be emphasized due to its greater relevance compared to a numerical zero.

Note 3 In general, and as previously mentioned, a quadratic solution must have two solutions or answers. This is a principle that cannot be ignored.

Another set of examples to try.

Example 9

Using the factorisation method, determine the roots of the following quadratic equations.

a) $(x-3)^2 - 16 = 0$ **b)** $x^4 - 5x^2 + 4 = 0$

What did you get? Find the solution below to double-check your answer.

Solution to Example 9

a) $(x-3)^2 - 16 = 0$
Solution
Let's first write the equation in the standard form as:

$$(x-3)^2 - 16 = 0$$
$$x^2 - 6x + 9 - 16 = 0$$
$$x^2 - 6x - 7 = 0$$
$$(x \quad)(x \quad) = 0$$

For this case, $b = -6$ and $c = -7$. Next, we need two numbers whose sum is -6 and product is -7. The answer is -7 and 1. Hence,

$$x^2 - 6x - 7 = 0$$
$$(x-7)(x+1) = 0$$

Now that we have completed factorisation, let's proceed to solve the equation as:

$$x - 7 = 0 \quad \textbf{OR} \quad x + 1 = 0$$
$$\therefore x = 7 \qquad\qquad \therefore x = -1$$

---

b) $x^4 - 5x^2 + 4 = 0$
Solution
Let's first write the equation in the standard form as:

$$x^4 - 5x^2 + 4 = 0$$
$$\left(x^2\right)^2 - 5\left(x^2\right) + 4 = 0$$
$$\left(x^2 \quad\right)\left(x^2 \quad\right) = 0$$

Notice how we've placed x^2 in brackets, which will be treated like x. For this case, $b = -5$ and $c = 4$. Next, we need two numbers whose sum is -5 and product is 4. The answer is -4 and -1. Hence,

$$x^4 - 5x^2 + 4$$
$$\left(x^2 - 1\right)\left(x^2 - 4\right) = 0$$

Now that we have completed factorisation, let's proceed to solve the equation as:

$$x^2 - 1 = 0 \quad \textbf{OR} \quad x^2 - 4 = 0$$
$$x^2 = 1 \qquad\qquad x^2 = 4$$
$$x = \sqrt{1} \qquad\qquad x = \sqrt{4}$$
$$\therefore x = \pm 1 \qquad\qquad \therefore x = \pm 2$$

NOTE

Technically, this is a quartic equation, but we've modelled it as a quadratic equation. It is therefore not surprising that we obtained four solutions.

Let's try another example.

Example 10

Show that $(x + 2)^2 + x(7x - 8) = (1 - 3x)^2$ can be written as $ax^2 + bx + c = 0$, where a, b, and c are integers. Hence, solve for x using the factorisation method.

What did you get? Find the solution below to double-check your answer.

Solution to Example 10

HINT

This does not look like a quadratic equation, but let's work through it.

Solution

$$(x + 2)^2 + x(7x - 8) = (1 - 3x)^2$$
$$(x + 2)^2 + x(7x - 8) - (1 - 3x)^2 = 0$$
$$(x^2 + 4x + 4) + 7x^2 - 8x - (1 - 6x + 9x^2) = 0$$
$$x^2 + 4x + 4 + 7x^2 - 8x - 1 + 6x - 9x^2 = 0$$
$$x^2 + 7x^2 - 9x^2 + 4x - 8x + 6x + 4 - 1 = 0$$
$$-x^2 + 2x + 3 = 0$$

The root is when $f(x) = 0$, thus we have:

$$-\left(x^2 - 2x - 3\right) = 0$$
$$x^2 - 2x - 3 = 0$$
$$(x \quad)(x \quad) = 0$$

For this case, $b = -2$ and $c = -3$. Next, we need two numbers whose sum is -2 and product is -3. The answer is -3 and 1. Hence,

$$x^2 - 2x - 3 = 0$$
$$(x - 3)(x + 1) = 0$$

Now that we have completed factorisation, let's proceed to solve the equation as:

$$x - 3 = 0 \quad \textbf{OR} \quad x + 1 = 0$$
$$\therefore x = 3 \qquad\qquad \therefore x = -1$$

6.4.4.2 Case 2: $a \neq 1$

An example of this case is $3x^2 - 4x - 5 = 0$. To factorise a quadratic equation where $a \neq 1$ can be challenging and requires more practice and trials. We will try to explain a couple of methods here in addition to the one we showed in Chapter 3 for factorising quadratic expressions. We should mention that these methods can also be used where $a = 1$.

Method 1

To use this method, follow these steps:

Step 1: If not already, write the equation in the standard form as:

$$ax^2 + bx + c = 0$$

Step 2: If applicable, attempt to factorise the expression. If after factorisation, you end up with $a = 1$, follow the steps for solving the case of $a = 1$ as previously demonstrated, otherwise go to the next step.

Step 3: Draw two vertical lines to create a grid with two columns and two rows. This looks like a 2 by 2 matrix with four unknown elements indicated by '$\cdots$'.

$$\begin{vmatrix} \cdots & \cdots \\ \cdots & \cdots \end{vmatrix}$$

Step 4: For ease of reference, we have labelled them. $C1$ and $C2$ are column 1 and column 2 respectively, and $R1$ and $R2$ represent row 1 and row 2 respectively. Columns will be designated for identifying the four elements, while rows will be used to determine the factors and the solutions of the equation.

$$\begin{array}{c} \quad C1 \quad C2 \\ \begin{array}{c} R1 \\ \\ R2 \end{array}\begin{vmatrix} \cdots & \cdots \\ \cdots & \cdots \end{vmatrix} \end{array}$$

Step 5: Remove the labels $C1$ and $C2$ and replace them with ax and c respectively. Do not worry that b or bx has not appeared, they will shortly.

$$\begin{array}{c} \quad (ax \quad c) \\ \begin{array}{c} R1 \\ \\ R2 \end{array}\begin{vmatrix} \cdots & \cdots \\ \cdots & \cdots \end{vmatrix} \end{array}$$

Step 6: Choose two numbers m and n such that $mn = a$ and place them in the first column.

$$
\begin{array}{c}
(ax \quad c) \\
R1 \begin{vmatrix} m & \cdots \\ \vdots & \vdots \\ \end{vmatrix} \\
R2 \begin{vmatrix} n & \cdots \end{vmatrix}
\end{array}
$$

Step 7: Choose another pair of numbers λ and μ such that $\lambda m + \mu n = b$ and place these in the second column. This step requires trial and error, so don't give up.

$$
\begin{array}{c}
(ax \quad c) \\
R1 \begin{vmatrix} m & \lambda \\ & \\ \end{vmatrix} \\
R2 \begin{vmatrix} n & \mu \end{vmatrix}
\end{array}
$$

Step 8: Finally, we are there. The $R1$ and $R2$ are the factors of the quadratic equation written as:

$$(mx + \lambda)\,(nx + \mu) = 0$$

$$
\begin{array}{c}
(ax \quad c) \\
\begin{vmatrix} m & \lambda \\ n & \mu \end{vmatrix}
\begin{array}{l} R1 \to mx + \lambda \\ R2 \to (nx + \mu) \end{array}
\end{array}
$$

This is a long one, but with practice you will find it easy and quick. The first column will be quick to fill, but the second column will come with practice. Sometimes, all we need is to change the sign or swap the positions of λ and μ.

Before we get into examples, let's explain the second method.

Method 2

Follow these steps to use this method:

Step 1: If written differently, write the equation in the standard form as:

$$ax^2 + bx + c = 0$$

Step 2: If applicable, attempt to factorise the expression. If after factorisation, you end up with $a = 1$, follow the steps for solving this case as previously demonstrated, otherwise go to the next step.

Step 3: Write the two brackets and put x in each, with coefficients whose product is equal to a (i.e., $mn = a$) as:

$$(mx \quad)(nx \quad) = 0$$

Although it rarely happens, you may have to change the choice if the next step proves impossible. It is also possible that the factorisation method is not the right option, try another method.

Step 4: This is a tricky one. We need to think about two numbers whose product is equal to c and at the same time when you multiply one of these numbers with m and the other with n and add them together, the result is b.

Let's say the two numbers are λ and μ, then $\lambda\mu = c$ and $m\mu + n\lambda = b$.

Step 5: Put each number along with its sign into each bracket as:

$$(mx + \lambda)(nx + \mu) = 0$$

Ensure that the numbers you multiply together in step 4 above are not in the same brackets.

These are the two methods that we would like to cover here. If we look carefully, we will realise that these two methods are essentially the same or at least similar.

Let's try some examples.

Example 11

Solve the following quadratic equations by factorisation method.

a) $3x^2 - x - 2 = 0$ **b)** $2s^2 - 11s + 14 = 0$

c) $12v^2 + 2v - 70 = 0$ **d)** $33 + 5z - 2z^2 = 0$

What did you get? Find the solution below to double-check your answer.

Solution to Example 11

a) $3x^2 - x - 2 = 0$
Solution
For this case, $a = 3$, $b = -1$, and $c = -2$. Now we need to set the scene as:

$$\begin{Vmatrix} 3 & 2 \\ 1 & -1 \end{Vmatrix}$$

We've used double vertical lines so as not to confuse this with a matrix. It can be observed from the above that:

    1) Column 1

$$3 \times 1 = 3 = a$$

    2) Column 2

$$2 \times -1 = -2 = c$$

The addition of the cross-multiplication

$$(3 \times -1) + (1 \times 2) = -3 + 2 = -1 = b$$

Hence, we can write the factors using rows 1 and 2 as:

$$3x^2 - x - 2 = 0$$
$$(3x + 2)(x - 1) = 0$$

We have now completed factorisation; let's proceed to solve the equation as:

$$(3x + 2) = 0 \qquad \textbf{OR} \qquad (x - 1) = 0$$
$$3x + 2 = 0 \qquad\qquad x - 1 = 0$$
$$3x = -2 \qquad\qquad \therefore x = 1$$
$$\therefore x = -\frac{2}{3}$$

b) $2s^2 - 11s + 14 = 0$

Solution

For this case, $a = 2$, $b = -11$, and $c = 14$. Let's set the scene as:

$$\begin{Vmatrix} 2 & -7 \\ 1 & -2 \end{Vmatrix}$$

It can be seen from the above that:

1) Column 1

$$2 \times 1 = 2 = a$$

2) Column 2

$$-7 \times -2 = 14 = c$$

The addition of the cross-multiplication

$$(2 \times -2) + (1 \times -7) = -4 - 7 = -11 = b$$

Hence, we can write the factors using rows 1 and 2 as:

$$2s^2 - 11s + 14 = 0$$
$$(2s - 7)(s - 2) = 0$$

We have now completed factorisation; let's proceed to solve the equation as:

$$(2s - 7) = 0 \qquad \textbf{OR} \qquad (s - 2) = 0$$
$$2s - 7 = 0 \qquad\qquad s - 2 = 0$$
$$2s = 7 \qquad\qquad \therefore s = 2$$
$$\therefore s = \frac{7}{2}$$

c) $12v^2 + 2v - 70 = 0$

Solution

To begin, we need to factorise this equation (since 2 is a common factor to all the terms) as:

$$12v^2 + 2v - 70 = 0$$
$$2\left(6v^2 + v - 35\right) = 0$$
$$6v^2 + v - 35 = 0$$

For the last expression $a = 6$, $b = 1$, and $c = -35$. Now we need to set the scene as:

$$\begin{Vmatrix} 2 & 5 \\ 3 & -7 \end{Vmatrix}$$

It can be seen from the above that:

1) Column 1

$$2 \times 3 = 6 = a$$

2) Column 2

$$5 \times -7 = -35 = c$$

The addition of the cross-multiplication

$$(2 \times -7) + (3 \times 5) = -14 + 15 = 1 = b$$

Hence, we can write the factors using rows 1 and 2 as:

$$6v^2 + v - 35 = 0$$

Remember that

$$2\left(6v^2 + v - 35\right) = 0$$

So we have

$$2\left(2v + 5\right)\left(3v - 7\right) = 0$$

We have now completed factorisation; let's proceed to solve the equation as:

$$
\begin{array}{ccc}
2\left(2v + 5\right) = 0 & \textbf{OR} & \left(3v - 7\right) = 0 \\
2v + 5 = 0 & & 3v - 7 = 0 \\
2v = -5 & & 3v = 7 \\
\therefore v = -\dfrac{5}{2} & & \therefore v = \dfrac{7}{3}
\end{array}
$$

ALTERNATIVE METHOD

We could have used $12v^2 + 2v - 70 = 0$ as:

$$\left\| \begin{array}{cc} 2 & 5 \\ 6 & -14 \end{array} \right\|$$

which implies

$$12v^2 + 2v - 70 = 0$$
$$(2v + 5)(6v - 14) = 0$$

OR

$$\left\| \begin{array}{cc} 4 & 10 \\ 3 & -7 \end{array} \right\|$$

which implies

$$12v^2 + 2v - 70 = 0$$
$$(4v + 10)(3v - 7) = 0$$

That's

$$\therefore v = -\frac{5}{2} \quad \textbf{OR} \quad \therefore v = \frac{7}{3}$$

The two options above will result in the same answers.

---

d) $33 + 5z - 2z^2 = 0$
Solution
Let us first write the equation in the standard form as:

$$33 + 5z - 2z^2 = 0$$

which implies that

$$-2z^2 + 5z + 33 = 0$$

For this case, $a = -2$, $b = 5$ and $c = 33$. Now we need to set the scene as:

$$\left\| \begin{matrix} -2 & 11 \\ 1 & 3 \end{matrix} \right\|$$

It can be seen from the above that

1) Column 1

$$-2 \times 1 = -2 = a$$

2) Column 2

$$11 \times 3 = 33 = c$$

The addition of the cross-multiplication

$$(-2 \times 3) + (1 \times 11) = -6 + 11 = 5 = b$$

Hence, we can write the factors using rows 1 and 2 as:

$$-2z^2 + 5z + 33 = 0$$
$$(-2z + 11)(z + 3) = 0$$

We have now completed factorisation; let's proceed to solve the equation as:

$$
\begin{array}{ccc}
(-2z + 11) = 0 & \textbf{OR} & (z + 3) = 0 \\
-2z + 11 = 0 & & z + 3 = 0 \\
-2z = -11 & & \therefore z = -3 \\
2z = 11 & & \\
\therefore z = \frac{11}{2} & &
\end{array}
$$

ALTERNATIVE METHOD

Starting with the standard form

$$-2z^2 + 5z + 33 = 0$$

multiply both sides with -1 or transfer all the terms to the right-hand side, we will have:

$$2z^2 - 5z - 33 = 0$$

For this case, $a = 2$, $b = -5$, and $c = -33$. Now we need to set the scene as:

$$\begin{Vmatrix} 2 & -11 \\ 1 & 3 \end{Vmatrix}$$

Following the same method, we will have

$$2z^2 - 5z - 33 = 0$$
$$(2z - 11)(z + 3) = 0$$

Solving this will produce

$$\therefore z = \frac{11}{2} \quad \textbf{OR} \quad \therefore z = -3$$

NOTE

We can observe that both equations yield the same answer, right? This is because the answers correspond to where the graph crosses the x-axis and is the same for an equation and its mirror graph. In other words, $-2z^2 + 5z + 33 = 0$ and $2z^2 - 5z - 33 = 0$ are mirrors of each other and they should therefore cross the x-axis at the same point(s). However, they are not the same; the first equation has an inverted u-shape, and the second equation has a u-shaped curve. A similar explanation can be given regarding Example 11(c) such that the solutions for $12v^2 + 2v - 70 = 0$ and $6v^2 + v - 35 = 0$ are the same.

Another set of examples to try.

Example 12

Determine the x coordinates of the point(s) at which the following functions meet the x-axis.

a) $f(x) = 2(x - 3)(x - 2) - x(x + 3)$ **b)** $g(x) = (2x - 1)^2 - (x + 2)^2$

What did you get? Find the solution below to double-check your answer.

Solution to Example 12

a) $f(x) = 2(x-3)(x-2) - x(x+3)$

Solution

We need to open the brackets and simplify the expression as:

$$\begin{aligned} f(x) &= 2(x-3)(x-2) - x(x+3) \\ &= 2(x^2 - 5x + 6) - x^2 - 3x \\ &= 2x^2 - 10x + 12 - x^2 - 3x \\ &= x^2 - 13x + 12 \end{aligned}$$

This has now been written in standard form. Thus, $a = 1$, $b = -13$, and $c = 12$. Next, we need to set the scene as:

$$\begin{Vmatrix} 1 & -12 \\ 1 & -1 \end{Vmatrix}$$

Although this is a case of $a = 1$, we've decided to use this method to show that it is applicable to all cases. It can be seen from the above that:

1) Column 1

$$1 \times 1 = 1 = a$$

2) Column 2

$$-12 \times -1 = 12 = c$$

The addition of the cross-multiplication is:

$$(1 \times -1) + (1 \times -12) = -1 - 12 = -13 = b$$

Hence, we can write the factors using rows 1 and 2 as:

$$x^2 - 13x + 12 = 0$$
$$(x - 12)(x - 1) = 0$$

We have now completed factorisation; let's proceed to solve the equation as:

$$\begin{array}{ccc} (x - 12) = 0 & \textbf{OR} & (x - 1) = 0 \\ x - 12 = 0 & & x - 1 = 0 \\ \therefore x = 12 & & \therefore x = 1 \end{array}$$

For this type of complex expression, it is always good to check if the answers satisfy the initial question. How? We just need to substitute each value.

- When $x = 1$, we write it as $f(1)$:

$$\begin{aligned} f(x) &= 2(x-3)(x-2) - x(x+3) \\ f(1) &= 2(1-3)(1-2) - 1(1+3) \\ &= 2(-2)(-1) - (4) \\ &= 2(2) - 4 = 0 \end{aligned}$$

- When $x = 12$, we write it as $f(12)$:

$$f(x) = 2(x-3)(x-2) - x(x+3)$$
$$f(12) = 2(12-3)(12-2) - 12(12+3)$$
$$= 2(9)(10) - 12(15)$$
$$= 180 - 180 = 0$$

Both look alright. Good!

---

b) $g(x) = (2x-1)^2 - (x+2)^2$

Solution

We need to open the brackets and simplify the expression as:

$$f(x) = (2x-1)^2 - (x+2)^2$$
$$= (4x^2 - 4x + 1) - (x^2 + 4x + 4)$$
$$= 4x^2 - 4x + 1 - x^2 - 4x - 4$$
$$= 3x^2 - 8x - 3$$

We are now able to write it in standard form. Thus, $a = 3$, $b = -8$, and $c = -3$. Next, we need to set the scene as:

$$\begin{Vmatrix} 1 & -3 \\ 3 & 1 \end{Vmatrix}$$

It can be seen from the above that:

1) Column 1

$$1 \times 3 = 3 = a$$

2) Column 2

$$-3 \times 1 = -3 = c$$

The addition of the cross-multiplication is:

$$(1 \times 1) + (3 \times -3) = 1 - 9 = -8 = b$$

Hence, we can write the factors using rows 1 and 2 as:

$$3x^2 - 8x - 3 = 0$$
$$(x-3)(3x+1) = 0$$

We have now completed factorisation; let's proceed to solve the equation as:

$$\begin{array}{lll} (x-3) = 0 & \textbf{OR} & (3x+1) = 0 \\ x - 3 = 0 & & 3x + 1 = 0 \\ \therefore x = 3 & & 3x = -1 \\ & & \therefore x = -\dfrac{1}{3} \end{array}$$

Another set of examples to try.

Example 13

Determine the equation whose roots are given below. Write the equation in the form $ax^2+bx+c = 0$, such that a, b, and c are integers.

a) $x = -3$ and $x = 2$ 	**b)** $x = 0.5$ and $x = -1$ 	**c)** $x = -\frac{2}{3}$ and $x = \frac{3}{7}$

What did you get? Find the solution below to double-check your answer.

Solution to Example 13

HINT

Use the fact that if α and β are the roots of a quadratic equation, then

$$(x - \alpha)(x - \beta) = 0$$

a) $x = -3$ and $x = 2$
Solution
If $x = -3$ and $x = 2$ are the roots of the equation $f(x)$, then:

$$f(x) = (x + 3)(x - 2)$$
$$= x^2 - 2x + 3x - 6$$
$$= x^2 + x - 6$$
$$\therefore x^2 + x - 6 = 0$$

NOTE

- $x^2+x-6 = 0$ is not the only equation that has $x = -3$ and $x = 2$ as its roots. $2x^2+2x-12 = 0$ and $10x^2 + 10x - 60 = 0$ are two others. In fact, there are infinite numbers of them, as they are produced by transformation in the vertical axis.

- The general expression for this family is $\lambda(x^2 + x - 6) = 0$, where λ is any real number.

- Unless we are given the vertex or turning point, there will always be many equations that will satisfy this condition.

b) $x = 0.5$ and $x = -1$

Solution

If $x = 0.5$ and $x = -1$ are the roots of the equation $f(x)$, then:

$$f(x) = (x - 0.5)(x + 1)$$
$$= x^2 + x - 0.5x - 0.5$$
$$= x^2 + 0.5x - 0.5$$
$$\therefore 2x^2 + x - 1 = 0$$

---

c) $x = -\dfrac{2}{3}$ and $x = \dfrac{3}{7}$

Solution

If $x = -\dfrac{2}{3}$ and $x = \dfrac{3}{7}$ are the roots of the equation $f(x)$, then:

$$f(x) = \left(x + \frac{2}{3}\right)\left(x - \frac{3}{7}\right)$$
$$= x^2 - \frac{3}{7}x + \frac{2}{3}x - \left(\frac{2}{3}\right)\left(\frac{3}{7}\right)$$
$$= x^2 + \frac{5}{21}x - \frac{2}{7}$$
$$= 21x^2 + 5x - 6$$
$$\therefore 21x^2 + 5x - 6 = 0$$

6.5 CHAPTER SUMMARY

1) An algebraic term (or simply a term) is the simplest algebraic unit (like cells in biology, atoms in chemistry, bits in computing, etc.) and may consist of:

  • A constant (number) or constants

  • A variable or variables

  • A combination of constant(s) and variable(s)

2) The key difference between an **expression** and an **equation** is the presence of the equal sign (=) in the latter.

3) If an equation is valid for all values of the variable, this is technically called an identity and is represented with a three-line symbol $\equiv$.

4) Solving an equation is a process of finding the value or values of the variable for which the expression on the LHS equals that on the RHS. The values are also called **solutions** or **roots** of the equation. You can always verify if these values are correct by replacing the variables with these values and see if the RHS equals the LHS.

5) A linear expression in one variable is an algebraic expression which takes the form:

$$\boxed{ax + b}$$

6) A linear equation (or sometimes called a simple equation) will generally take the form:

$$ax + b = 0$$

- where x is the unknown quantity to be found.

7) In another technical jargon, a linear equation is referred to as a **polynomial of degree 1**.

8) The graph of a linear equation is always a straight line.

9) A quadratic expression is an algebraic expression that has the form:

$$ax^2 + bx + c$$

- where a, b, and c are real numbers. a and b are also called the coefficients of x^2 and x term respectively, and c is the constant term.

10) A function is written as $f(x)$ and is read as '**f of x**'.

11) A quadratic equation is a quadratic expression that is equated to zero, a constant, or an expression. In another technical term, a quadratic equation is referred to as a **polynomial of degree 2**.

12) A quadratic equation in one variable takes a general form:

$$ax^2 + bx + c = 0$$

13) Every quadratic equation must have two answers (roots or solutions), both real values or both complex values.

14) Factorisation is one of the methods used to solve a quadratic equation analytically and it is usually used if all that is needed is the roots of the equation.

15) By factorisation, it implies that the quadratic equation is in (or should be placed into) two brackets, with each part called a **factor**. Thus, the term 'factorisation' is used to describe the method.

16) The procedure for factorising $ax^2 + bx + c = 0$ depends on whether $a = \pm 1$ or $a \neq \pm 1$.

****

6.6 FURTHER PRACTICE

To access complementary contents, including additional exercises, please go to www.dszak.com.

7 Algebraic Equations II

Learning Outcomes

Once you have studied the content of this chapter, you should be able to:

- Solve quadratic equations by completing the square
- Solve quadratic equations by using the quadratic formula
- Determine the discriminant in order to determine the number and nature of roots
- Sketch the graph of a quadratic function
- Solve simultaneous linear and non-linear equations

7.1 INTRODUCTION

In this chapter, we will continue our discussion on solving equations, covering three other methods of solving quadratic equations. We will also discuss simultaneous equations.

7.2 SOLVING QUADRATIC EQUATIONS

Factorisation is a very common method of solving a quadratic equation, but there are other methods that prove indispensable. We will discuss these one after another.

7.2.1 COMPLETING THE SQUARE METHOD

This method is particularly used when we need to sketch the graph of a quadratic equation. It gives not only the x-intercepts (or the roots of the equation), but it additionally provides us with the turning point (or vertex: maximum or minimum) and the line of symmetry.

In this method, a quadratic equation $ax^2 + bx + c = 0$ is reduced to a two-term form, consisting of a square and a constant term as:

$$ax^2 + bx + c = 0 \qquad a(x + \lambda)^2 + \mu = 0$$
$$\downarrow$$
$$a(x + \lambda)^2 + \mu = 0 \qquad \text{1. Square term} \qquad \text{2. Constant term}$$

where both λ and μ are new constants to be found. $a(x + \lambda)^2$ and μ are the square and constant terms, respectively. This format is also called the **vertex form**.

DOI:10.1201/9781003027928-7

Let's make the following notes about the expression above:

Note 1 Roots (or x-intercepts)

To find the roots of the quadratic equation, i.e., where the graph crosses the x-axis, we need to solve $a(x + \lambda)^2 + \mu = 0$. The coordinates of these points are given as $(x_1, 0)$ and $(x_2, 0)$. There are three further points to make here:

  a) If a and μ have **different signs**, then there will be **two distinct real roots**, i.e., $x_1 \neq x_2$.

  b) If $\mu = 0$, then there will be **one single real root**, i.e., $x_1 = x_2$.

  c) If a and μ have the **same signs**, then there will be **no real roots**.

Note 2 Vertex (minimum or maximum point)

The coordinates of the turning point or vertex are $(-\lambda, \mu)$.

Note 3 Line of symmetry

It is a line that divides the graph into two equal and identical parts. It occurs when the expression in the brackets is equal to zero; in other words, $x + \lambda = 0$ or $x = -\lambda$. The line passes through the vertex or coordinates $(-\lambda, \mu)$.

Note 4 y-intercept

The y-intercept is where the curve cuts (or passes) through the y-axis. This is easily identifiable in the quadratic expression $ax^2 + bx + c$ as c (or the constant term) and so the coordinates at this point are $(0, c)$. However, in the vertex form $a(x + \lambda)^2 + \mu = 0$, the coordinates of y-intercept are not $(0, \mu)$. We can, however, substitute $x = 0$ into the equation to find the corresponding y-value at this point as:

$$a(x + \lambda)^2 + \mu = y$$
$$a(0 + \lambda)^2 + \mu = y$$
$$a\lambda^2 + \mu = y$$

Hence, the coordinates of the y-intercept are $(0, a\lambda^2 + \mu)$.

Note 5 μ and λ

These were chosen arbitrarily and the above formula is not to be memorised. It is always easier to follow the procedure of arriving at these terms instead of looking for formulas to determine the new constants.

Like with factorisation, there are two scenarios to consider for this method.

7.2.1.1 Case 1: $a = 1$

For this case, $ax^2 + bx + c = 0$ becomes $x^2 + bx + c = 0$, e.g., $x^2 + 6x + 5 = 0$. We will use this to illustrate the method in the following steps:

Step 1: If applicable, write the equation in the standard form as:

$$x^2 + bx + c = 0$$

Step 2: Find half of the constant b. In this case, $b = 6$ and half of this is 3. This is our λ in the above general expression.

Step 3: We can now write the **square term** as:

$$(x + 3)^2$$

Step 4: Subtract half of b squared or $\left(\frac{1}{2}b\right)^2$ from the square term and add c. For this case, $\left(\frac{1}{2}b\right)^2 = \left(\frac{1}{2} \times 6\right)^2 = 9$ and $c = 5$. Here we go:

$$(x + 3)^2 - 9 + 5 = (x + 3)^2 - 4$$

Step 5: This represents and is equivalent to the left-hand side of the equation $x^2 + 6x + 5 = 0$. Let's add zero to make our completed square an equation.

$$(x + 3)^2 - 4 = 0$$

Step 6: Finally, we need to simplify the above as follows:

$$(x + 3)^2 = 4$$

$$\sqrt{(x + 3)^2} = \pm\sqrt{4}$$

$$x + 3 = \pm 2$$

$$x = -3 \pm 2$$

$$
\begin{array}{ccc}
x = -3 + 2 & & x = -3 - 2 \\
\therefore x = -1 & \textbf{OR} & \therefore x = -5
\end{array}
$$

Let's try this using $x^2 + bx + c = 0$. Here we go:

$$x^2 + bx + c = 0$$

$$\left(x + \frac{b}{2}\right)^2 - \left(\frac{b}{2}\right)^2 + c = 0$$

$$\left(x + \frac{b}{2}\right)^2 - \frac{b^2}{4} + c = 0$$

$$\left(x + \frac{b}{2}\right)^2 - \frac{b^2}{4} + \frac{4c}{4} = 0$$

$$\left(x + \frac{b}{2}\right)^2 + \frac{4c}{4} - \frac{b^2}{4} = 0$$

$$\left(x + \frac{b}{2}\right)^2 + \frac{4c - b^2}{4} = 0$$

Therefore, when compared with our format given before, we have that

$$\lambda = \frac{b}{2}; \mu = \frac{4c - b^2}{4}$$

Let's try some examples to show how to complete the square of a quadratic expression.

Example 1

Complete the square for the following quadratic expressions.

a) $x^2 + 2x$ b) $x^2 - 3x$ c) $x^2 - 4x + 1$ d) $x^2 + x + 5$

What did you get? Find the solution below to double-check your answer.

Solution to Example 1

a) $x^2 + 2x$
Solution

$$x^2 + 2x = (x + 1)^2 - 1^2$$
$$= (x + 1)^2 - 1$$

b) $x^2 - 3x$
Solution

$$x^2 - 3x = \left(x - \frac{3}{2}\right)^2 - \left(\frac{3}{2}\right)^2$$
$$= \left(x - \frac{3}{2}\right)^2 - \frac{9}{4}$$

c) $x^2 - 4x + 1$
Solution

$$x^2 - 4x + 1 = (x - 2)^2 - 2^2 + 1$$
$$= (x - 2)^2 - 4 + 1$$
$$= (x - 2)^2 - 3$$

d) $x^2 + x + 5$
Solution

$$x^2 + x + 5 = \left(x + \frac{1}{2}\right)^2 - \left(\frac{1}{2}\right)^2 + 5$$
$$= \left(x + \frac{1}{2}\right)^2 - \frac{1}{4} + 5$$
$$= \left(x + \frac{1}{2}\right)^2 + \frac{19}{4}$$

Let's try another set of examples. Our focus here is to be able to identify the vertex of a quadratic function.

Example 2

Determine the coordinates of the vertex of the following equations.

a) $(x-1)^2 - 3 = 0$

b) $2\left(x - \frac{2}{3}\right)^2 = -9$

What did you get? Find the solution below to double-check your answer.

Solution to Example 2

a) $(x-1)^2 - 3 = 0$
Solution
For $a(x+\lambda)^2 + \mu = 0$, the vertex is $(-\lambda, \mu)$. Comparing $(x-1)^2 - 3 = 0$ with this, hence the vertex is

$$\therefore (1, -3)$$

ALTERNATIVE METHOD

We can equally say that the vertex happens when

$$(x-1)^2 = 0 \text{ or } x - 1 = 0$$

which implies that

$$x = 1$$

At this point,

$$y = (1-1)^2 - 3 = -3$$

The vertex is therefore

$$(1, -3)$$

b) $2\left(x - \frac{2}{3}\right)^2 = -9$
Solution
Let us first write it in standard form as:

$$2\left(x - \frac{2}{3}\right)^2 = -9$$
$$2\left(x - \frac{2}{3}\right)^2 + 9 = 0$$

For $a(x + \lambda)^2 + \mu = 0$, the vertex is $(-\lambda, \mu)$. Comparing $2\left(x - \dfrac{2}{3}\right)^2 + 9$ with this, hence the vertex is

$$\left(\frac{2}{3}, 9\right)$$

We have now illustrated how to complete the square of quadratic expressions. Solving a quadratic equation using this method is essentially the same, though requires a few more steps.

Let's try some examples.

Example 3

Solve the following quadratic equations by completing the square method. Present your answers correct to 2 significant figures.

a) $x^2 - 14x + 33 = 0$

b) $m^2 + 2m + \dfrac{1}{5} = 0$

c) $\alpha^2 + 0.2\alpha - 1.2 = 0$

d) $\dfrac{3}{4} - \dfrac{1}{2}k - k^2 = 0$

What did you get? Find the solution below to double-check your answer.

Solution to Example 3

a) $x^2 - 14x + 33 = 0$
Solution

$$x^2 - 14x + 33 = 0$$
$$(x - 7)^2 - 7^2 + 33 = 0$$
$$(x - 7)^2 - 49 + 33 = 0$$
$$(x - 7)^2 - 16 = 0$$
$$(x - 7)^2 = 16$$
$$\sqrt{(x - 7)^2} = \pm\sqrt{16}$$
$$x - 7 = \pm 4$$

Therefore:

$$
\begin{array}{ccc}
x - 7 = 4 & & x - 7 = -4 \\
x = 7 + 4 & \text{OR} & x = 7 - 4 \\
\therefore x = 11 & & \therefore x = 3.0
\end{array}
$$

NOTE
When solving equations, you may find it easier to move the constant term to the RHS before beginning the process of completing the square, as shown in the alternative method. They are essentially the same, except their initial steps.

ALTERNATIVE METHOD

We can equally say that the vertex happens when

$$x^2 - 14x + 33 = 0$$
$$(x - 7)^2 - 7^2 = -33$$
$$(x - 7)^2 - 49 = -33$$
$$(x - 7)^2 = 49 - 33$$
$$(x - 7)^2 = 16$$
$$\sqrt{(x - 7)^2} = \pm\sqrt{16}$$
$$x - 7 = \pm 4$$

Therefore:

$$x - 7 = 4 \qquad\qquad x - 7 = -4$$
$$x = 7 + 4 \quad \textbf{OR} \quad x = 7 - 4$$
$$\therefore x = 11 \qquad\qquad \therefore x = 3.0$$

---

b) $m^2 + 2m + \frac{1}{5} = 0$

Solution

$$m^2 + 2m + \frac{1}{5} = 0$$
$$(m + 1)^2 - 1^2 + \frac{1}{5} = 0$$
$$(m + 1)^2 - 1 + \frac{1}{5} = 0$$
$$(m + 1)^2 - \frac{4}{5} = 0$$
$$(m + 1)^2 = \frac{4}{5}$$
$$\sqrt{(m + 1)^2} = \pm\sqrt{\frac{4}{5}}$$
$$m + 1 = \pm\frac{2}{\sqrt{5}}$$

Therefore:

$$m + 1 = \frac{2}{\sqrt{5}} \qquad\qquad m + 1 = -\frac{2}{\sqrt{5}}$$
$$m = -1 + \frac{2}{\sqrt{5}} \quad \textbf{OR} \quad m = -1 - \frac{2}{\sqrt{5}}$$
$$= \frac{-5 + 2\sqrt{5}}{5} \qquad\qquad = -\frac{5 + 2\sqrt{5}}{5}$$
$$\therefore m = -0.11 \qquad\qquad \therefore m = -1.9$$

c) $\alpha^2 + 0.2\alpha - 1.2 = 0$

Solution

$$\alpha^2 + 0.2\alpha - 1.2 = 0$$
$$(\alpha + 0.1)^2 - 0.1^2 - 1.2 = 0$$
$$(\alpha + 0.1)^2 - 0.01 - 1.2 = 0$$
$$(\alpha + 0.1)^2 - 1.21 = 0$$
$$(\alpha + 0.1)^2 = 1.21$$
$$\sqrt{(\alpha + 0.1)^2} = \pm\sqrt{1.21} = \pm\sqrt{\frac{121}{100}}$$
$$\alpha + 0.1 = \pm\frac{11}{10}$$

Therefore:

$$\alpha + 0.1 = \frac{11}{10} = 1.1 \qquad\qquad \alpha + 0.1 = -\frac{11}{10} = -1.1$$
$$\alpha = -0.1 + 1.1 \qquad \textbf{OR} \qquad \alpha = -0.1 - 1.1$$
$$\therefore \alpha = \mathbf{1.0} \qquad\qquad\qquad \therefore \alpha = \mathbf{-1.2}$$

---

d) $\frac{3}{4} - \frac{1}{2}k - k^2 = 0$

Solution

There are a few ways to handle this question including:

1) Multiply through by -1 so that it is a case of $a = 1$.

2) Consider it under the case of $a \neq 1$.

Either way, the roots remain unchanged, but the curve and vertex will be flipped. We will be using the first approach here, but the second one will be used later so that we can compare the answers. Let us first write the equation in the standard form as:

$$-k^2 - \frac{1}{2}k + \frac{3}{4} = 0$$

Now multiply through by -1.

$$k^2 + \frac{1}{2}k - \frac{3}{4} = 0$$
$$\left(k + \frac{1}{4}\right)^2 - \left(\frac{1}{4}\right)^2 - \frac{3}{4} = 0$$
$$\left(k + \frac{1}{4}\right)^2 - \frac{13}{16} = 0$$
$$\left(k + \frac{1}{4}\right)^2 = \frac{13}{16}$$
$$\sqrt{\left(k + \frac{1}{4}\right)^2} = \pm\sqrt{\frac{13}{16}}$$
$$k + \frac{1}{4} = \pm\frac{\sqrt{13}}{4}$$

Therefore:

$$k + \frac{1}{4} = \frac{\sqrt{13}}{4} \qquad\qquad k + \frac{1}{4} = -\frac{\sqrt{13}}{4}$$

$$k = -\frac{1}{4} + \frac{\sqrt{13}}{4} \qquad \textbf{OR} \qquad k = -\frac{1}{4} - \frac{\sqrt{13}}{4}$$

$$= \frac{-1+\sqrt{13}}{4} \qquad\qquad = -\frac{1+\sqrt{13}}{4}$$

$$\therefore k = 0.65 \qquad\qquad \therefore k = -1.2$$

All answers have been given correct to 2 s.f.

7.2.1.2 Case 2: $a \neq 1$

This case is like what we've just covered, but there is a need to tweak things due to a not being equal to 1. An example of this case is $2x^2 - x - 15 = 0$, which we will use to illustrate.

Step 1: If not already, write the equation in the standard form as:

$$ax^2 + bx + c = 0$$

Step 2: We will focus on the first two terms $ax^2 + bx$ and take out the constant a, which is 2 here.

$$2(x^2 - \frac{1}{2}x) - 15 = 0$$

Note that the expression in the brackets (i.e., $x^2 - \frac{1}{2}x$) is quadratic where $a = 1$, $b = -\frac{1}{2}$, and $c = 0$. Let's call this a '**minor quadratic expression**'.

Step 3: The steps we applied to the case of $a = 1$ will be applied to the minor quadratic expression.

$$2\left[\left(x - \frac{1}{4}\right)^2 - \left(\frac{1}{4}\right)^2\right] - 15 = 0$$

$$2\left[\left(x - \frac{1}{4}\right)^2 - \frac{1}{16}\right] - 15 = 0$$

Step 4: From here, it is a matter of simplification. Open the square brackets [] as:

$$2\left(x - \frac{1}{4}\right)^2 - 2 \times \frac{1}{16} - 15 = 0$$

$$2\left(x - \frac{1}{4}\right)^2 - \frac{1}{8} - 15 = 0$$

$$2\left(x - \frac{1}{4}\right)^2 - \frac{121}{8} = 0$$

Step 5: Let's finish the simplification and then solve as follows:

$$2\left(x - \frac{1}{4}\right)^2 = \frac{121}{8}$$

$$\left(x - \frac{1}{4}\right)^2 = \frac{121}{16}$$

$$\sqrt{\left(x - \frac{1}{4}\right)^2} = \pm\sqrt{\frac{121}{16}}$$

$$x - \frac{1}{4} = \pm\frac{11}{4}$$

$$x = \frac{1}{4} \pm \frac{11}{4}$$

$x = \frac{1}{4} + \frac{11}{4}$	$x = \frac{1}{4} - \frac{11}{4}$
$= \frac{1+11}{4} = \frac{12}{4}$ **OR**	$\frac{1-11}{4} = \frac{-10}{4}$
$\therefore x = 3$	$\therefore x = -2.5$

Let's try this using $ax^2 + bx + c = 0$. Here we go:

$$ax^2 + bx + c = 0$$

$$a\left(x^2 + \frac{b}{a}x\right) + c = 0$$

$$a\left[\left(x + \frac{b}{2a}\right)^2 - \left(\frac{b}{2a}\right)^2\right] + c = 0$$

$$a\left[\left(x + \frac{b}{2a}\right)^2 - \frac{b^2}{4a^2}\right] + c = 0$$

$$a\left(x + \frac{b}{2a}\right)^2 - \frac{b^2}{4a} + c = 0$$

$$a\left(x + \frac{b}{2a}\right)^2 - \frac{b^2}{4a} + \frac{4ac}{4a} = 0$$

$$a\left(x + \frac{b}{2a}\right)^2 + \frac{4ac}{4a} - \frac{b^2}{4a} = 0$$

$$a\left(x + \frac{b}{2a}\right)^2 + \frac{4ac - b^2}{4a} = 0$$

Therefore, when compared with our format given before, we have that

$$\lambda = \frac{b}{2a}; \mu = \frac{4ac - b^2}{4a}$$

We'll try some examples now.

Example 4

Complete the square for the following expressions.

a) $3x^2 + 5x$ **b)** $2x^2 - x$ **c)** $2x^2 + 8x + 3$ **d)** $5x^2 - 4x - 1$

What did you get? Find the solution below to double-check your answer.

Solution to Example 4

a) $3x^2 + 5x$
Solution

$$3x^2 + 5x = 3\left(x^2 + \frac{5}{3}x\right)$$
$$= 3\left[\left(x + \frac{5}{6}\right)^2 - \left(\frac{5}{6}\right)^2\right]$$
$$= 3\left[\left(x + \frac{5}{6}\right)^2 - \frac{25}{36}\right]$$
$$= 3\left(x + \frac{5}{6}\right)^2 - 3 \times \frac{25}{36}$$
$$\therefore 3x^2 + 5x = 3\left(x + \frac{5}{6}\right)^2 - \frac{25}{12}$$

b) $2x^2 - x$
Solution

$$2x^2 - x = 2\left(x^2 - \frac{1}{2}x\right)$$
$$= 2\left[\left(x - \frac{1}{4}\right)^2 - \left(\frac{1}{4}\right)^2\right]$$
$$= 2\left[\left(x - \frac{1}{4}\right)^2 - \frac{1}{16}\right]$$
$$= 2\left(x - \frac{1}{4}\right)^2 - 2 \times \frac{1}{16}$$
$$\therefore 2x^2 - x = 2\left(x - \frac{1}{4}\right)^2 - \frac{1}{8}$$

c) $2x^2 + 8x + 3$
Solution

$$2x^2 + 8x + 3 = (2x^2 + 8x) + 3$$
$$= 2(x^2 + 4x) + 3$$
$$= 2\left[(x + 2)^2 - (2)^2\right] + 3$$
$$= 2\left[(x + 2)^2 - 4\right] + 3$$
$$= 2(x + 2)^2 - 8 + 3$$
$$\therefore 2x^2 + 8x + 3 = 2(x + 2)^2 - 5$$

---

d) $5x^2 - 4x - 1$
Solution

$$5x^2 - 4x - 1 = (5x^2 - 4x) - 1$$
$$= 5\left(x^2 - \frac{4}{5}x\right) - 1$$
$$= 5\left[\left(x - \frac{2}{5}\right)^2 - \left(\frac{2}{5}\right)^2\right] - 1$$
$$= 5\left[\left(x - \frac{2}{5}\right)^2 - \frac{4}{25}\right] - 1$$
$$= 5\left(x - \frac{2}{5}\right)^2 - \frac{4}{5} - 1$$
$$\therefore 5x^2 - 4x - 1 = 5\left(x - \frac{2}{5}\right)^2 - \frac{9}{5}$$

Let's try some examples.

Example 5

Complete the square of the following quadratic equations to determine the number and nature of their roots.

a) $2x^2 - 4x + 5 = 0$ **b)** $3 - 2\omega - 5\omega^2 = 0$

c) $4\alpha^2 - \alpha + \frac{1}{16} = 0$ **d)** $0.1 + 0.8p - p^2 = 0$

What did you get? Find the solution below to double-check your answer.

Solution to Example 5

a) $2x^2 - 4x + 5 = 0$
Solution

$$2x^2 - 4x + 5 = 0$$
$$2(x^2 - 2x) + 5 = 0$$
$$2\left[(x-1)^2 - 1^2\right] + 5 = 0$$
$$2\left[(x-1)^2 - 1\right] + 5 = 0$$
$$2(x-1)^2 - 2 + 5 = 0$$
$$\mathbf{2(x-1)^2 + 3 = 0}$$

For this case, $a = 2$ and $\mu = 5$. Since both have the same sign, there will be no real roots.

b) $3 - 2\omega - 5\omega^2 = 0$
Solution
Let us first write the equation in the standard form as:

$$3 - 2\omega - 5\omega^2 = 0$$
$$-5\omega^2 - 2\omega + 3 = 0$$
$$-5\left(\omega^2 + \frac{2}{5}\omega\right) + 3 = 0$$
$$-5\left[\left(\omega + \frac{1}{5}\right)^2 - \left(\frac{1}{5}\right)^2\right] + 3 = 0$$
$$-5\left[\left(\omega + \frac{1}{5}\right)^2 - \frac{1}{25}\right] + 3 = 0$$
$$-5\left(\omega + \frac{1}{5}\right)^2 + \frac{1}{5} + 3 = 0$$
$$\mathbf{-5\left(\omega + \frac{1}{5}\right)^2 + \frac{16}{5} = 0}$$

For this case, $a = -5$ and $\mu = \frac{16}{5}$. Since both have different signs, there will be two distinct real roots.

c) $4\alpha^2 - \alpha + \frac{1}{16} = 0$

Solution

$$4\alpha^2 - \alpha + \frac{1}{16} = 0$$

$$4\left(\alpha^2 - \frac{1}{4}\alpha\right) + \frac{1}{16} = 0$$

$$4\left[\left(\alpha - \frac{1}{8}\right)^2 - \left(\frac{1}{8}\right)^2\right] + \frac{1}{16} = 0$$

$$4\left[\left(\alpha - \frac{1}{8}\right)^2 - \frac{1}{64}\right] + \frac{1}{16} = 0$$

$$4\left(\alpha - \frac{1}{8}\right)^2 - \frac{1}{16} + \frac{1}{16} = 0$$

$$4\left(\alpha - \frac{1}{8}\right)^2 + 0 = 0$$

For this case, $a = 4$ and $\mu = 0$. Since $\mu = 0$, there will be one single root, which is repeated.

d) $0.1 + 0.8p - p^2 = 0$

Solution

Let us first write the equation in the standard form as:

$$-p^2 + 0.8p + 0.1 = 0$$

$$\left[-p^2 + 0.8p + 0.1\right] = 0$$

$$-\left[p^2 - 0.8p - 0.1\right] = 0$$

$$-\left[(p - 0.4)^2 - (0.4)^2 - 0.1\right] = 0$$

$$-\left[(p - 0.4)^2 - 0.16 - 0.1\right] = 0$$

$$-\left[(p - 0.4)^2 - 0.26\right] = 0$$

$$-(p - 0.4)^2 + 0.26 = 0$$

For this case, $a = -1$ and $\mu = 0.26$. Since both have different signs, there will be two distinct real roots.

Let's try some examples on solving quadratic equations.

Example 6

Solve the following quadratic equations by completing the square method. Present your answers correct to 3 significant figures.

a) $2x^2 + 4x - 9 = 0$ **b)** $3\omega^2 + \omega - \frac{1}{2} = 0$ **c)** $\frac{1}{6}\beta^2 - 5\beta + \frac{3}{2} = 0$ **d)** $\frac{3}{4} - \frac{1}{2}k - k^2 = 0$

What did you get? Find the solution below to double-check your answer.

Solution to Example 6

a) $2x^2 + 4x - 9 = 0$
Solution

$$2x^2 + 4x - 9 = 0$$
$$2(x^2 + 2x) - 9 = 0$$
$$2\left[(x+1)^2 - 1\right] - 9 = 0$$
$$2(x+1)^2 - 2 - 9 = 0$$
$$2(x+1)^2 - 11 = 0$$
$$2(x+1)^2 = 11$$
$$(x+1)^2 = \frac{11}{2}$$
$$\sqrt{(x+1)^2} = \pm\sqrt{\frac{11}{2}}$$
$$x + 1 = \pm\sqrt{\frac{11}{2}}$$

Therefore:

$$x + 1 = \sqrt{\frac{11}{2}} \qquad \textbf{OR} \qquad x + 1 = -\sqrt{\frac{11}{2}}$$
$$x = -1 + \sqrt{\frac{11}{2}} \qquad\qquad x = -1 - \sqrt{\frac{11}{2}}$$
$$= \frac{-2 + \sqrt{22}}{2} \qquad\qquad = -\frac{2 + \sqrt{22}}{2}$$
$$\therefore x = \mathbf{1.35} \qquad\qquad \therefore x = \mathbf{-3.35}$$

b) $3\omega^2 + \omega - \frac{1}{2} = 0$
Solution

$$3\omega^2 + \omega - \frac{1}{2} = 0$$
$$3\left(\omega^2 + \frac{1}{3}\omega\right) - \frac{1}{2} = 0$$
$$3\left[\left(\omega + \frac{1}{6}\right)^2 - \left(\frac{1}{6}\right)^2\right] - \frac{1}{2} = 0$$
$$3\left[\left(\omega + \frac{1}{6}\right)^2 - \frac{1}{36}\right] - \frac{1}{2} = 0$$
$$3\left(\omega + \frac{1}{6}\right)^2 - \frac{3}{36} - \frac{1}{2} = 0$$
$$3\left(\omega + \frac{1}{6}\right)^2 - \frac{7}{12} = 0$$

$$3\left(\omega + \frac{1}{6}\right)^2 = \frac{7}{12}$$

$$\left(\omega + \frac{1}{6}\right)^2 = \frac{7}{36}$$

$$\sqrt{\left(\omega + \frac{1}{6}\right)^2} = \pm\sqrt{\frac{7}{36}}$$

$$\omega + \frac{1}{6} = \pm\frac{\sqrt{7}}{6}$$

Therefore:

$$\omega + \frac{1}{6} = \frac{\sqrt{7}}{6} \quad \textbf{OR} \quad \omega + \frac{1}{6} = -\frac{\sqrt{7}}{6}$$

$$\omega = -\frac{1}{6} + \frac{\sqrt{7}}{6} \qquad\qquad \omega = -\frac{1}{6} - \frac{\sqrt{7}}{6}$$

$$= \frac{-1 + \sqrt{7}}{6} \qquad\qquad = -\frac{1 + \sqrt{7}}{6}$$

$$\therefore \omega = 0.274 \qquad\qquad \therefore \omega = -0.608$$

---

c) $\frac{1}{6}\beta^2 - 5\beta + \frac{3}{2} = 0$

Solution

$$\frac{1}{6}\beta^2 - 5\beta + \frac{3}{2} = 0$$

$$\frac{1}{6}\left(\beta^2 - 30\beta\right) + \frac{3}{2} = 0$$

$$\frac{1}{6}\left[(\beta - 15)^2 - (15)^2\right] + \frac{3}{2} = 0$$

$$\frac{1}{6}\left[(\beta - 15)^2 - 225\right] + \frac{3}{2} = 0$$

$$\frac{1}{6}(\beta - 15)^2 - \frac{225}{6} + \frac{3}{2} = 0$$

$$\frac{1}{6}(\beta - 15)^2 - 36 = 0$$

$$\frac{1}{6}(\beta - 15)^2 = 36$$

$$(\beta - 15)^2 = 36 \times 6$$

$$\sqrt{(\beta - 15)^2} = \pm\sqrt{216}$$

$$\beta - 15 = \pm 6\sqrt{6}$$

Therefore:

$$\beta - 15 = 6\sqrt{6} \qquad\qquad \beta - 15 = -6\sqrt{6}$$

$$\beta = 15 + 6\sqrt{6} \quad \textbf{OR} \quad \beta = 15 - 6\sqrt{6}$$

$$\therefore \beta = 29.7 \qquad\qquad \therefore \beta = 0.303$$

d) $\frac{3}{4} - \frac{1}{2}k - k^2 = 0$

Solution

Here we are repeating Example 3(d) that came before under $a = 1$. Notice the slight difference in workings, but more importantly the same answers.

Let us first write the equation in the standard form as:

$$-k^2 - \frac{1}{2}k + \frac{3}{4} = 0$$

$$\left[-k^2 - \frac{1}{2}k\right] + \frac{3}{4} = 0$$

$$-\left[k^2 + \frac{1}{2}k\right] + \frac{3}{4} = 0$$

$$-\left[\left(k + \frac{1}{4}\right)^2 - \left(\frac{1}{4}\right)^2\right] + \frac{3}{4} = 0$$

$$-\left[\left(k + \frac{1}{4}\right)^2 - \frac{1}{16}\right] + \frac{3}{4} = 0$$

$$-\left(k + \frac{1}{4}\right)^2 + \frac{1}{16} + \frac{3}{4} = 0$$

From here, the steps are the same in both approaches.

$$-\left(k + \frac{1}{4}\right)^2 + \frac{13}{16} = 0$$

$$-\left(k + \frac{1}{4}\right)^2 = -\frac{13}{16}$$

$$\left(k + \frac{1}{4}\right)^2 = \frac{13}{16}$$

$$\sqrt{\left(k + \frac{1}{4}\right)^2} = \pm\sqrt{\frac{13}{16}}$$

$$k + \frac{1}{4} = \pm\frac{\sqrt{13}}{4}$$

Therefore:

$$k + \frac{1}{4} = \frac{\sqrt{13}}{4} \qquad \textbf{OR} \qquad k + \frac{1}{4} = -\frac{\sqrt{13}}{4}$$

$$k = -\frac{1}{4} + \frac{\sqrt{13}}{4} \qquad\qquad\qquad k = -\frac{1}{4} - \frac{\sqrt{13}}{4}$$

$$= \frac{-1 + \sqrt{13}}{4} \qquad\qquad\qquad\qquad = -\frac{1 + \sqrt{13}}{4}$$

$$\therefore k = 0.65 \qquad\qquad\qquad\qquad\quad \therefore k = -1.2$$

Another set of examples to try.

Example 7

x_1 and x_2 are the roots of $ax^2 + bx + c = 0$, where a, b, and c are constants. Using completing the square method, show that:

$$x_1 = \frac{-b + \sqrt{b^2 - 4ac}}{2a}, \quad x_2 = \frac{-b - \sqrt{b^2 - 4ac}}{2a}$$

What did you get? Find the solution below to double-check your answer.

Solution to Example 7

Solution

$$ax^2 + bx + c = 0$$

$$a\left(x^2 + \frac{b}{a}x\right) + c = 0$$

$$a\left[\left(x + \frac{b}{2a}\right)^2 - \left(\frac{b}{2a}\right)^2\right] + c = 0$$

$$a\left[\left(x + \frac{b}{2a}\right)^2 - \frac{b^2}{4a^2}\right] + c = 0$$

$$\left[a\left(x + \frac{b}{2a}\right)^2 - a\frac{b^2}{4a^2}\right] + c = 0$$

$$a\left(x + \frac{b}{2a}\right)^2 - \frac{b^2}{4a} + c = 0$$

$$a\left(x + \frac{b}{2a}\right)^2 = \frac{b^2}{4a} - c$$

$$a\left(x + \frac{b}{2a}\right)^2 = \frac{b^2 - 4ac}{4a}$$

$$\left(x + \frac{b}{2a}\right)^2 = \frac{b^2 - 4ac}{4a^2}$$

$$\sqrt{\left(x + \frac{b}{2a}\right)^2} = \pm\sqrt{\frac{b^2 - 4ac}{4a^2}}$$

$$x + \frac{b}{2a} = \pm\frac{\sqrt{b^2 - 4ac}}{\sqrt{4a^2}}$$

$$x = -\frac{b}{2a} \pm \frac{\sqrt{b^2 - 4ac}}{2a}$$

We can therefore write the roots as

$$x_1 = \frac{-b + \sqrt{b^2 - 4ac}}{2a}, \quad x_2 = \frac{-b - \sqrt{b^2 - 4ac}}{2a}$$

The results obtained are the basis of our next method of solving quadratic equations.

7.2.2 QUADRATIC FORMULA METHOD

The two methods we just introduced are great for and essential to solving quadratic equations. However, sometimes, we may find it challenging to factorise because the possible combinations are too many, especially since fractions or irrational numbers are potential combinations. Furthermore, the information we obtain when using completing the square method may not be needed, particularly if we are not sketching the graph.

However, we are lucky to have another method that can be used to solve any quadratic equation with ease. It only requires identifying a, b and c of a quadratic equation. Yes, it is as simple as that, because your calculator will do the rest for you. This method is called the **Quadratic Formula** method and uses the formula:

$$\boxed{x = \frac{-b \pm \sqrt{b^2 - 4ac}}{2a}} \tag{7.1}$$

where a, b and c are the constants in the standard form $ax^2 + bx + c = 0$.

Notice the $\pm$ sign in the formula, which implies that there are two solutions for x. Let's denote these solutions as x_1 and x_2, then we have:

$$\boxed{x_1 = \frac{-b + \sqrt{b^2 - 4ac}}{2a}} \text{ OR } \boxed{x_2 = \frac{-b - \sqrt{b^2 - 4ac}}{2a}} \tag{7.2}$$

We always leave this splitting until the end, as will be shown in the worked examples to follow.

Let's try some examples.

Example 8

Solve the following equations using the quadratic formula.

a) $x^2 + 6x + 8 = 0$ **b)** $2y^2 - y - 15 = 0$ **c)** $1 - 4m^2 = 0$

What did you get? Find the solution below to double-check your answer.

Solution to Example 8

HINT

All we need in this method is to arrange the given equation in the standard form $ax^2 + bx + c = 0$, note the coefficients and apply the formula. That's all.

a) $x^2 + 6x + 8 = 0$

Solution

Comparing with the standard form $ax^2 + bx + c = 0$, we note that $a = 1$, $b = 6$, and $c = 8$. Hence:

$$x = \frac{-b \pm \sqrt{b^2 - 4ac}}{2a}$$

$$= \frac{-6 \pm \sqrt{(6)^2 - 4(1)(8)}}{2 \times 1}$$

$$= \frac{-6 \pm \sqrt{36 - 32}}{2}$$

$$= \frac{-6 \pm \sqrt{4}}{2} = \frac{-6 \pm 2}{2}$$

The roots are:

$$x = \frac{-6 + 2}{2} = -\frac{4}{2} \quad \textbf{OR} \quad x = \frac{-6 - 2}{2} = -\frac{8}{2}$$

$$\therefore x = -2 \qquad\qquad\qquad \therefore x = -4$$

---

b) $2y^2 - y - 15 = 0$

Solution

Comparing with the standard form $ax^2 + bx + c = 0$, we note that $a = 2$, $b = -1$, and $c = -15$. Hence:

$$y = \frac{-b \pm \sqrt{b^2 - 4ac}}{2a}$$

$$= \frac{-(-1) \pm \sqrt{(-1)^2 - 4(2)(-15)}}{2 \times 2}$$

$$= \frac{1 \pm \sqrt{1 + 120}}{4}$$

$$= \frac{1 \pm \sqrt{121}}{4} = \frac{1 \pm 11}{4}$$

The roots are:

$$y = \frac{1 + 11}{4} = \frac{12}{4} \quad \textbf{OR} \quad y = \frac{1 - 11}{4} = \frac{-10}{4}$$

$$\therefore y = 3 \qquad\qquad\qquad \therefore y = -\frac{5}{2}$$

---

c) $1 - 4m^2 = 0$

Solution

Let's re-write this (though we don't have to, provided we can spot the coefficients, but take caution with negative coefficients) as:

$$1 - 4m^2 = 0$$

$$-4m^2 + 1 = 0$$

$$-4m^2 + 0m + 1 = 0$$

Comparing with the standard form $ax^2 + bx + c = 0$, we note that $a = -4$, $b = 0$, and $c = 1$. Hence:

$$m = \frac{-b \pm \sqrt{b^2 - 4ac}}{2a}$$

$$= \frac{-0 \pm \sqrt{(0)^2 - 4(-4)(1)}}{2 \times -4}$$

$$= \frac{0 \pm \sqrt{0 + 16}}{-8}$$

$$= \frac{\pm\sqrt{16}}{-8} = \frac{\pm 4}{-8}$$

The roots are:

$$m = \frac{4}{-8} = -\frac{1}{2} \qquad \text{OR} \qquad m = \frac{-4}{-8} = \frac{1}{2}$$
$$\therefore m = -0.5 \qquad\qquad\qquad\qquad \therefore m = 0.5$$

ALTERNATIVE METHOD

This equation can be solved quicker using factorisation method. This is a difference of two squares (Section 3.5.2), so we have:

$$1 - 4m^2 = 0$$
$$1^2 - (2m)^2 = 0$$
$$(1 - 2m)(1 + 2m) = 0$$

Therefore:

$$1 - 2m = 0 \qquad\qquad 1 + 2m = 0$$
$$-2m = -1 \qquad \text{OR} \qquad 2m = -1$$
$$\therefore m = 0.5 \qquad\qquad \therefore m = -0.5$$

Let's try another set of examples.

Example 9

Solve the following equations using the formula. Present your answer correct to 2 d.p. where applicable.

a) $7 + x - x^2 = 2(1 - x)$ b) $3x^2 - 0.2x - 0.1 = 0$ c) $4x^6 - 5x^3 - 6 = 0$

d) $x = 3\sqrt{x} - 2$ e) $5x^2 - 7x + 11 = 0$ f) $x^2 + x + 3 = 0$

What did you get? Find the solution below to double-check your answer.

Solution to Example 9

a) $7 + x - x^2 = 2(1 - x)$

Solution

Let's simplify the equation as:

$$7 + x - x^2 = 2(1 - x)$$
$$7 + x - x^2 = 2 - 2x$$
$$5 + 3x - x^2 = 0$$
$$-x^2 + 3x + 5 = 0$$

Now comparing with the standard form $ax^2 + bx + c = 0$, we note that $a = -1$, $b = 3$, and $c = 5$. Hence:

$$x = \frac{-b \pm \sqrt{b^2 - 4ac}}{2a}$$
$$= \frac{-3 \pm \sqrt{(3)^2 - 4(-1)(5)}}{2 \times -1}$$
$$= \frac{-3 \pm \sqrt{9 + 20}}{-2}$$
$$= \frac{-3 \pm \sqrt{29}}{-2} = \frac{3 \mp \sqrt{29}}{2}$$

The roots are:

$$x = \frac{3 - \sqrt{29}}{2} = \frac{3 - 5.3852}{2} \quad \textbf{OR} \quad x = \frac{3 + \sqrt{29}}{2} = \frac{3 + 5.3852}{2}$$
$$\therefore x = -1.19 \text{ (2 d.p.)} \qquad\qquad \therefore x = 4.19 \text{ (2 d.p.)}$$

NOTE

- We can write the equation as $x^2 - 3x - 5 = 0$ either by multiplying through with -1 or transferring all the terms to the RHS, this will not affect our solutions.

- Notice that we changed the sign from $\pm$ to $\mp$. This is because we divided with $-$, which also changed -3 to 3 and -2 to 2.

- It is also advisable to evaluate at the end (or at a go). For example, enter $\frac{3 - \sqrt{21}}{2}$ once on your calculator to minimise errors. If this is not feasible, try to retain as many digits as possible. Here, when we evaluated $\sqrt{21}$, we gave the answer in 4 d.p. since the answer is required to be rounded to 2 d.p. This ensures that we are not far away from the expected answer, though on this occasion we are spot on.

b) $3x^2 - 0.2x - 0.1 = 0$

Solution

It is already in the standard form, comparing with the standard form $ax^2 + bx + c = 0$, we note that $a = 3$, $b = -0.2$, and $c = -0.1$. Hence:

$$x = \frac{-b \pm \sqrt{b^2 - 4ac}}{2a}$$

$$= \frac{-(-0.2) \pm \sqrt{(-0.2)^2 - 4\,(3)\,(-0.1)}}{2 \times 3}$$

$$= \frac{0.2 \pm \sqrt{0.04 + 1.2}}{6}$$

$$= \frac{0.2 \pm \sqrt{1.24}}{6}$$

The roots are:

$$x = \frac{0.2 + \sqrt{1.24}}{6} = \frac{0.2 + 1.1136}{6} \qquad \textbf{OR} \qquad x = \frac{0.2 - \sqrt{1.24}}{6} = \frac{0.2 - 1.1136}{6}$$

$$\therefore x = \mathbf{0.22}\ \textbf{(2 d.p.)} \qquad\qquad\qquad\qquad \therefore x = \mathbf{-0.15}\ \textbf{(2 d.p.)}$$

NOTE

This shows that the quadratic can be used for any real values of a, b, and c. In other words, we do not have to worry about simplifying to ensure that a, b, and c are integers.

---

c) $4x^6 - 5x^3 - 6 = 0$

Solution

This is not a quadratic but let's re-write it to look like it, so we can proceed using the formula.

$$4x^6 - 5x^3 - 6 = 0$$

$$4\left(x^3\right)^2 - 5\left(x^3\right) - 6 = 0$$

Let $y = x^3$, we can then write the equation as

$$4y^2 - 5y - 6 = 0$$

This is now in the standard form, comparing with the standard form $ax^2 + bx + c = 0$, we note that $a = 4$, $b = -5$, and $c = -6$. Hence:

$$y = \frac{-b \pm \sqrt{b^2 - 4ac}}{2a}$$

$$= \frac{-(-5) \pm \sqrt{(-5)^2 - 4\,(4)\,(-6)}}{2 \times 4}$$

$$= \frac{5 \pm \sqrt{25 + 96}}{8}$$

$$= \frac{5 \pm \sqrt{121}}{8}$$

The roots are:

$$y = \frac{5 + \sqrt{121}}{8} = \frac{5 + 11}{8} \qquad \text{OR} \qquad y = \frac{5 - \sqrt{121}}{8} = \frac{5 - 11}{8}$$

$$\therefore y = 2 \qquad\qquad\qquad\qquad \therefore y = -\frac{3}{4}$$

Now, let's solve for x as follows.

When $y = 2$	When $y = -\frac{3}{4}$
$x^3 = 2$	$x^3 = -\frac{3}{4}$
$x = \sqrt[3]{2}$	$x = \sqrt[3]{-\frac{3}{4}}$
$\therefore x = 1.26$ (2 d.p.)	$\therefore x = -0.91$ (2 d.p.)

NOTE

The solution obtained are the real values. We expect six roots, since the equation is a polynomial of degree 6.

---

d) $x = 3\sqrt{x} - 2$

Solution

This is not a quadratic but let's re-write to look like it, so we can proceed using the formula.

$$x = 3\sqrt{x} - 2$$

$$x - 3\sqrt{x} + 2 = 0$$

$$\left(x^{\frac{1}{2}}\right)^2 - 3\sqrt{x} + 2 = 0$$

$$\left(\sqrt{x}\right)^2 - 3\sqrt{x} + 2 = 0$$

Let $y = \sqrt{x}$, we can then write the equation as

$$y^2 - 3y + 2 = 0$$

This is now in the standard form, comparing with the standard form $ax^2 + bx + c = 0$, we note that $a = 1$, $b = -3$, and $c = 2$. Hence:

$$y = \frac{-b \pm \sqrt{b^2 - 4ac}}{2a}$$

$$= \frac{-(-3) \pm \sqrt{(-3)^2 - 4(1)(2)}}{2 \times 1}$$

$$= \frac{3 \pm \sqrt{9 - 8}}{2}$$

$$= \frac{3 \pm \sqrt{1}}{2}$$

The roots are:

$$y = \frac{3 + \sqrt{1}}{2} = \frac{3 + 1}{2} \qquad \text{OR} \qquad y = \frac{3 - \sqrt{1}}{2} = \frac{3 - 1}{2}$$

$$\therefore y = 2 \qquad\qquad\qquad\qquad \therefore y = 1$$

Now, let's solve for x as follows.

When $y = 2$	When $y = 1$
$\sqrt{x} = 2$	$\sqrt{x} = 1$
$x = 2^2$	$x = 1^2$
$\therefore x = 4$	$\therefore x = 1$

NOTE

- We can equally solve this problem by squaring first, but we should ensure that the terms are arranged so that the root can be eliminated. For this, we can proceed as:

$$x = 3\sqrt{x} - 2$$
$$x + 2 = 3\sqrt{x}$$
$$(x + 2)^2 = \left(3\sqrt{x}\right)^2$$
$$x^2 + 4x + 4 = 9x$$
$$x^2 - 5x + 4 = 0$$
$$(x - 1)(x - 4) = 0$$
$$\therefore x = 1, \ x = 4$$

- It is advisable to check that the solutions obtained still satisfy the original equation.

---

e) $5x^2 - 7x + 11 = 0$

Solution

It is already in the standard form, comparing with the standard form $ax^2 + bx + c = 0$, we note that $a = 5, b = -7$, and $c = 11$. Hence:

$$x = \frac{-b \pm \sqrt{b^2 - 4ac}}{2a}$$
$$= \frac{-(-7) \pm \sqrt{(-7)^2 - 4(5)(11)}}{2 \times 5}$$
$$= \frac{7 \pm \sqrt{49 - 220}}{10}$$
$$= \frac{7 \pm \sqrt{-171}}{10}$$

Unfortunately, we can't find $\sqrt{-171}$ because the radicand is negative. So, we say there are no roots for this equation, more precisely no **real** roots.

---

f) $x^2 + x + 3 = 0$

Solution

It is already in the standard form, comparing with the standard form $ax^2 + bx + c = 0$, we note that $a = 1$, $b = 1$, and $c = 3$. Hence:

$$x = \frac{-b \pm \sqrt{b^2 - 4ac}}{2a}$$

$$= \frac{-1 \pm \sqrt{(1)^2 - 4(1)(3)}}{2 \times 1}$$

$$= \frac{-1 \pm \sqrt{1 - 12}}{2}$$

$$= \frac{-1 \pm \sqrt{-11}}{2}$$

Again, we can't find $\sqrt{-11}$ because the radicand is negative. So, we say there are no real roots for this equation. This equation looks simple but still can be solved. Is there any way to know that an equation cannot be solved? The answer to this is in the next section.

7.2.3 DISCRIMINANT

With reference to the quadratic formula, we may notice that the solutions of the equation depend on the value of the expression in the root, i.e., $b^2 - 4ac$. This expression is called **discriminant** and is generally denoted by the letter D. Thus:

$$\boxed{D = b^2 - 4ac}$$
(7.3)

The value of D provides us with three categories of roots as follows:

Case 1 When $D > 0$ or $b^2 > 4ac$, the quadratic equation has two distinct roots, and the graph (Figure 7.1) crosses the x-axis at two different points. The roots will be rational numbers if $b^2 - 4ac$ is a perfect square, otherwise it will be irrational numbers.

Case 2 When $D = 0$ or $b^2 = 4ac$, the quadratic equation has only one root. Technically speaking, there are two roots but they have the same value and overlap each other on the graph, which is commonly said to be repeated (or happened twice). The value of this root is $x_1 = x_2 = -\frac{b}{2a}$.

The quadratic equation can be written as $\left(x + \frac{b}{2a}\right)^2$, and it touches the x-axis at one point. This point also represents the turning point (minimum or maximum) or vertex, as shown in Figure 7.2.

Case 3 When $D < 0$ or $b^2 < 4ac$, the quadratic equation has no (real) roots and the graph does not cross the x-axis at all (Figure 7.3).

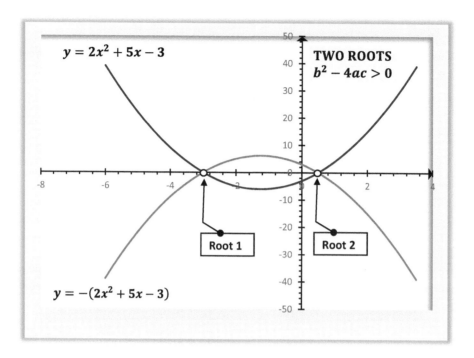

FIGURE 7.1　When $b^2 - 4ac > 0$ illustrated.

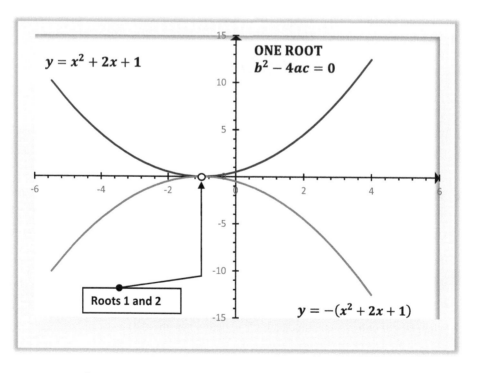

FIGURE 7.2　When $b^2 - 4ac = 0$ illustrated.

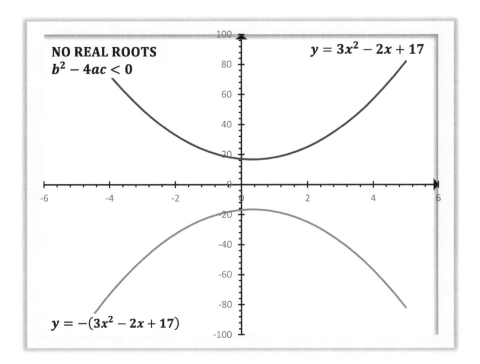

FIGURE 7.3   When $b^2 - 4ac < 0$ illustrated.

Let's try some examples to illustrate.

Example 10

Determine the discriminant of the following quadratic equations and comment on the nature of their roots.

a) $\phi^2 - 7\phi + 11 = 0$ **b)** $2\beta^2 + 50 = 0$ **c)** $3s^2 - 4s = 0$ **d)** $6x - \dfrac{9}{5} - 5x^2 = 0$

What did you get? Find the solution below to double-check your answer.

Solution to Example 10

a) $\phi^2 - 7\phi + 11 = 0$
Solution
Comparing $f(\phi)$ with the standard form $ax^2 + bx + c = 0$, we note that $a = 1$, $b = -7$, and $c = 11$, hence

$$D = b^2 - 4ac$$
$$= (-7)^2 - 4(1)(11)$$
$$= 49 - 44$$
$$= 5$$
$$\therefore D > 0$$

Thus, $f(\phi)$ has two distinct roots that are irrational numbers, and the graph crosses the x-axis at two points. The roots are irrational because D is not a perfect square.

---

b) $2\beta^2 + 50 = 0$

Solution

Let us write $f(\beta)$ in the standard form $ax^2 + bx + c = 0$ as:

$$2\beta^2 + 0\beta + 50 = 0$$

We note from the above that $a = 2$, $b = 0$, and $c = 50$, hence

$$D = b^2 - 4ac$$
$$= (0)^2 - 4\,(2)\,(50)$$
$$= 0 - 400$$
$$= -400$$
$$\therefore D < 0$$

Thus, $f(\beta)$ has no real roots and the graph does not cross the x-axis at all.

---

c) $3s^2 - 4s = 0$

Solution

Let us write $f(s)$ in the standard form $ax^2 + bx + c = 0$ as:

$$3s^2 - 4s + 0 = 0$$

We note from above that $a = 3$, $b = -4$, and $c = 0$, hence

$$D = b^2 - 4ac$$
$$= (-4)^2 - 4\,(3)\,(0)$$
$$= 16 - 0$$
$$= 16$$
$$\therefore D > 0$$

Thus, $f(s)$ has two distinct roots that are rational numbers, because 16 is a perfect square. The graph crosses the x-axis at two points.

---

d) $6x - \dfrac{9}{5} - 5x^2 = 0$

Solution

Let us write $f(x)$ in the standard form $ax^2 + bx + c = 0$ as:

$$-5x^2 + 6x - \frac{9}{5} = 0$$

We note from the above that $a = -5$, $b = 6$, and $c = -\dfrac{9}{5}$, hence

$$D = b^2 - 4ac$$
$$= (6)^2 - 4\,(-5)\left(-\frac{9}{5}\right)$$
$$= 36 - 36$$
$$= 0$$
$$\therefore D = 0$$

Thus, $f(x)$ has one root, which is repeated. The graph touches the x-axis at one single point.

Let's try another example.

Example 11

Determine the values of constant k if $f(x) = 5x^2 + kx + 2$ has only one root. Present the answers correct to 2 decimal places.

What did you get? Find the solution below to double-check your answer.

Solution to Example 11

Comparing $f(x)$ with the standard form $ax^2 + bx + c = 0$, we note that $a = 5$, $b = k$, and $c = 2$. If $f(x)$ has one root, then

$$b^2 - 4ac = 0$$

Thus

$$k^2 - 4(5)(2) = 0$$
$$k^2 - 40 = 0$$
$$k^2 = 40$$
$$k = \pm\sqrt{40} = \pm2\sqrt{10} = 6.32$$
$$\therefore k = \pm 6.32$$

Let's try one final example.

Example 12

Determine the range of values of α for which $f(\theta) = 3\theta^2 - \theta + \alpha$ has two distinct roots. Present the answers correct to 2 decimal places.

What did you get? Find the solution below to double-check your answer.

Solution to Example 12

Comparing $f(\theta)$ with the standard form $ax^2 + bx + c = 0$, we can see that $a = 3$, $b = -1$, and $c = \alpha$. If $f(\theta)$ has two distinct roots, then

$$b^2 - 4ac > 0$$

Thus

$$(-1)^2 - 4(3)(\alpha) > 0$$
$$1 - 12\alpha > 0$$
$$1 > 12\alpha$$
$$\frac{1}{12} > \alpha$$
$$\therefore \alpha < \frac{1}{12}$$

7.2.4 SKETCHING THE GRAPH OF $f(x) = ax^2 + bx + c$

Sketching and being able to use the information on a graph of a quadratic equation is a helpful skill. We've covered most of what is required to do this earlier in this chapter, but we will summarise the three main features, illustrated in Figure 7.4, needed to achieve this. They are:

1) Shape

The graph of quadratic equations has the general shape of a parabola and can be either of these two:

a) A U-like shape if the coefficient of the x^2 term is positive i.e. $a > 0$.

b) An inverted U (or ∩-like) shape if the coefficient of the x^2 term is negative i.e. $a < 0$.

This feature can easily be determined by visual inspection of the equation.

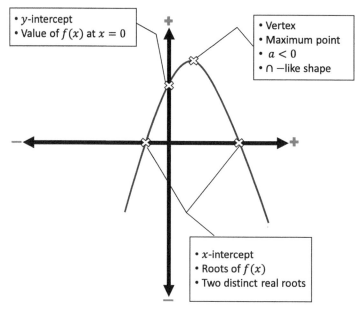

FIGURE 7.4 Graph of a quadratic equation showing intercepts and a vertex.

2) Intercepts

There are two intercepts, namely:

a) x-intercepts: They are the values of x when $f(x) = 0$. Use any suitable method to determine these.

b) y-intercept: It is the value of c when the equation is expressed as $ax^2 + bx + c = 0$. Alternatively, it can be found by substituting $x = 0$ in the equation.

Note that the discriminant or completed square form can help in determining the number and nature of roots.

3) Vertex

Vertex (also called the turning point) is an additional requirement, as it completely describes the shape of a given equation. Note that when the discriminant is negative there are no real roots, and the graph does not cross the x-axis. In this case, knowing the vertex will be helpful in sketching the graph.

The vertex is maximum when $a < 0$ and minimum when $a > 0$. Both can be obtained when a quadratic equation is expressed in the completed square form.

Note that the line through the vertex divides the graph into two equal and identical parts; it is called a **line of symmetry**. For a one-root equation, the vertex is the same as the point at which the graph touches the x-axis. If the only root is β, the vertex is $(\beta, 0)$.

Let's illustrate with examples.

Example 13

Sketch the graph of each of the following quadratic functions.

a) $f(x) = x^2 - x - 6$ **b)** $g(x) = (4 - x)(2x - 3)$ **c)** $h(x) = \frac{3}{5}x^2 + 4x - 15$

What did you get? Find the solution below to double-check your answer.

Solution to Example 13

HINT

For these questions, we need to:

- Obtain the coordinates of the turning point, x-intercepts, and y-intercept.
- Determine whether the parabola is a U-like shape or ∩-like shape.
- Locate the above three/four points on the graph and connect them together with a smooth curve.

a) $f(x) = x^2 - x - 6$
Solution
We need the following information to be able sketch the graph (Figure 7.5).

- **Shape**: It is a U-like shape as $a > 0$.
- **y-intercept**: $(0, -6)$.
- **x-intercept**: Solving the equation, we have the following points: $(-2, 0)$, $(3, 0)$.
- **Vertex (or turning point)**: Using the complete square method, we have $\left(x - \frac{1}{2}\right)^2 - \frac{25}{4}$, so the vertex is $\left(\frac{1}{2}, -\frac{25}{4}\right)$.

---

b) $g(x) = (4 - x)(2x - 3)$
Solution
Expand the function as:

$$g(x) = (4 - x)(2x - 3)$$
$$= -2x^2 + 11x - 12$$

We need the following information to be able sketch the graph (Figure 7.6).

- **Shape**: It is an inverted U shape as $a < 0$.
- **y-intercept**: $(0, -12)$.
- **x-intercept**: Solving the equation, we have the following points$(1.5, 0)$, $(4, 0)$.

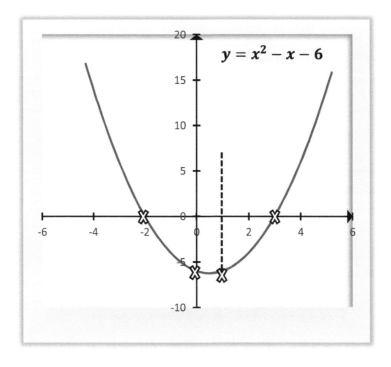

FIGURE 7.5 Solution to Example 13(a).

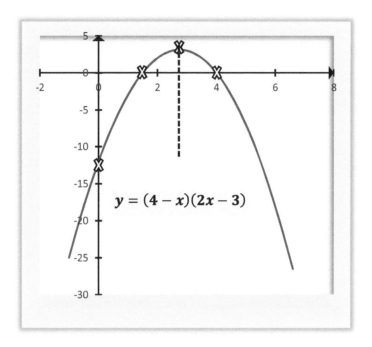

$$y = (4 - x)(2x - 3)$$

FIGURE 7.6 Solution to Example 13(b).

- **Vertex (or turning point)**: Using the complete square method, we have $-2\left(x - \frac{11}{4}\right)^2 + \frac{25}{8}$, so the vertex is $\left(\frac{11}{4}, \frac{25}{8}\right)$.

---

c) $h(x) = \frac{3}{5}x^2 + 4x - 15$

Solution

We need the following information to be able sketch the graph (Figure 7.7).

- **Shape**: It is a U shape as $a > 0$.
- **y-intercept**: $(0, -15)$.
- **x-intercept**: Solving the equation, we have the following points: $(-\frac{10}{3} + \frac{5}{3}\sqrt{13}, 0)$, $(-\frac{10}{3} - \frac{5}{3}\sqrt{13}, 0)$.

- **Vertex (or turning point)**: Using the complete square method, we have $\frac{3}{5}\left(x + \frac{10}{3}\right)^2 - \frac{65}{3}$, so the vertex is $\left(-\frac{10}{3}, -\frac{65}{3}\right)$.

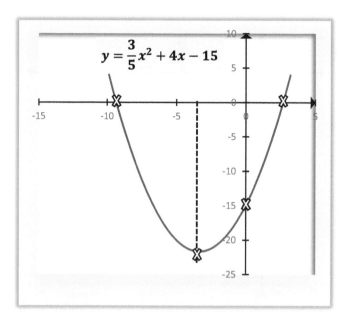

FIGURE 7.7 Solution to Example 13(c).

Let's have another set of examples.

Example 14

Use the graph of $f(x) = x^2 - x - 6$ obtained in Example 13(a) above to find the roots of the following.

a) $g(x) = x^2 - x - 2$ b) $g(x) = x^2 - x - 12$

c) $g(x) = 3x^2 - 3x - 18$ d) $g(x) = x^2 - 2x - 24$

What did you get? Find the solution below to double-check your answer.

Solution to Example 14

a) $g(x) = x^2 - x - 2$
Solution

$$g(x) = x^2 - x - 2$$

implies

$$x^2 - x - 2 = 0$$

Now let's subtract 4 from both sides, thus we have:

$$x^2 - x - 2 - 4 = -4$$
$$x^2 - x - 6 = -4$$

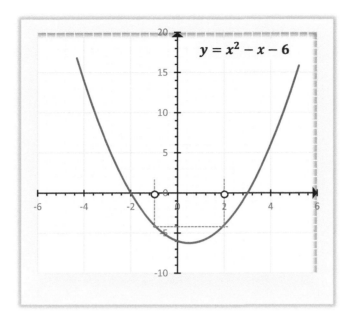

FIGURE 7.8 Solution to Example 14(a).

What this implies is that

$$f(x) = -4$$

Locate where $f(x) = -4$ or $y = -4$, draw a horizontal line until it touches the curve and then draw vertical lines until they touch the x-axis. You now need to read the values at these points, as shown in Figure 7.8.

From Figure 7.8, the solutions of $g(x) = x^2 - x - 2$ are:

$$x = -1 \quad \textbf{OR} \quad x = 2$$

b) $g(x) = x^2 - x - 12$
Solution

$$g(x) = x^2 - x - 12$$

implies

$$x^2 - x - 12 = 0$$

Now let's add 6 to both sides, thus we have:

$$x^2 - x - 12 + 6 = 6$$
$$x^2 - x - 6 = 6$$

What this implies is that

$$f(x) = 6$$

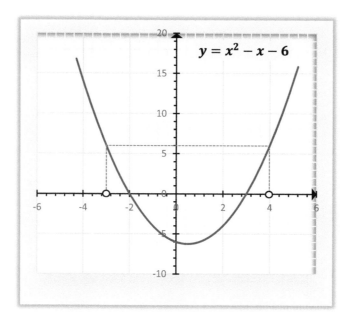

FIGURE 7.9   Solution to Example 14(b).

Locate where $f(x) = 6$ or $y = 6$, draw a horizontal line until it touches the curve and then draw vertical lines until they touch the x-axis. You now need to read the values at these points, as shown in Figure 7.9.

From Figure 7.9, the solutions of $g(x) = x^2 - x - 12$ are:

$$x = -3 \quad \textbf{OR} \quad x = 4$$

---

c) $g(x) = 3x^2 - 3x - 18$
Solution

$$g(x) = 3x^2 - 3x - 18$$

implies

$$3x^2 - 3x - 18 = 0$$

Here, we will try to factorise the expression on the LHS, thus we have:

$$3\left(x^2 - x - 6\right) = 0$$

What this implies is that

$$f(x) = \frac{0}{3}$$

Or simply

$$f(x) = 0$$

In transformation, we say that this is a stretch in the vertical direction or the y-axis. Therefore, the zeros of $3\left(x^2 - x - 6\right) = 0$ and $x^2 - x - 6 = 0$ are the same.

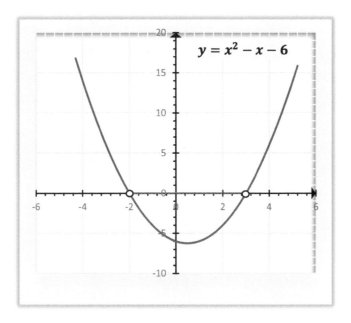

FIGURE 7.10 Solution to Example 14(c).

Now we need to locate where $f(x) = 0$ or $y = 0$, draw a horizontal line until it touches the curve and read the values at these points, as shown in Figure 7.10.

From Figure 7.10, the solutions of $g(x) = 3x^2 - 3x - 18$ are:

$$x = -2 \quad \textbf{OR} \quad x = 3$$

d) $g(x) = x^2 - 2x - 24$
Solution

$$g(x) = x^2 - 2x - 24$$

implies

$$x^2 - 2x - 24 = 0$$

Here we will try to divide each term by 4, thus we have:

$$\frac{x^2}{4} - \frac{2x}{4} - \frac{24}{4} = \frac{0}{4}$$

Simplifying this, we have:

$$\frac{x^2}{2^2} - \frac{x}{2} - 6 = 0$$
$$\left(\frac{x}{2}\right)^2 - \left(\frac{x}{2}\right) - 6 = 0$$

This doesn't look like $x^2 - x - 6$. With a closer look, however, we can see that it does but instead of x we have $\frac{x}{2}$. In transformation, we say that this is a stretch in the horizontal direction or x-axis.

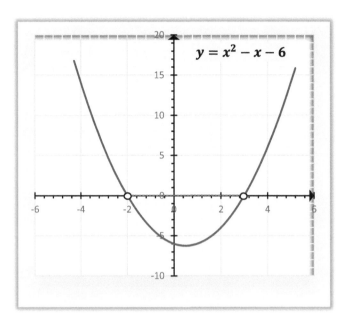

FIGURE 7.11   Solution to Example 14(d).

The stretching factor here is 2. Therefore, the zeros of $x^2 - x - 6 = 0$ are half of the zeros of $\left(\frac{x}{2}\right)^2 - \left(\frac{x}{2}\right) - 6 = 0$.

Now we need to locate where $f(x) = 0$ or $y = 0$, draw a horizontal line until it touches the curve, and read the values at these points, as shown in Figure 7.11.

From Figure 7.11, these give us $x = -2$ or $x = 3$. We need to double these values to obtain the solutions of $\left(\frac{x}{2}\right)^2 - \left(\frac{x}{2}\right) - 6 = 0$, which is the same as zeros of $g(x) = x^2 - 2x - 24$. Thus, we have:

$$x = -4 \quad \textbf{OR} \quad x = 6$$

The above examples are used specifically for sketching, which requires us to show only the key parts of the shape. Let's see when we are asked to plot. Here we go.

Example 15

Draw the graph of $v(t) = t^2 + 3t - 4$ in the interval $-6 \le t \le 4$. Write down the coordinates of the vertex and the equation of the line of symmetry.

What did you get? Find the solution below to double-check your answer.

TABLE 7.1
Solution to Example 15

t	t^2	$3t$	-4	$v(t)$	Working Out	Coordinates: (t, v)
-6	36	-18	-4	**14**	$v(t) = (-6)^2 + 3(-6) - 4 = 14$	**(−6, 14)**
-5	25	-15	-4	**6**	$v(t) = (-5)^2 + 3(-5) - 4 = 6$	**(−5, 6)**
-4	16	-12	-4	**0**	$v(t) = (-4)^2 + 3(-4) - 4 = 0$	**(−4, 0)**
-3	9	-9	-4	**−4**	$v(t) = (-3)^2 + 3(-3) - 4 = -4$	**(−3, −4)**
-2	4	-6	-4	**−6**	$v(t) = (-2)^2 + 3(-2) - 4 = -6$	**(−2, −6)**
-1	1	-3	-4	**−6**	$v(t) = (-1)^2 + 3(-1) - 4 = -6$	**(−1, −6)**
0	0	0	-4	**−4**	$v(t) = (0)^2 + 3(0) - 4 = -4$	**(0, −4)**
1	1	3	-4	**0**	$v(t) = (1)^2 + 3(1) - 4 = 0$	**(1, 0)**
2	4	6	-4	**6**	$v(t) = (2)^2 + 3(2) - 4 = 6$	**(2, 6)**
3	9	9	-4	**14**	$v(t) = (3)^2 + 3(3) - = 14$	**(3, 14)**
4	16	12	-4	**24**	$v(t) = (4)^2 + 3(4) - 4 = 24$	**(4, 24)**

Solution to Example 15

We will go through the steps required to plot a graph of a quadratic function, but these are the same (or very similar at least) to any other function.

Step 1: Using the range of values of t, produce Table 7.1.

Let's illustrate how to fill the table by using the first row, where $t = -6$, as a guide. Thus, the second column is $t^2 = (-6)^2 = 36$ and the third column is $3t = 3(-6) = -18$. The fourth column is a constant and it is therefore -4, which is the same for all the rows. $v(t)$ or the fifth column is the addition of the second, third, and fourth columns, i.e., $36 + (-18) + (-4) = 14$. The 'working out' column shows the full working using the function of t, as explained in Chapter 6. The remaining rows are obtained in the same way.

For this case, the graph is $v(t)$ versus t, so the former will be on the y-axis and the latter on the x-axis. We need to get used to using the exact physical quantities instead of x and y, which are default variables.

Step 2: Choose appropriate scales for the vertical and horizontal axes. The range of t and $v(t)$ will serve as a guide.

Step 3: Using (t, v) coordinates in the table above, locate the 11 points on the graph. For example, the first point for this case is $(-6, 14)$, the second is $(-5, 6)$, and so on, as shown in the last column of Table 7.1.

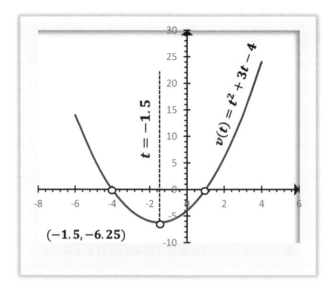

FIGURE 7.12 Solution to Example 15.

Step 4: Join the points together using a smooth curve. Note that the shape is U since $a > 0$.

Step 5: The vertex is a minimum point. It is possible to see this on the table or by looking at the smooth curve. Alternatively, it is a point mid-way between the points at which the graph crosses the x-axis. Since the graph crosses at -4 and 1, the x-coordinate at the mid-point is

$$t_m = \frac{-4 + 1}{2} = -1.5$$

v_m can be found either from the graph or by substituting t_m as follows:

$$v(t) = t^2 + 3t - 4$$
$$v(-1.5) = (-1.5)^2 + 3(-1.5) - 4$$
$$= -6.25$$

Hence, vertex

$$(t_m, v_m) = (-1.5, -6.25)$$

Step 6: The equation of the line of symmetry is

$$t = t_m = -1.5$$

All information is indicated in Figure 7.12.

NOTE

The above is like what we can use to plot an experimental data with an already obtained dependent (y) and independent (x). For this case, we do not need step 1 above, and steps 4–6 are also not essentially required.

7.3 SOLVING SIMULTANEOUS EQUATIONS

We've looked at equations (linear and quadratic), where there is only one unknown quantity in them. For example, t (time) is the only unknown variable in $2t - 3 = 0$ and $t^2 - 5t - 11 = 0$. Sometimes equations may contain two or more unknown quantities. Examples include $x - y = 3$, $x + 2y - z = 0$, and $x^2 y = 4$. In each of these, it is not possible to completely solve for the unknown quantities. However, if we are given two or more equations that are **consistent**, we should be able to solve the equations. This is called **simultaneous equations**.

Consistency of equations implies that there is a value or values which satisfy all the equations simultaneously. For example, $3x + y = 1$ and $x + y = -1$ are consistent as $x = 1$ and $y = -2$ will satisfy both equations. On the other hand, $x - 3y = 7$ and $2x - 6y = 10$ are inconsistent as there are no values of x and y that will satisfy both equations at the same time; we usually say they have **no solution**.

A clue to why this is the case is when the two are written as $y = f(x)$, we will realise that they have the same gradient and are distinct. Since their y-intercept is different, their graphs will be parallel when drawn on the same scale and will never cross each, which is a condition for consistency.

In general, the minimum number of equations required is equal to the number of unknowns to be determined. In other words, if there is one unknown variable, we need one equation and for two variables, at least two equations are required and so on for three, four, and up to n variables.

In this chapter, we will restrict ourselves to simultaneous equations involving: (i) two unknown quantities, and (ii) linear and quadratic. Thus, we will consider:

a) Two linear equations with two unknown quantities (one set of solutions).

b) One linear equation and one quadratic equation with two unknown quantities (one or two sets of solutions).

c) Two quadratic equations with two unknown quantities (one or two sets of solutions).

For the above cases, we will apply three different methods. These methods can also be applied to cases where the unknown quantities are more than two and the power of the unknown is more than two, though with different levels of difficulty. For these latter cases, matrices will be suitable, which is beyond the scope of this book but covered in *Advanced Mathematics for Engineers and Scientists with Worked Examples* by the same author.

7.3.1 ELIMINATION METHOD

To explain this method, we will use the following simultaneous linear equations in x and y.

$$2x - y = 4$$
$$2y + 3x = -1$$

Here are the steps to follow to solve the above equations simultaneously.

Step 1: For ease of reference, label the original equations and all subsequently derived equations accordingly.

$$2x - y = 4 \qquad ------(i)$$
$$2y + 3x = -1 \qquad ------(ii)$$

Step 2: Ensure that the equations have an identical format such that the unknown variables are vertically aligned. For this case, we can assume $ax + by = c$, where x and y are the unknown variables to be found and a, b, and c are constants or numbers.

$$2x - y = 4$$
$$3x + 2y = -1$$

Step 3: Decide which variable to eliminate (or remove) first, and this is the source of the method's name. Let's start by eliminating x. To do this, the coefficient of x in both equations must be the same. This is achieved by using the LCM of the two coefficients. As a result, multiply equation (i) by 3 and equation (ii) by 2, we then have:

$$2x - y = 4 \quad \times 3 \qquad ------(i)$$
$$3x + 2y = -1 \quad \times 2 \qquad -----(ii)$$

We therefore obtain

$$6x - 3y = 12 \qquad ------(iii)$$
$$6x + 4y = -2 \qquad ------(iv)$$

The new equations are numbered (iii) and (iv) and the coefficient of x in both is 6.

Step 4: We now need to remove the x term by subtracting the corresponding term in equation (iv) from equation (iii). We use 'subtraction' here because it achieves the goal of elimination; addition could be appropriate in another case.

$$
\begin{array}{rrclc}
6x & -3y & = & 12 & ------(iii) \\
- & - & & - & \\
6x & +4y & = & -2 & ------(iv) \\
\hline
0 & -7y & = & 14 & ------(v) \\
\hline
\end{array}
$$

Step 5: Now simplify equation (v) as:

$$-7y = 14$$
$$y = \frac{14}{-7}$$
$$\therefore y = -2$$

We now need to find x using elimination, though substitution is more appropriate here to find the second unknown quantity because we have obtained the value of y.

Step 6: To eliminate y, we need to multiply equation (i) by 2 and keep equation (ii) as it is or say multiply it by 1.

$$2x - y = 4 \quad \times 2 \qquad ------(i)$$
$$3x + 2y = -1 \quad \times 1 \qquad ------(ii)$$

We therefore obtain

$$4x - 2y = 8 \qquad ------(\text{vi})$$
$$3x + 2y = -1 \qquad ------(\text{ii})$$

Step 7: We now need to remove the y term by adding the corresponding term in equation (ii) from equation (vi). We use 'addition' here because it achieves the goal of elimination.

$4x$	$-2y$	$=$	8	$------(\text{vi})$
$+$	$+$		$+$	
$3x$	$+2y$	$=$	-1	$------(\text{ii})$
$7x$	0	$=$	7	$------(\text{vii})$

NOTE

In step 6, the equation could have been multiplied by -2 to ensure that the coefficient of the variable to be eliminated has the same sign in both equations. We therefore need to apply subtraction in step 7.

In general, if the coefficient of a term is the same but the sign is different, use 'addition'. However, if the sign is the same, use 'subtraction'.

Step 8: Now simplify equation (vii)

$$7x = 7$$
$$x = \frac{7}{7}$$
$$\therefore x = 1$$

All done!

Before we move on, notice that 'to obtain one variable you will need to eliminate the other'. This is important to note because sometimes you're not required to find both variables. Just eliminate the one that is not needed and keep the one that is required.

7.3.2 SUBSTITUTION METHOD

To explain this method, we will use the following simultaneous linear equations with variables a and b.

$$5a - b = 4$$
$$3b + a = -4$$

Step 1: For ease of reference, label the original equations and all subsequently derived equations accordingly.

$$5a - b = 4 \qquad - - - - - - \text{(i)}$$
$$3b + a = -4 \qquad - - - - - - \text{(ii)}$$

This method does not require the terms to be aligned.

Step 2: Choose one of the equations and make one of the two variables the subject, i.e., standing alone on the left-hand side. For this, we will make b in equation (i) the subject and call it equation (iii) as:

$$5a - b = 4$$
$$5a - 4 = b$$
$$\therefore b = 5a - 4 \qquad - - - - - - \text{(iii)}$$

Step 3: Substitute equation (iii) in equation (ii). Ensure that you do not substitute it in the equation you derived it from, i.e., equation (i). The word 'substitute' here (source of the method's name) means that whenever we see b in equation (ii) we will use its equivalent $(5a - 4)$ found in equation (iii) as:

$$3b + a = -4 \qquad - - - - - - \text{(ii)}$$
$$3(5a - 4) + a = -4 \qquad - - - - - - \text{(iv)}$$

Step 4: Simplify equation (iv) as:

$$3(5a - 4) + a = -4$$
$$15a - 12 + a = -4$$
$$16a = 12 - 4$$
$$16a = 8$$
$$a = \frac{8}{16}$$
$$\therefore a = \frac{1}{2}$$

We've obtained a and we are left with b.

Step 5: Substitute a in equation (iii) as:

$$b = 5a - 4 \qquad - - - - - - \text{(iii)}$$
$$b = 5\left(\frac{1}{2}\right) - 4 = \frac{5}{2} - 4$$
$$\therefore b = -\frac{3}{2}$$

Before we move to examples, it is important to mention the following:

Note 1 The above two methods are not used in isolation, they rather complement each other and can both be used in a single problem. For example, once a variable has been found by elimination, you can apply substitution to find the second variable.

Note 2 Elimination usually works well with linear equations.

Note 3 If a simultaneous equation has one linear and one quadratic, start with the linear to make one variable the subject and substitute it in the quadratic equation.

Note 4 We can verify our answers by substituting them into either of the original equations.

Let's try some examples.

Example 16

Using appropriate methods, solve the following simultaneous equations.

a) $2s - t = 4$ **and** $3s + t = 1$ b) $7x + 3y = 0.6$ **and** $5x - 2y = 2.5$

c) $x - y = 2$ **and** $3x^2 + 2y = 1$ d) $5u + v = 7$ **and** $uv = -6$

e) $pq = -2$ **and** $2p^2 - q^2 = -2$ f) $m - n = -2$ **and** $m^2 + 3mn - n^2 = 1$

What did you get? Find the solution below to double-check your answer.

Solution to Example 16

a) $2s - t = 4$ and $3s + t = 1$
Solution
We have

$$2s - t = 4 \quad ----(i)$$
$$3s + t = 1 \quad ----(ii)$$

Let's eliminate t by adding (i) and (ii), we have

$$5s = 5$$
$$\therefore s = 1$$

Now let's substitute the answer to either equation (i) or (ii). Let's go for equation (ii), thus we have:

$$3(1) + t = 1$$
$$3 + t = 1$$
$$t = 1 - 3$$
$$\therefore t = -2$$

Hence, the solutions to the simultaneous equation are

$$\therefore s = 1, t = -2$$

ALTERNATIVE METHOD

We could have used the elimination method to find t. By multiplying equation (i) by 3 and equation (ii) by 2, we have:

$$6s - 3t = 12 \quad ---- \text{(iii)}$$
$$6s + 2t = 2 \quad ---- \text{(iv)}$$

(iii)–(iv), we have:

$$-5t = 10$$
$$\therefore t = -2$$

---

b) $7x + 3y = 0.6$ **and** $5x - 2y = 2.5$
Solution
We have:

$$7x + 3y = 0.6 \quad ---- \text{(i)}$$
$$5x - 2y = 2.5 \quad ---- \text{(ii)}$$

Let's start by eliminating y, so multiply (i) by 2 and (ii) by 3, we have:

$$14x + 6y = 1.2 \quad ---- \text{(iii)}$$
$$15x - 6y = 7.5 \quad ---- \text{(iv)}$$

Adding (iii) and (iv) together, we have:

$$29x = 8.7$$
$$x = \frac{8.7}{29}$$
$$\therefore x = 0.3$$

Now, let's substitute the answer to either equation (i) or (ii). Let's go for equation (i), thus we have:

$$7x + 3y = 0.6$$
$$3y = 0.6 - 7x = 0.6 - 7(0.3) = 0.6 - 2.1$$

which implies that

$$3y = -1.5$$
$$y = -\frac{1.5}{3}$$
$$\therefore y = -0.5$$

---

c) $x - y = 2$ **and** $3x^2 + 2y = 1$
Solution
We have

$$x - y = 2 \quad ---- \text{(i)}$$
$$3x^2 + 2y = 1 \quad ---- \text{(ii)}$$

We will use the substitution method here. Let's obtain y from (i) as:

$$x - y = 2$$
$$y = x - 2 \quad ----\text{(iii)}$$

Substitute (iii) in (ii) as:

$$3x^2 + 2y = 1$$
$$3x^2 + 2(x - 2) = 1$$
$$3x^2 + 2x - 4 = 1$$
$$3x^2 + 2x - 5 = 0$$
$$(3x + 5)(x - 1) = 0$$

We therefore have:

$$3x + 5 = 0$$
$$3x = -5 \qquad \textbf{OR} \qquad x - 1 = 0$$
$$\therefore x = -\frac{5}{3} \qquad\qquad\qquad \therefore x = \mathbf{1}$$

Now we need to find the value of y using (iii) as:

- When $x = -\dfrac{5}{3}$

$$y = x - 2 = -\frac{5}{3} - 2$$
$$\therefore y = -\frac{11}{3}$$

- When $x = 1$

$$y = x - 2 = 1 - 2$$
$$\therefore y = -\mathbf{1}$$

ALTERNATIVE METHOD

For this current pair of equations, we can make y the subject in both (i) and (ii) as follows:

- From equation (i), we have:

$$x - y = 2$$
$$y = x - 2 \quad ----\text{(v)}$$

- From equation (ii), we have:

$$3x^2 + 2y = 1$$
$$2y = 1 - 3x^2$$
$$y = \frac{1}{2}(1 - 3x^2) \quad ----\text{(vi)}$$

Now we will equate (v) and (vi) as:

$$x - 2 = \frac{1}{2}\left(1 - 3x^2\right)$$
$$2x - 4 = 1 - 3x^2$$
$$3x^2 + 2x - 5 = 0$$

From here, what's left is the same as obtained above.

---

d) $5u + v = 7$ **and** $uv = -6$
Solution
We have:

$$5u + v = 7 \qquad ----\text{(i)}$$
$$uv = -6 \qquad ----\text{(ii)}$$

Using equation (i), we can make v the subject as:

$$v = 7 - 5u \qquad ----\text{(iii)}$$

Now, let's substitute equation (iii) in equation (ii), we have:

$$u\left(7 - 5u\right) = -6$$
$$7u - 5u^2 = -6$$
$$5u^2 - 7u - 6 = 0$$

This is a quadratic equation and can be solved using any of the known methods. We will use factorisation for this. $5u^2 - 7u - 6 = 0$ is already in the standard form $ax^2 + bx + c = 0$. The product of ac

$$ac = 5 \times -6 = -30$$

Two numbers whose product is -30 and sum is -7 are:

$$\mathbf{-10, 3}$$

Re-write the expression and factorise as

$$5u^2 - 10u + 3u - 6 = 0$$
$$5u\left(u - 2\right) + 3\left(u - 2\right) = 0$$
$$\left(u - 2\right)\left(5u + 3\right) = 0$$

Thus

$$\begin{array}{ccc} & & 5u + 3 = 0 \\ u - 2 = 0 & \textbf{OR} & 5u = -3 \\ \therefore \boldsymbol{u = 2} & & \therefore \boldsymbol{u = -\dfrac{3}{5}} \end{array}$$

Let's find v using equation (iii).

When $u = 2$

$v = 7 - 5u$
$= 7 - 5(2)$ **AND**
$= 7 - 10$
$\therefore v = -3$

When $u = -\frac{3}{5}$

$v = 7 - 5u$
$= 7 - 5\left(-\frac{3}{5}\right)$
$= 7 + 3$
$\therefore v = 10$

Hence, the solutions to the simultaneous equation are

$$\therefore u = 2, \, v = -3 \, ; \, u = -\frac{3}{5}, \, v = 10$$

---

e) $pq = -2$ **and** $2p^2 - q^2 = -2$
Solution
We have:

$$pq = -2 \quad ----(i)$$
$$2p^2 - q^2 = -2 \quad ----(ii)$$

Using equation (i), we can make p the subject as:

$$p = -\frac{2}{q} \quad ----(iii)$$

Now, let's substitute equation (iii) in equation (ii), we have:

$$2p^2 - q^2 = -2$$
$$2\left(-\frac{2}{q}\right)^2 - q^2 = -2$$
$$2\left(\frac{4}{q^2}\right) - q^2 = -2$$
$$\frac{8}{q^2} - q^2 = -2$$
$$8 - q^4 = -2q^2$$
$$q^4 - 2q^2 - 8 = 0$$

This is a quartic equation but can be modelled as a quadratic equation by re-writing as

$$(q^2)^2 - 2(q^2) - 8 = 0 \quad ----(iv)$$

where $a = 1, b = -2, c = -8$, and q^2 is the variable to be found. We can factorise this as

$$(q^2 - 4)(q^2 + 2) = 0$$

Thus

$q^2 - 4 = 0$
$q^2 = 4$
$q = \sqrt{4} = \pm 2$ **OR**
$\therefore q = \pm 2$

$q^2 + 2 = 0$
$q^2 = -2$
$q = \sqrt{-2}$
$\therefore q = $ **Not a real root**

Let's find v using equation (iii).

$$\text{When } q = 2$$
$$p = -\frac{2}{q} = -\frac{2}{2} = -1 \quad \textbf{AND} \quad p = -\frac{2}{q} = -\frac{2}{-2} = 1$$
$$\therefore \boldsymbol{p = -1} \qquad\qquad\qquad \therefore \boldsymbol{p = 1}$$

Hence, the solutions to the simultaneous equation are

$$\therefore \boldsymbol{p = \pm 1, \; q = \mp 2}$$

NOTE

- It is customary to re-write equation (iv) by saying, for example, let $y = q^2$. We can then write it as $y^2 - 2y - 8 = 0$, which is clearly a quadratic equation.
- We could not find solutions for $q = \sqrt{-2}$ as we are concerned with real values, otherwise there are two answers for this.
- Since the original equation is of degree four (quartic), it is not a surprise that there are four possible solutions for the unknown variable q.

f) $m - n = -2$ **and** $m^2 + 3mn - n^2 = 1$
Solution
We have:

$$m - n = -2 \qquad ----(i)$$
$$m^2 + 3mn - n^2 = 1 \qquad ----(ii)$$

Using equation (i), we can make n the subject as:

$$n = m + 2 \qquad ----(iii)$$

Now, let's substitute equation (iii) in equation (ii), we have:

$$m^2 + 3mn - n^2 = 1$$
$$m^2 + 3m(m+2) - (m+2)^2 = 1$$
$$m^2 + 3m^2 + 6m - (m^2 + 4m + 4) - 1 = 0$$
$$m^2 + 3m^2 + 6m - m^2 - 4m - 4 - 1 = 0$$
$$3m^2 + 2m - 5 = 0$$

Let's use the formula to solve this. Comparing with the standard form $ax^2 + bx + c$, we note that $a = 3$, $b = 2$, and $c = -5$. Hence,:

$$m = \frac{-b \pm \sqrt{b^2 - 4ac}}{2a}$$
$$= \frac{-2 \pm \sqrt{(2)^2 - 4(3)(-5)}}{2 \times 3}$$
$$= \frac{-2 \pm \sqrt{4 + 60}}{6}$$
$$= \frac{-2 \pm \sqrt{64}}{6} = \frac{-2 \pm 8}{6}$$

The roots are:

$$m = \frac{-2+8}{6} = \frac{6}{6} \qquad\qquad m = \frac{-2-8}{6} = \frac{-10}{6}$$
$$\therefore m = 1 \qquad \textbf{OR} \qquad \therefore m = -\frac{5}{3}$$

Let's find n using equation (iii).

$$\text{When } m = 1 \qquad\qquad \textbf{AND} \qquad \text{When } m = -\frac{5}{3}$$

$$n = m + 2 = 1 + 2 \qquad\qquad n = m + 2 = -\frac{5}{3} + 2$$
$$\therefore n = 3 \qquad\qquad\qquad \therefore n = \frac{1}{3}$$

Hence, the solutions to the simultaneous equation are

$$\therefore m = 1, \ n = 3 \text{ AND } m = -\frac{5}{3}, \ n = \frac{1}{3}$$

7.3.3 GRAPHICAL METHOD

A graph is always an indispensable method of solving problems. We've looked at simultaneous linear equations in Chapter 5 (Coordinate Geometry II) and sketching the graph of a quadratic equation has been covered earlier in this chapter.

Let's quickly set out the steps for solving simultaneous equations graphically as follows:

Step 1: Draw or sketch the graphs of the equations on a graph paper using the same scale.
Step 2: Extend the curves of the functions to ensure that they cross each other. For simultaneous linear equations, the lines only cross at one point, otherwise you should expect two or more points of intersections of the curves.
Step 3: Find the coordinates of the intersections. These are the values of the unknown quantities.

An example is given for each of the three cases in Figure 7.13.

What does it mean if the two lines in Figure 7.13(a) do not meet no matter how much they are extended? This simply implies that the lines are parallel and there is no solution to the simultaneous

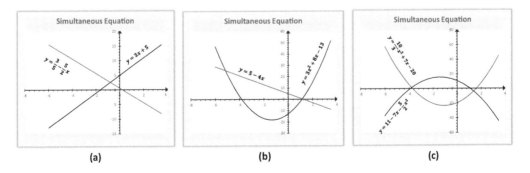

(a)   **(b)**   **(c)**

FIGURE 7.13 Graphical solutions of simultaneous equations illustrated: (a) two linear equations, (b) a linear equation and a quadratic equation, and (c) two quadratic equations.

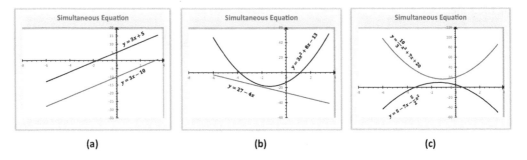

FIGURE 7.14 Graphical illustrations of when there are no solutions to a pair of equations: (a) two linear equations, (b) a linear equation and a quadratic equation, and (c) two quadratic equations.

equations. In other words, no single value of x satisfies both equations at the same time. The same can be said about the two other cases in Figure 7.13(b) and 7.13(c), i.e., no solutions to the simultaneous equations as shown in Figure 7.14.

7.4 CHAPTER SUMMARY

1) Factorisation is a very common method of solving a quadratic equation but there are other methods that prove indispensable, namely:

   - completing the square,
   - the quadratic formula, and
   - graphical method.

2) When using completing the square method, a quadratic equation $ax + bx + c = 0$ is reduced to a 2-term form, consisting of a square and a constant term as:

$$a(x + \lambda)^2 + \mu = 0$$

where both λ and μ are new constants to be found. $a(x + \lambda)^2$ and μ are the square and constant terms respectively. This format is also called the **vertex form**.

3) The **quadratic formula method** uses the formula:

$$x = \frac{-b \pm \sqrt{b^2 - 4ac}}{2a}$$

where a, b, and c are the constants when the equation is written as $y = ax^2 + bx + c$. The sign $\pm$ in the formula implies that there are two solutions for x, which can be given as

$$x_1 = \frac{-b + \sqrt{b^2 - 4ac}}{2a} \quad \text{OR} \quad x_2 = \frac{-b - \sqrt{b^2 - 4ac}}{2a}$$

4) The solutions of a quadratic equation depend on the value of the expression in the root, i.e., $b^2 - 4ac$. This expression is called **discriminant**. It is generally denoted by the letter D, where:

$$D = b^2 - 4ac$$

5) The value of D results in three categories of roots as follows:

- When $D > 0$ or $b^2 > 4ac$, the quadratic equation has two distinct roots, and the graph crosses the x-axis at two different points. The roots will be rational numbers if $b^2 - 4ac$ is a perfect square, otherwise they will be irrational numbers.

- When $D = 0$ or $b^2 = 4ac$, the quadratic equation has only one root. Technically speaking, there are two roots here, but they have the same value and overlap with each other on the graph, which is commonly said to be repeated. The value of this root is $x_1 = x_2 = -\dfrac{b}{2a}$.

- When $D < 0$ or $b^2 < 4ac$, the quadratic equation has no (real) roots and the graph does not cross the x-axis at all.

6) To completely sketch the graph of $f(x) = ax^2 + bx + c$, we need to know the following:

- **Shape**: A U shape if the coefficient of the x^2 term is positive i.e. $a > 0$. An inverted U shape if the coefficient of the x^2 term is negative i.e. $a < 0$.

- **x-intercepts**: These are the values of x when $f(x) = 0$. Use any suitable method to determine these.

- **y-intercept**: It is the value of c when the equation is expressed as $ax^2 + bx + c = 0$. Alternatively, it can be found by substituting $x = 0$ in the equation.

- **Vertex**: The vertex (also called the turning point) is maximum when $a < 0$ and minimum when $a > 0$.

7) Simultaneous equations can be solved using the elimination method, the substitution method, or the combination.

8) In general, the minimum number of equations required is equal to the number of the unknown to be determined. If there is one unknown variable, we need one equation and for two unknown quantities, at least two equations are required and so on for three, four and up to n unknown quantities.

9) Two or more equations are said to be **consistent** if there is a value or values which satisfies all the equations.

****

7.5 FURTHER PRACTICE

To access complementary contents, including additional exercises, please go to www.dszak.com.

8 Surds

Once you have studied the content of this chapter, you should be able to:

- Simplify expressions involving surds
- Carry out addition, subtraction, and multiplication of surds
- Discuss conjugates of surds
- Carry out division of surds using rationalising the denominator
- Determine the square root of a surd

8.1 INTRODUCTION

Presenting numbers in surd forms is common and highly essential in science and engineering, including when solving a quadratic equation using the formula method or when working with trigonometric angles. In this chapter, we will explain this concept, covering its meaning, types, rules, and applications.

8.2 WHAT IS A SURD?

A surd is a number or an expression written in the form $\pm k\sqrt[n]{x}$, which cannot be expressed exactly as a fraction; in other words, it's an irrational number. It's as simple as that. Let's state the conditions required for a number, given as $\pm k\sqrt[n]{x}$, to be regarded as a surd:

Condition 1 : $n \in \mathbb{Z}^+ - \{0, 1\}$

This means that the value of n can be any integer except 0 (zero) and 1 (one). However, the most common case is $n = 2$, i.e., $\sqrt[2]{x}$ (or $\sqrt{x}$). This is an example of **quadratic surds** or **quadratic irrationals**. Similarly, $\sqrt[3]{x}$ is an example of **cubic surds**.

Condition 2 : $x \in \mathbb{Q}^+$

This implies that the value of x can be any positive rational number. A rational number is a number that can be expressed as a ratio of two non-zero integers, such as $\frac{a}{b}$, where a and b are integers and $b \neq 0$. This has been covered in Chapter 1.

Condition 3 : $\pm k$ where $k \in \mathbb{R} - \{0\}$

The constant k is any real number excluding 0 (zero); it can be regarded as a coefficient of a surd, just like 3 is the coefficient of the term $3x$. This shows that a surd can either be positive or negative and has a value between $-\infty$ and ∞.

Condition 4 : $\pm\sqrt[n]{x} \neq \frac{a}{b}$ where $a, b \in \mathbb{Z}$ and $b \neq 0$

This implies that if the first three conditions are met but the expression can be simplified further to obtain a fraction or whole number, then it is not a surd. For example, $\sqrt{4}$ is not a surd since $\sqrt{4}$ equals 2.

The square root of a prime number is a typical example of a surd. It is also worth noting that whilst a calculator can provide the answer to a surd, it is usually an approximation.

Let's try some examples.

Example 1

Sort out the following numbers into surds and non-surds and provide a brief comment in each case.

a) $\sqrt{2}$ b) $\sqrt{3}$ c) $\sqrt[3]{4}$ d) $\sqrt[5]{7}$

e) $\sqrt{36}$ f) $\sqrt{\frac{1}{4}}$ g) $\sqrt[3]{27}$ h) $\sqrt[5]{32}$

What did you get? Find the solution below to double-check your answer.

Solution to Example 1

HINT

- Apply the four conditions above to these numbers. The answers are provided in Table 8.1 with comments.
- A calculator will give an answer to surd, but this is only an approximation, correct to a given significant figure or decimal places.
- Check the answers using a calculator to verify.
- If a number is required to be accurate, it is better to present it in surd form. This is particularly important if it will be used in a subsequent calculation.

8.3 FUNDAMENTALS OF SURDS

8.3.1 MIXED SURDS

Recall that an improper fraction such as $\frac{7}{3}$ can be written as a mixed fraction $2\frac{1}{3}$. Similarly, when a number is made up of rational and irrational parts (surd), it is called a **mixed surd**. The general expression for a mixed surd is

$$\boxed{a \pm b\sqrt{c}}$$ (8.1)

TABLE 8.1
Solution to Example 1

		Type	Notes
a)	$\sqrt{2}$	Surd	
b)	$\sqrt{3}$	Surd	They meet the conditions above as they cannot be expressed as a fraction.
c)	$\sqrt[3]{4}$	Surd	
d)	$\sqrt[5]{7}$	Surd	
e)	$\sqrt{36}$	Non-surd	This is a rational number as $\sqrt{36} = 6$.
f)	$\sqrt{\frac{1}{4}}$	Non-surd	This is a rational number as $\sqrt{\frac{1}{4}} = \frac{1}{2}$.
g)	$\sqrt[3]{27}$	Non-surd	This is a rational number as $\sqrt[3]{27} = 3$.
h)	$\sqrt[5]{32}$	Non-surd	This is a rational number as $\sqrt[5]{32} = 2$.

where a, b, and c are rational numbers but $\sqrt{c}$ is an irrational number, i.e., a surd. The above is also an example of a **quadratic mixed surd**. In general, a mixed surd can be written as:

$$\boxed{a \pm b\sqrt[n]{c}} \tag{8.2}$$

Again, n is any positive integer excluding 1 and 0.

8.3.2 COMPOUND SURDS

A **compound surd** is made up of two irrational parts and takes the general form

$$\boxed{\sqrt[n]{a} \pm \sqrt[n]{b}} \tag{8.3}$$

If $n = 2$, the above expression can be written as $\sqrt{a} \pm \sqrt{b}$. This is an example of a quadratic **compound surd**. Sometimes, the terms 'mixed surd' and 'compound surd' are used interchangeably and are therefore the same.

8.3.3 SIMILAR SURDS

Two quadratic surds $\pm\sqrt{a}$ and $\pm\sqrt{b}$ are said to be similar if their radicands (i.e., the number in the square root) are the same, i.e., $a = b$.

The following pairs are similar surds:

a) $\sqrt{2}$ **and** $-\sqrt{2}$ **b)** $3\sqrt{y}$ **and** $-2\sqrt{y}$ **c)** $6\sqrt{7}$ **and** $5\sqrt{7}$

On the other hand, the following pairs are not similar surds:

a) $2\sqrt{2}$ **and** $\sqrt{3}$ **b)** $2\sqrt{3}$ **and** $3\sqrt{2}$ **c)** $5\sqrt{x}$ **and** $5\sqrt{2x}$ **d)** $\sqrt{2}$ **and** $\sqrt[3]{2}$

Note that, though $\sqrt{2}$ and $\sqrt[3]{2}$ have the same radicand and their coefficients are also the same, they are however not similar surds. This is because one is a square root while the other is a cube root. When surds are similar, they can be combined algebraically.

In general, two surds (quadratic or otherwise) $\pm\sqrt[n]{a}$ and $\pm\sqrt[m]{b}$ are said to be similar if:

1) their radicands are equal, i.e., $a = b$, and

2) $m = n$.

8.3.4 RULES OF SURDS

Essentially, there are three fundamental rules which will be covered here.

Rule 1 Multiplication

Two or more surds can be multiplied together as shown below.

$$\boxed{k_1\sqrt{a} \times k_2\sqrt{b} = k_1k_2\sqrt{a \times b}} \tag{8.4}$$

If $k_1 = k_2 = 1$, the above can be reduced to:

$$\boxed{\sqrt{a} \times \sqrt{b} = \sqrt{a \times b}} \tag{8.5}$$

Similarly,

$$\boxed{\sqrt[3]{a} \times \sqrt[3]{b} = \sqrt[3]{a \times b}}$$

In general, we have:

$$\boxed{k_1\sqrt[n]{a} \times k_2\sqrt[n]{b} = k_1k_2\sqrt[n]{a \times b}} \tag{8.6}$$

However, if $m \neq n$, then

$$\boxed{\sqrt[m]{a} \times \sqrt[n]{b} \neq \sqrt[m]{a \times b}} \;\; \textbf{OR} \;\; \boxed{\sqrt[m]{a} \times \sqrt[n]{b} \neq \sqrt[n]{a \times b}} \tag{8.7}$$

Let's try some examples.

Example 2

Evaluate each of the following:

a) $\sqrt{2} \times \sqrt{2}$ **b)** $\sqrt{24} \times \sqrt{24}$ **c)** $\sqrt{-1} \times \sqrt{-1}$ **d)** $\sqrt[3]{2} \times \sqrt[3]{4}$

e) $\sqrt{a} \times \sqrt{b} \times \sqrt{c}$ **f)** $\sqrt{5x} \times \sqrt{20x}$ **g)** $\sqrt{2} \times \sqrt{3}$ **h)** $\sqrt{x} \times \sqrt{y}$

i) $\sqrt{5} \times \sqrt{x}$ **j)** $\sqrt{2} \times \sqrt{8}$ **k)** $\sqrt{27} \times \sqrt{3}$ **l)** $-\sqrt{3} \times \sqrt{12}$

m) $\sqrt[3]{4} \times \sqrt[3]{4} \times \sqrt[3]{4}$ **n)** $\sqrt[5]{16} \times \sqrt[5]{16} \times \sqrt[5]{16} \times \sqrt[5]{16} \times \sqrt[5]{16}$

What did you get? Find the solution below to double-check your answer.

Solution to Example 2

HINT

In this example, we will be applying the fact that since $x > 0$

$$\sqrt{x} \times \sqrt{x} = x \text{ and } \sqrt[3]{x} \times \sqrt[3]{x} \times \sqrt[3]{x} = x$$

In general,

$$\sqrt[n]{x_1} \times \sqrt[n]{x_2} \times \sqrt[n]{x_2} \times \cdots \times \sqrt[n]{x_{n-1}} \times \sqrt[n]{x_n}$$
$$= x$$
$$for\, x_1 = x_2 = x_3 \cdots = x_{n-1} = x_n$$

This is a special case of multiplication rule where the radicands are equal.

a) $\sqrt{2} \times \sqrt{2}$
Solution

$$\sqrt{2} \times \sqrt{2} = \left(\sqrt{2}\right)^2 = 2$$
$$\therefore \sqrt{2} \times \sqrt{2} = 2$$

b) $\sqrt{24} \times \sqrt{24}$
Solution

$$\sqrt{24} \times \sqrt{24} = \left(\sqrt{24}\right)^2 = 24$$
$$\therefore \sqrt{24} \times \sqrt{24} = 24$$

c) $\sqrt{-1} \times \sqrt{-1}$
Solution

$$\sqrt{-1} \times \sqrt{-1} = \left(\sqrt{-1}\right)^2 = -1$$
$$\therefore \sqrt{-1} \times \sqrt{-1} = -1$$

NOTE
It is also correct to say that $\sqrt{-1} \times \sqrt{-1} = \sqrt{-1 \times -1} = \sqrt{1} = 1$ as per Equation 8.5. However, $\sqrt{-1}$ has a special case in mathematics. In complex numbers, $\sqrt{-1} = j$. This topic is beyond the scope of this book, but it's covered in *Advanced Mathematics for Engineers and Scientists with Worked Examples* by the same author.

d) $\sqrt[3]{2} \times \sqrt[3]{4}$

Solution

$$\sqrt[3]{2} \times \sqrt[3]{4} = \sqrt[3]{2 \times 4}$$
$$= \sqrt[3]{8} = 2$$
$$\therefore \sqrt[3]{2} \times \sqrt[3]{4} = \mathbf{2}$$

e) $\sqrt{a} \times \sqrt{b} \times \sqrt{c}$
Solution

$$\sqrt{a} \times \sqrt{b} \times \sqrt{c} = \sqrt{a \times b \times c}$$
$$= \sqrt{abc}$$
$$\therefore \sqrt{a} \times \sqrt{b} \times \sqrt{c} = \sqrt{abc}$$

f) $\sqrt{5x} \times \sqrt{20x}$
Solution

$$\sqrt{5x} \times \sqrt{20x} = \sqrt{(5x) \times (20x)}$$
$$= \sqrt{5 \times x \times 20 \times x}$$
$$= \sqrt{100x^2} = \sqrt{100 \times x^2}$$
$$= \sqrt{10^2} \times \sqrt{x^2}$$
$$= 10 \times x = 10x$$
$$\therefore \sqrt{5x} \times \sqrt{20x} = \mathbf{10x}$$

g) $\sqrt{2} \times \sqrt{3}$
Solution

$$\sqrt{2} \times \sqrt{3} = \sqrt{2 \times 3} = \sqrt{6}$$
$$\therefore \sqrt{2} \times \sqrt{3} = \sqrt{6}$$

h) $\sqrt{x} \times \sqrt{y}$
Solution

$$\sqrt{x} \times \sqrt{y} = \sqrt{x \times y} = \sqrt{xy}$$
$$\therefore \sqrt{x} \times \sqrt{y} = \sqrt{xy}$$

i) $\sqrt{5} \times \sqrt{x}$
Solution

$$\sqrt{5} \times \sqrt{x} = \sqrt{5 \times x} = \sqrt{5x}$$

$$\therefore \sqrt{5} \times \sqrt{x} = \sqrt{5x}$$

j) $\sqrt{2} \times \sqrt{8}$

Solution

$$\sqrt{2} \times \sqrt{8} = \sqrt{2 \times 8} = \sqrt{16}$$
$$= \sqrt{4^2} = 4$$
$$\therefore \sqrt{2} \times \sqrt{8} = 4$$

k) $\sqrt{27} \times \sqrt{3}$

Solution

$$\sqrt{27} \times \sqrt{3} = \sqrt{3 \times 27}$$
$$= \sqrt{81} = \sqrt{9^2} = 9$$
$$\therefore \sqrt{27} \times \sqrt{3} = 9$$

l) $-\sqrt{3} \times \sqrt{12}$

Solution

$$-\sqrt{3} \times \sqrt{12} = -\sqrt{3 \times 12}$$
$$= -\sqrt{36} = -\sqrt{6^2} = -6$$
$$\therefore -\sqrt{3} \times \sqrt{12} = -6$$

m) $\sqrt[3]{4} \times \sqrt[3]{4} \times \sqrt[3]{4}$

Solution

$$\sqrt[3]{4} \times \sqrt[3]{4} \times \sqrt[3]{4} = \left(\sqrt[3]{4}\right)^3 = 4$$
$$\therefore \sqrt[3]{4} \times \sqrt[3]{4} \times \sqrt[3]{4} = 4$$

n) $\sqrt[5]{16} \times \sqrt[5]{16} \times \sqrt[5]{16} \times \sqrt[5]{16} \times \sqrt[5]{16}$

Solution

$$\sqrt[5]{16} \times \sqrt[5]{16} \times \sqrt[5]{16} \times \sqrt[5]{16} \times \sqrt[5]{16}$$
$$= \left(\sqrt[5]{16}\right)^5 = 16$$
$$\therefore \sqrt[5]{16} \times \sqrt[5]{16} \times \sqrt[5]{16} \times \sqrt[5]{16} \times \sqrt[5]{16} = 16$$

Another set of examples to try.

Example 3

Simplify each of the following:

a) $3\sqrt{5} \times \sqrt{7}$ **b)** $3\sqrt{2} \times 5\sqrt{8}$ **c)** $2\sqrt{6} \times 5\sqrt{7}$ **d)** $4\sqrt{10} \times 3\sqrt{8}$ **e)** $\sqrt{3} + \sqrt{2}$

What did you get? Find the solution below to double-check your answer.

Solution to Example 3

HINT

$a\sqrt{b}$ should be viewed as a product of a and $\sqrt{b}$ as you would do with ab.

a) $3\sqrt{5} \times \sqrt{7}$
Solution

$$3\sqrt{5} \times \sqrt{7} = 3 \times \sqrt{5} \times \sqrt{7}$$
$$= 3 \times \sqrt{5 \times 7} = 3\sqrt{35}$$
$$\therefore 3\sqrt{5} \times \sqrt{7} = 3\sqrt{35}$$

b) $3\sqrt{2} \times 5\sqrt{8}$
Solution

$$3\sqrt{2} \times 5\sqrt{8} = 3 \times 5 \times \sqrt{2} \times \sqrt{8}$$
$$= 15 \times \sqrt{2 \times 8}$$
$$= 15 \times \sqrt{16}$$
$$= 15 \times 4 = 60$$
$$\therefore 3\sqrt{2} \times 5\sqrt{8} = 60$$

c) $2\sqrt{6} \times 5\sqrt{7}$
Solution

$$2\sqrt{6} \times 5\sqrt{7} = 2 \times 5 \times \sqrt{6} \times \sqrt{7}$$
$$= 10 \times \sqrt{6 \times 7} = 10 \times \sqrt{42}$$
$$= 10\sqrt{42}$$
$$\therefore 2\sqrt{6} \times 5\sqrt{7} = 10\sqrt{42}$$

d) $4\sqrt{10} \times 3\sqrt{8}$
Solution

$$4\sqrt{10} \times 3\sqrt{8} = 4 \times 3 \times \sqrt{10} \times \sqrt{8}$$
$$= 12 \times \sqrt{10 \times 8} = 12\sqrt{80}$$
$$= 12 \times \sqrt{16 \times 5} = 12 \times 4\sqrt{5}$$
$$= 48\sqrt{5}$$
$$\therefore 4\sqrt{10} \times 3\sqrt{8} = 48\sqrt{5}$$

e) $\sqrt{3} + \sqrt{2}$

Solution

$$\sqrt{3} + \sqrt{2} = \sqrt{3} + \sqrt{2}$$

$$\therefore \sqrt{3} + \sqrt{2} = \sqrt{3} + \sqrt{2}$$

NOTE

The rule above does not apply to addition. Hence, we cannot simplify this.

As a result of Example 3(e) above, we can now introduce the next rule.

Rule 2 Addition and subtraction

It states that

$$\boxed{m\sqrt{a} + n\sqrt{a} = (m + n)\sqrt{a}} \text{ AND } \boxed{m\sqrt{b} - n\sqrt{b} = (m - n)\sqrt{b}} \tag{8.8}$$

On the contrary, the following are not valid:

$$\boxed{\sqrt{a} + \sqrt{b} \neq \sqrt{a + b}} \text{ AND } \boxed{\sqrt{a} - \sqrt{b} \neq \sqrt{a - b}} \tag{8.9}$$

In other words, the addition of surds is not equal to the surd of an addition. Similarly, the subtraction of surds is not equal to the surd of a subtraction.

Rule 3 Division

A surd can be divided by another surd

$$\boxed{\frac{\sqrt{a}}{\sqrt{b}} = \sqrt{\frac{a}{b}}} \text{ OR } \boxed{\sqrt{a} \div \sqrt{b} = \sqrt{\frac{a}{b}}} \tag{8.10}$$

This third rule of division will be useful when we simplify expressions involving surds.

Let's try some examples.

Example 4

Evaluate each of the following, giving the final answer in the form $a\sqrt{b}$, where a and b are real numbers.

a) $\sqrt{14} \div \sqrt{7}$ **b)** $2\sqrt{6x} \div \sqrt{3x}$ **c)** $\dfrac{\sqrt{8}}{\sqrt{2}}$

d) $\dfrac{\sqrt{63}}{\sqrt{7}}$ **e)** $\dfrac{\sqrt{125}}{\sqrt{5}}$ **f)** $\dfrac{5\sqrt{35}}{15\sqrt{7}}$

What did you get? Find the solution below to double-check your answer.

Solution to Example 4

a) $\sqrt{14} \div \sqrt{7}$
Solution

$$\sqrt{14} \div \sqrt{7} = \frac{\sqrt{14}}{\sqrt{7}}$$

$$= \sqrt{\frac{14}{7}} = \sqrt{2}$$

$$\therefore \sqrt{14} \div \sqrt{7} = \sqrt{2}$$

b) $2\sqrt{6x} \div \sqrt{3x}$
Solution

$$2\sqrt{6x} \div \sqrt{3x} = \frac{2 \times \sqrt{6x}}{\sqrt{3x}}$$

$$= 2 \times \sqrt{\frac{6x}{3x}} = 2 \times \sqrt{2}$$

$$= 2\sqrt{2}$$

$$\therefore 2\sqrt{6x} \div \sqrt{3x} = 2\sqrt{2}$$

c) $\frac{\sqrt{8}}{\sqrt{2}}$
Solution

$$\frac{\sqrt{8}}{\sqrt{2}} = \sqrt{\frac{8}{2}}$$

$$= \sqrt{4} = 2$$

$$\therefore \frac{\sqrt{8}}{\sqrt{2}} = 2$$

d) $\frac{\sqrt{63}}{\sqrt{7}}$
Solution

$$\frac{\sqrt{63}}{\sqrt{7}} = \sqrt{\frac{63}{7}} = \sqrt{9} = 3$$

$$\therefore \frac{\sqrt{63}}{\sqrt{7}} = 3$$

e) $\dfrac{\sqrt{125}}{\sqrt{5}}$

Solution

$$\frac{\sqrt{125}}{\sqrt{5}} = \sqrt{\frac{125}{5}} = \sqrt{25} = 5$$

$$\therefore \frac{\sqrt{125}}{\sqrt{5}} = 5$$

---

f) $\dfrac{5\sqrt{35}}{15\sqrt{7}}$

Solution

$$\frac{5\sqrt{35}}{15\sqrt{7}} = \frac{5}{15} \times \frac{\sqrt{35}}{\sqrt{7}}$$

$$= \frac{1}{3} \times \sqrt{\frac{35}{7}} = \frac{1}{3} \times \sqrt{5}$$

$$\therefore \frac{5\sqrt{35}}{15\sqrt{7}} = \frac{1}{3}\sqrt{5}$$

8.4 SIMPLIFICATION OF SURDS

Simplifying surds implies reducing the value of the radicands where possible. Reverse multiplication rule will be used for whole numbers whilst reverse division rule will be relevant for fractions.

To simplify a surd, the radicand must be expressed in terms of a product of its largest perfect square, which is sometimes difficult to know. This can be overcome if the number is written as a product of its prime factors (see Chapter 1). For example:

$$54 = 2 \times 3 \times 3 \times 3$$

So, to find the square root of 54, we then have:

$$\sqrt{54} = \sqrt{2 \times 3 \times 3 \times 3}$$

$$= \sqrt{2 \times 3 \times 3^2}$$

$$= \sqrt{2 \times 3} \times \sqrt{3^2}$$

$$= \sqrt{6} \times 3$$

$$= 3\sqrt{6}$$

Let's try some examples.

Example 5

Express the following in the simplest possible form, i.e., $a\sqrt{b}$.

a) $\sqrt{20}$ b) $\sqrt{96}$ c) $\sqrt{243}$ d) $2\sqrt{27}$

e) $\sqrt{\frac{1}{4}}$ f) $\sqrt{\frac{3}{4}}$ g) $\sqrt{\frac{18}{50}}$ h) $\sqrt{0.04}$

What did you get? Find the solution below to double-check your answer.

Solution to Example 5

a) $\sqrt{20}$
Solution

$$\sqrt{20} = \sqrt{2 \times 2 \times 5}$$
$$= \sqrt{2^2 \times 5} = \sqrt{4 \times 5}$$
$$= \sqrt{4} \times \sqrt{5} = 2 \times \sqrt{5}$$
$$= 2\sqrt{5}$$
$$\therefore \sqrt{20} = 2\sqrt{5}$$

b) $\sqrt{96}$
Solution

$$\sqrt{96} = \sqrt{2 \times 2 \times 2 \times 2 \times 2 \times 3}$$
$$= \sqrt{2^4 \times 2 \times 3}$$
$$= \sqrt{16 \times 6}$$
$$= \sqrt{16} \times \sqrt{6} = \sqrt{4^2} \times \sqrt{6}$$
$$= 4 \times \sqrt{6} = 4\sqrt{6}$$
$$\therefore \sqrt{96} = 4\sqrt{6}$$

c) $\sqrt{243}$
Solution

$$\sqrt{243} = \sqrt{3 \times 3 \times 3 \times 3 \times 3}$$
$$= \sqrt{3^4 \times 3} = \sqrt{81 \times 3} = \sqrt{81} \times \sqrt{3}$$
$$= \sqrt{9^2} \times \sqrt{3} = 9 \times \sqrt{3} = 9\sqrt{3}$$
$$\therefore \sqrt{243} = 9\sqrt{3}$$

d) $2\sqrt{27}$
Solution

$$2\sqrt{27} = 2 \times \sqrt{3 \times 3 \times 3}$$
$$= 2 \times \sqrt{3^2 \times 3}$$
$$= 2 \times \sqrt{9} \times \sqrt{3}$$
$$= 2 \times 3 \times \sqrt{3} = 6\sqrt{3}$$
$$\therefore 2\sqrt{27} = 6\sqrt{3}$$

e) $\sqrt{\dfrac{1}{4}}$
Solution

$$\sqrt{\frac{1}{4}} = \frac{\sqrt{1}}{\sqrt{4}} = \frac{\sqrt{1}}{\sqrt{2^2}} = \frac{1}{2}$$
$$\therefore \sqrt{\frac{1}{4}} = \frac{1}{2}$$

f) $\sqrt{\dfrac{3}{4}}$
Solution

$$\sqrt{\frac{3}{4}} = \frac{\sqrt{3}}{\sqrt{4}} = \frac{\sqrt{3}}{\sqrt{2^2}} = \frac{\sqrt{3}}{2} = \frac{1}{2}\sqrt{3}$$
$$\therefore \sqrt{\frac{3}{4}} = \frac{1}{2}\sqrt{3}$$

g) $\sqrt{\dfrac{18}{50}}$
Solution

$$\sqrt{\frac{18}{50}} = \sqrt{\frac{9}{25}} = \frac{\sqrt{9}}{\sqrt{25}} = \frac{\sqrt{3^2}}{\sqrt{5^2}} = \frac{3}{5}$$
$$\therefore \sqrt{\frac{18}{50}} = \frac{3}{5}$$

h) $\sqrt{0.04}$
Solution

$$\sqrt{0.04} = \sqrt{\frac{4}{100}} = \frac{\sqrt{4}}{\sqrt{100}}$$
$$= \frac{\sqrt{2^2}}{\sqrt{10^2}} = \frac{2}{10} = \frac{1}{5}$$
$$\therefore \sqrt{0.04} = \frac{1}{5}$$

Another set of examples to try.

Example 6

Given that $\sqrt{2} = 1.414$, $\sqrt{3} = 1.732$, $\sqrt{5} = 2.236$, and $\sqrt{7} = 2.646$, evaluate each of the following. Present your answers correct to 4 significant figures.

a) $\sqrt{20}$ b) $\sqrt{112}$ c) $\sqrt{72}$

d) $\sqrt{175}$ e) $\sqrt{48} - 10$ f) $2\sqrt{27} + 3\sqrt{80}$

What did you get? Find the solution below to double-check your answer.

Solution to Example 6

HINT

In this case, we need to simplify first, ensuring that the radicand is **2, 3, 5,** or **7** and then substitute.

a) $\sqrt{20}$
Solution

$$\sqrt{20} = \sqrt{4 \times 5} = 2\sqrt{5}$$
$$= 2\,(2.236) = 4.472$$
$$\therefore \sqrt{20} = \textbf{4.472 (4 s.f.)}$$

b) $\sqrt{112}$
Solution

$$\sqrt{112} = \sqrt{16 \times 7} = 4\sqrt{7}$$
$$= 4\,(2.646) = 10.58$$
$$\therefore \sqrt{112} = \textbf{10.58 (4 s.f.)}$$

c) $\sqrt{72}$
Solution

$$\sqrt{72} = \sqrt{36 \times 2} = \sqrt{36} \times \sqrt{2}$$
$$= \sqrt{6^2} \times \sqrt{2} = 6 \times \sqrt{2}$$
$$= 6\,(1.414) = 8.484$$
$$\therefore \sqrt{72} = \textbf{8.484 (4 s.f.)}$$

d) $\sqrt{175}$
Solution

$$\sqrt{175} = \sqrt{25 \times 7} = \sqrt{25} \times \sqrt{7}$$
$$= \sqrt{5^2} \times \sqrt{7} = 5 \times \sqrt{7}$$
$$= 5\,(2.646) = 13.23$$
$$\therefore \sqrt{175} = \mathbf{13.23}\ \textbf{(4 s.f.)}$$

e) $\sqrt{48} - 10$
Solution

$$\sqrt{48} - 10 = \sqrt{16 \times 3} - 10 = 4\sqrt{3} - 10$$
$$= 4\,(1.732) - 10 = -3.072$$
$$\therefore \sqrt{48} - 10 = \mathbf{-3.072}\ \textbf{(4 s.f.)}$$

f) $2\sqrt{27} + 3\sqrt{80}$
Solution

$$2\sqrt{27} + 3\sqrt{80} = 2\sqrt{9 \times 3} + 3\sqrt{16 \times 5}$$
$$= \left(2 \times 3\sqrt{3}\right) + \left(3 \times 4\sqrt{5}\right)$$
$$= 6\sqrt{3} + 12\sqrt{5} = 6\,(1.732) + 12\,(2.236)$$
$$= 10.392 + 26.832 = 37.224 = 37.22$$
$$\therefore 2\sqrt{27} + 3\sqrt{80} = \mathbf{37.22}\ \textbf{(4 s.f.)}$$

Sometimes while carrying out simplification, it will become necessary to change a surd from $k\sqrt{a}$ to $\sqrt{b}$ form such that $k\sqrt{a} = \sqrt{b}$. This is easy to do since $k = \sqrt{k^2}$, thus we can say that

$$k\sqrt{a} = \sqrt{k^2} \times \sqrt{a} = \sqrt{k^2 a}$$
$$\therefore b = k^2 a$$

Let's try some examples to illustrate this.

Example 7

Write each of the following surds in the form of $\sqrt{a}$.

a) $10\sqrt{2}$ **b)** $7\sqrt{3}$ **c)** $5\sqrt{6}$ **d)** $3\sqrt{11}$ **e)** $2\sqrt{12}$

What did you get? Find the solution below to double-check your answer.

Solution to Example 7

a) $10\sqrt{2}$
Solution

$$10\sqrt{2} = \left(\sqrt{10}\right)^2 \times \sqrt{2}$$
$$= \sqrt{100} \times \sqrt{2} = \sqrt{100 \times 2} = \sqrt{200}$$
$$\therefore 10\sqrt{2} = \sqrt{200}$$

b) $7\sqrt{3}$
Solution

$$7\sqrt{3} = \left(\sqrt{7}\right)^2 \times \sqrt{3}$$
$$= \sqrt{49} \times \sqrt{3} = \sqrt{49 \times 3} = \sqrt{147}$$
$$\therefore 7\sqrt{3} = \sqrt{147}$$

c) $5\sqrt{6}$
Solution

$$5\sqrt{6} = \left(\sqrt{5}\right)^2 \times \sqrt{6}$$
$$= \sqrt{25} \times \sqrt{6} = \sqrt{25 \times 6} = \sqrt{150}$$
$$\therefore 5\sqrt{6} = \sqrt{150}$$

d) $3\sqrt{11}$
Solution

$$3\sqrt{11} = \left(\sqrt{3}\right)^2 \times \sqrt{11}$$
$$= \sqrt{9} \times \sqrt{11} = \sqrt{9 \times 11} = \sqrt{99}$$
$$\therefore 3\sqrt{11} = \sqrt{99}$$

e) $2\sqrt{12}$
Solution

$$2\sqrt{12} = \left(\sqrt{2}\right)^2 \times \sqrt{12}$$
$$= \sqrt{4} \times \sqrt{12} = \sqrt{4 \times 12} = \sqrt{48}$$
$$\therefore 2\sqrt{12} = \sqrt{48}$$

Another set of examples to try.

Example 8

Simplify the following, giving the answer in the form of $a\sqrt[n]{b}$, where a is any real number, and b and n are positive integers.

a) $\sqrt[3]{40}$ b) $\sqrt[4]{2} - \sqrt[4]{162}$ c) $\sqrt[3]{40} + \sqrt[3]{5}$ d) $\sqrt[3]{24} + \sqrt[3]{3}$ e) $\sqrt[5]{192} + \sqrt[5]{6}$

What did you get? Find the solution below to double-check your answer.

Solution to Example 8

a) $\sqrt[3]{40}$
Solution

$$\sqrt[3]{40} = \sqrt[3]{8 \times 5} = \sqrt[3]{8} \times \sqrt[3]{5}$$
$$= \sqrt[3]{2^3} \times \sqrt[3]{5} = 2 \times \sqrt[3]{5} = 2\sqrt[3]{5}$$
$$\therefore \sqrt[3]{40} = 2\sqrt[3]{5}$$

b) $\sqrt[4]{2} - \sqrt[4]{162}$
Solution

$$\sqrt[4]{2} - \sqrt[4]{162} = \sqrt[4]{2} - \left(\sqrt[4]{81 \times 2}\right)$$
$$= \sqrt[4]{2} - \left(\sqrt[4]{81} \times \sqrt[4]{2}\right) = \sqrt[4]{2} - \left(\sqrt[4]{3^4} \times \sqrt[4]{2}\right)$$
$$= \sqrt[4]{2} - \left(3 \times \sqrt[4]{2}\right) = \sqrt[4]{2} - 3\sqrt[4]{2} = -2\sqrt[4]{2}$$
$$\therefore \sqrt[4]{2} - \sqrt[4]{162} = -2\sqrt[4]{2}$$

c) $\sqrt[3]{40} + \sqrt[3]{5}$
Solution

$$\sqrt[3]{40} + \sqrt[3]{5} = \sqrt[3]{8 \times 5} + \sqrt[3]{5}$$
$$= \left(\sqrt[3]{8} \times \sqrt[3]{5}\right) + \sqrt[3]{5} = \left(\sqrt[3]{2^3} \times \sqrt[3]{5}\right) + \sqrt[3]{5}$$
$$= \left(2 \times \sqrt[3]{5}\right) + \sqrt[3]{5} = 2\sqrt[3]{5} + \sqrt[3]{5} = 3\sqrt[3]{5}$$
$$\therefore \sqrt[3]{40} + \sqrt[3]{5} = 3\sqrt[3]{5}$$

d) $\sqrt[3]{24} + \sqrt[3]{3}$
Solution

$$\sqrt[3]{24} + \sqrt[3]{3} = \sqrt[3]{8 \times 3} + \sqrt[3]{3} = \left(\sqrt[3]{8} \times \sqrt[3]{3}\right) + \sqrt[3]{3}$$
$$= \left(2 \times \sqrt[3]{3}\right) + \sqrt[3]{3} = 2\sqrt[3]{3} + \sqrt[3]{3} = 3\sqrt[3]{3}$$
$$\therefore \sqrt[3]{24} + \sqrt[3]{3} = 3\sqrt[3]{3}$$

e) $\sqrt[5]{192} + \sqrt[5]{6}$

Solution

$$\sqrt[5]{192} + \sqrt[5]{6} = \sqrt[5]{32 \times 6} + \sqrt[5]{6}$$
$$= \left(\sqrt[5]{32} \times \sqrt[5]{6}\right) + \sqrt[5]{6} = \left(\sqrt[5]{2^5} \times \sqrt[5]{6}\right) + \sqrt[5]{6}$$
$$= \left(2 \times \sqrt[5]{6}\right) + \sqrt[5]{6} = 2\sqrt[5]{6} + \sqrt[5]{6} = 3\sqrt[5]{6}$$
$$\therefore \sqrt[5]{192} + \sqrt[5]{6} = 3\sqrt[5]{6}$$

8.5 FUNDAMENTAL OPERATIONS OF SURDS

The four main arithmetic operations can be applied to surds, singular or mixed, and these will be discussed in this section.

8.5.1 ADDITION AND SUBTRACTION

Addition of two or more surds can only be carried out if they are similar. If the surds are singular, the surds are added together as per the rule of addition, i.e. 'Rule 2 Addition and Subtraction'. For compound or mixed surds, the rational parts are added together and the same applies to the irrational parts. The procedure is also used when subtracting one surd from another. In summary, when surds are similar, they can be combined algebraically.

Let's try some examples.

Example 9

Simplify the following:

a) $2\sqrt{7} + 3\sqrt{7}$ **b)** $-2\sqrt{x} - 3\sqrt{x}$ **c)** $2\sqrt[3]{6} - 5\sqrt[3]{6}$

d) $\left(3 + 5\sqrt{3}\right) + \left(7 + \sqrt{3}\right)$ **e)** $3\sqrt{3} + 4\sqrt{3} + 5\sqrt{3}$ **f)** $3\sqrt{2} - 5\sqrt{3} - 2\sqrt{2} + 2\sqrt{3}$

What did you get? Find the solution below to double-check your answer.

Solution to Example 9

HINT

Consider $a\sqrt{x} + b\sqrt{x}$ as $2x + 3x$, factorise, and/or simplify.

a) $2\sqrt{7} + 3\sqrt{7}$

Solution

$$2\sqrt{7} + 3\sqrt{7} = \sqrt{7}\,(2 + 3)$$
$$= \sqrt{7}\,(5) = 5\sqrt{7}$$
$$\therefore 2\sqrt{7} + 3\sqrt{7} = 5\sqrt{7}$$

---

b) $-2\sqrt{x} - 3\sqrt{x}$

Solution

$$-2\sqrt{x} - 3\sqrt{x} = \sqrt{x}\,(-2 - 3)$$
$$= \sqrt{x}\,(-5) = -5\sqrt{x}$$
$$\therefore -2\sqrt{x} - 3\sqrt{x} = -5\sqrt{x}$$

---

c) $2\sqrt[3]{6} - 5\sqrt[3]{6}$

Solution

$$2\sqrt[3]{6} - 5\sqrt[3]{6} = \sqrt[3]{6}\,(2 - 5)$$
$$= \sqrt[3]{6}\,(-3) = -3\sqrt[3]{6}$$
$$\therefore 2\sqrt[3]{6} - 5\sqrt[3]{6} = -3\sqrt[3]{6}$$

---

d) $\left(3 + 5\sqrt{3}\right) + \left(7 + \sqrt{3}\right)$

Solution

Recognise that we have two mixed surds to be added. Combine like parts (i.e., rational with rational and irrational with irrational part).

$$\left(3 + 5\sqrt{3}\right) + \left(7 + \sqrt{3}\right) = (3 + 7) + \left(5\sqrt{3} + \sqrt{3}\right)$$
$$= (10) + \left(6\sqrt{3}\right)$$
$$= 10 + 6\sqrt{3} = 2\left(5 + 3\sqrt{3}\right)$$
$$\therefore \left(3 + 5\sqrt{3}\right) + \left(7 + \sqrt{3}\right) = 10 + 6\sqrt{3} = 2\left(5 + 3\sqrt{3}\right)$$

---

e) $3\sqrt{3} + 4\sqrt{3} + 5\sqrt{3}$

Solution

$$3\sqrt{3} + 4\sqrt{3} + 5\sqrt{3} = 12\sqrt{3}$$
$$\therefore 3\sqrt{3} + 4\sqrt{3} + 5\sqrt{3} = 12\sqrt{3}$$

---

f) $3\sqrt{2} - 5\sqrt{3} - 2\sqrt{2} + 2\sqrt{3}$

Solution

$$3\sqrt{2} - 5\sqrt{3} - 2\sqrt{2} + 2\sqrt{3} = 3\sqrt{2} - 2\sqrt{2} - 5\sqrt{3} + 2\sqrt{3}$$
$$= \sqrt{2} - 3\sqrt{3}$$
$$\therefore 3\sqrt{2} - 5\sqrt{3} - 2\sqrt{2} + 2\sqrt{3} = \sqrt{2} - 3\sqrt{3}$$

NOTE
We can only add similar surds. $\sqrt{2}$ is unlike $\sqrt{3}$, therefore no further simplification is possible in this case.

In the above set of examples, we should simply identify the like terms or similar surds and add them algebraically. It may be possible that simplification would require addition or subtraction as well as applying other rules. Let's try further examples.

Example 10

Simplify each of the following:

a) $\sqrt{8} + \sqrt{8}$ b) $\sqrt{75} - \sqrt{27}$ c) $\sqrt{45} - \sqrt{125}$

d) $\sqrt{48} + \sqrt{27} + \sqrt{12}$ e) $\sqrt{12} + \sqrt{75} - \sqrt{192}$ f) $\sqrt{1000} + \sqrt{1210} + \sqrt{1440}$

g) $10\sqrt{20} - 4\sqrt{45} - \sqrt{80}$ h) $4\sqrt{8} - 2\sqrt{75} + \sqrt{200} - 3\sqrt{48} + 5\sqrt{45}$

i) $2\sqrt{18} - 4\sqrt{72} - \sqrt{50} + 3\sqrt{98}$ j) $2\sqrt{48} + 5\sqrt{54} - \sqrt{75} - 2\sqrt{24}$

What did you get? Find the solution below to double-check your answer.

Solution to Example 10

a) $\sqrt{8} + \sqrt{8}$
Solution

$$\sqrt{8} + \sqrt{8} = \sqrt{4 \times 2} + \sqrt{4 \times 2}$$
$$= \left(\sqrt{4} \times \sqrt{2}\right) + \left(\sqrt{4} \times \sqrt{2}\right)$$
$$= \left(2 \times \sqrt{2}\right) + \left(2 \times \sqrt{2}\right)$$
$$= 2\sqrt{2} + 2\sqrt{2} = 4\sqrt{2}$$
$$\therefore \sqrt{8} + \sqrt{8} = 4\sqrt{2}$$

ALTERNATIVE METHOD

$$\sqrt{8} + \sqrt{8} = 2\sqrt{8}$$
$$= 2\sqrt{4 \times 2} = 2 \times \sqrt{4} \times \sqrt{2}$$
$$= 2 \times 2 \times \sqrt{2} = 4 \times \sqrt{2} = 4\sqrt{2}$$
$$\therefore \sqrt{8} + \sqrt{8} = 4\sqrt{2}$$

b) $\sqrt{75} - \sqrt{27}$

Solution

$$\sqrt{75} - \sqrt{27} = \sqrt{25 \times 3} - \sqrt{9 \times 3}$$
$$= \left(\sqrt{25} \times \sqrt{3}\right) - \left(\sqrt{9} \times \sqrt{3}\right)$$
$$= \left(5 \times \sqrt{3}\right) - \left(3 \times \sqrt{3}\right)$$
$$= 5\sqrt{3} - 3\sqrt{3} = 2\sqrt{3}$$
$$\therefore \sqrt{75} - \sqrt{27} = 2\sqrt{3}$$

---

c) $\sqrt{45} - \sqrt{125}$

Solution

$$\sqrt{45} - \sqrt{125} = \sqrt{9 \times 5} - \sqrt{25 \times 5}$$
$$= \left(\sqrt{9} \times \sqrt{5}\right) - \left(\sqrt{25} \times \sqrt{5}\right)$$
$$= \left(3 \times \sqrt{5}\right) - \left(5 \times \sqrt{5}\right)$$
$$= 3\sqrt{5} - 5\sqrt{5} = -2\sqrt{5}$$
$$\therefore \sqrt{45} - \sqrt{125} = -2\sqrt{5}$$

---

d) $\sqrt{48} + \sqrt{27} + \sqrt{12}$

Solution

$$\sqrt{48} + \sqrt{27} + \sqrt{12} = \sqrt{16 \times 3} + \sqrt{9 \times 3} + \sqrt{4 \times 3}$$
$$= \left(\sqrt{16} \times \sqrt{3}\right) + \left(\sqrt{9} \times \sqrt{3}\right) + \left(\sqrt{4} \times \sqrt{3}\right)$$
$$= \left(4 \times \sqrt{3}\right) + \left(3 \times \sqrt{3}\right) + \left(2 \times \sqrt{3}\right)$$
$$= 4\sqrt{3} + 3\sqrt{3} + 2\sqrt{3} = 9\sqrt{3}$$
$$\therefore \sqrt{48} + \sqrt{27} + \sqrt{12} = 9\sqrt{3}$$

---

e) $\sqrt{12} + \sqrt{75} - \sqrt{192}$

Solution

$$\sqrt{12} + \sqrt{75} - \sqrt{192} = \sqrt{4 \times 3} + \sqrt{25 \times 3} - \sqrt{64 \times 3}$$
$$= \left(\sqrt{4} \times \sqrt{3}\right) + \left(\sqrt{25} \times \sqrt{3}\right) - \left(\sqrt{64} \times \sqrt{3}\right)$$
$$= \left(2 \times \sqrt{3}\right) + \left(5 \times \sqrt{3}\right) - \left(8 \times \sqrt{3}\right)$$
$$= 2\sqrt{3} + 5\sqrt{3} - 8\sqrt{3} = -\sqrt{3}$$
$$\therefore \sqrt{12} + \sqrt{75} - \sqrt{192} = -\sqrt{3}$$

---

f) $\sqrt{1000} + \sqrt{1210} + \sqrt{1440}$

Solution

$$\sqrt{1000} + \sqrt{1210} + \sqrt{1440} = \sqrt{100 \times 10} + \sqrt{121 \times 10} + \sqrt{144 \times 10}$$
$$= \left(\sqrt{100} \times \sqrt{10}\right) + \left(\sqrt{121} \times \sqrt{10}\right) + \left(\sqrt{144} \times \sqrt{10}\right)$$
$$= \left(10 \times \sqrt{10}\right) + \left(11 \times \sqrt{10}\right) + \left(12 \times \sqrt{10}\right)$$
$$= 10\sqrt{10} + 11\sqrt{10} + 12\sqrt{10}$$
$$= 33\sqrt{10}$$
$$\therefore \sqrt{1000} + \sqrt{1210} + \sqrt{1440} = 33\sqrt{10}$$

---

g) $10\sqrt{20} - 4\sqrt{45} - \sqrt{80}$

Solution

$$10\sqrt{20} - 4\sqrt{45} - \sqrt{80} = 10\sqrt{4 \times 5} - 4\sqrt{9 \times 5} - \sqrt{16 \times 5}$$
$$= \left(10 \times \sqrt{4} \times \sqrt{5}\right) - \left(4 \times \sqrt{9} \times \sqrt{5}\right) - \left(\sqrt{16} \times \sqrt{5}\right)$$
$$= \left(10 \times 2\sqrt{5}\right) - \left(4 \times 3\sqrt{5}\right) - 4\sqrt{5}$$
$$= 20\sqrt{5} - 12\sqrt{5} - 4\sqrt{5}$$
$$= 4\sqrt{5}$$
$$\therefore 10\sqrt{20} - 4\sqrt{45} - \sqrt{80} = 4\sqrt{5}$$

---

h) $4\sqrt{8} - 2\sqrt{75} + \sqrt{200} - 3\sqrt{48} + 5\sqrt{45}$

Solution

$$4\sqrt{8} - 2\sqrt{75} + \sqrt{200} - 3\sqrt{48} + 5\sqrt{45}$$
$$= 4\sqrt{4 \times 2} - 2\sqrt{25 \times 3} + \sqrt{100 \times 2} - 3\sqrt{16 \times 3} + 5\sqrt{9 \times 5}$$
$$= \left(4 \times \sqrt{4} \times \sqrt{2}\right) - \left(2 \times \sqrt{25} \times \sqrt{3}\right) + \left(\sqrt{100} \times \sqrt{2}\right) - \left(3 \times \sqrt{16} \times \sqrt{3}\right) + \left(5 \times \sqrt{9} \times \sqrt{5}\right)$$
$$= \left(4 \times 2\sqrt{2}\right) - \left(2 \times 5\sqrt{3}\right) + 10\sqrt{2} - \left(3 \times 4\sqrt{3}\right) + \left(5 \times 3\sqrt{5}\right)$$
$$= 8\sqrt{2} - 10\sqrt{3} + 10\sqrt{2} - 12\sqrt{3} + 15\sqrt{5}$$
$$= 8\sqrt{2} + 10\sqrt{2} - 10\sqrt{3} - 12\sqrt{3} + 15\sqrt{5}$$
$$= 18\sqrt{2} - 22\sqrt{3} + 15\sqrt{5}$$
$$\therefore 4\sqrt{8} - 2\sqrt{75} + \sqrt{200} - 3\sqrt{48} + 5\sqrt{45} = 18\sqrt{2} - 22\sqrt{3} + 15\sqrt{5}$$

---

i) $2\sqrt{18} - 4\sqrt{72} - \sqrt{50} + 3\sqrt{98}$

Solution

$$
\begin{aligned}
2\sqrt{18} - 4\sqrt{72} - \sqrt{50} + 3\sqrt{98} &= 2\sqrt{9 \times 2} - 4\sqrt{36 \times 2} - \sqrt{25 \times 2} + 3\sqrt{49 \times 2} \\
&= \left(2 \times \sqrt{9} \times \sqrt{2}\right) - \left(4 \times \sqrt{36} \times \sqrt{2}\right) \\
&\quad - \left(\sqrt{25} \times \sqrt{2}\right) + \left(3 \times \sqrt{49} \times \sqrt{2}\right) \\
&= \left(2 \times 3\sqrt{2}\right) - \left(4 \times 6\sqrt{2}\right) - 5\sqrt{2} + \left(3 \times 7\sqrt{2}\right) \\
&= 6\sqrt{2} - 24\sqrt{2} - 5\sqrt{2} + 21\sqrt{2} \\
&= -2\sqrt{2}
\end{aligned}
$$

$$\therefore 2\sqrt{18} - 4\sqrt{72} - \sqrt{50} + 3\sqrt{98} = -2\sqrt{2}$$

---

j) $2\sqrt{48} + 5\sqrt{54} - \sqrt{75} - 2\sqrt{24}$

Solution

$$
\begin{aligned}
2\sqrt{48} + 5\sqrt{54} - \sqrt{75} - 2\sqrt{24} &= 2\sqrt{16 \times 3} + 5\sqrt{9 \times 6} - \sqrt{25 \times 3} - 2\sqrt{4 \times 6} \\
&= \left(2 \times \sqrt{16} \times \sqrt{3}\right) + \left(5 \times \sqrt{9} \times \sqrt{6}\right) \\
&\quad - \left(\sqrt{25} \times \sqrt{3}\right) - \left(2 \times \sqrt{4} \times \sqrt{6}\right) \\
&= \left(2 \times 4\sqrt{3}\right) + \left(5 \times 3\sqrt{6}\right) - \left(5\sqrt{3}\right) - \left(2 \times 2\sqrt{6}\right) \\
&= 8\sqrt{3} + 15\sqrt{6} - 5\sqrt{3} - 4\sqrt{6} \\
&= 3\sqrt{3} + 11\sqrt{6}
\end{aligned}
$$

$$\therefore 2\sqrt{48} + 5\sqrt{54} - \sqrt{75} - 2\sqrt{24} = 3\sqrt{3} + 11\sqrt{6}$$

8.5.2 MULTIPLICATION

Singular surds can be multiplied using the multiplication rule, i.e. 'Rule 1 Multiplication'. When multiplying a singular surd and a mixed (or compound) surd, we simply open the brackets. This is the same when the product of two mixed or compound surds are to be computed. It is as simple as that.

Let's try some examples.

Example 11

Simplify the following:

a) $\sqrt{2}\left(5 - \sqrt{7}\right)$

b) $\left(3 + \sqrt{2}\right)\left(5 + \sqrt{2}\right)$

c) $\left(\sqrt{2} - \sqrt{7}\right)\left(\sqrt{6} + \sqrt{3}\right)$

d) $\left(\sqrt{5} + \sqrt{2}\right)\left(\sqrt{2} - \sqrt{3}\right)$

What did you get? Find the solution below to double-check your answer.

Solution to Example 11

HINT

Essentially, we need the rule of opening the brackets covered in Chapter 3. We therefore treat $\sqrt{a}\left(b + \sqrt{c}\right)$ as $b\sqrt{a} + \sqrt{a} \times \sqrt{c}$ (and then $b\sqrt{a} + \sqrt{ac}$).

a) $\sqrt{2}\left(5 - \sqrt{7}\right)$

Solution

Multiply each term in the brackets by $\sqrt{2}$. Apply the rule of multiplication to $\sqrt{2}$ and $\sqrt{7}$.

$$\sqrt{2}\left(5 - \sqrt{7}\right) = \left(5 \times \sqrt{2}\right) - \left(\sqrt{2} \times \sqrt{7}\right)$$
$$= 5\sqrt{2} - \sqrt{2 \times 7}$$
$$= 5\sqrt{2} - \sqrt{14}$$

This cannot be simplified further because $\sqrt{2}$ and $\sqrt{14}$ are not similar.

$$\therefore \sqrt{2} \times \left(5 - \sqrt{7}\right) = 5\sqrt{2} - \sqrt{14}$$

b) $\left(3 + \sqrt{2}\right)\left(5 + \sqrt{2}\right)$

Solution

This is a multiplication of two mixed surds, so we have:

$$\left(3 + \sqrt{2}\right)\left(5 + \sqrt{2}\right) = 3\left(5 + \sqrt{2}\right) + \sqrt{2}\left(5 + \sqrt{2}\right)$$
$$= \left(3 \times 5 + 3 \times \sqrt{2}\right) + \left(5 \times \sqrt{2} + \sqrt{2} \times \sqrt{2}\right)$$
$$= \left(15 + 3\sqrt{2}\right) + \left(5\sqrt{2} + 2\right)$$
$$= (15 + 2) + \left(3\sqrt{2} + 5\sqrt{2}\right)$$
$$= 17 + 8\sqrt{2}$$
$$\therefore \left(3 + \sqrt{2}\right)\left(5 + \sqrt{2}\right) = 17 + 8\sqrt{2}$$

c) $\left(\sqrt{2} - \sqrt{7}\right)\left(\sqrt{6} + \sqrt{3}\right)$

Solution

This is a multiplication of two compound surds.

$$\left(\sqrt{2} - \sqrt{7}\right)\left(\sqrt{6} + \sqrt{3}\right) = \sqrt{2}\left(\sqrt{6} + \sqrt{3}\right) - \sqrt{7}\left(\sqrt{6} + \sqrt{3}\right)$$
$$= \left(\sqrt{2} \times \sqrt{6} + \sqrt{2} \times \sqrt{3}\right) - \left(\sqrt{7} \times \sqrt{6} + \sqrt{7} \times \sqrt{3}\right)$$
$$= \left(\sqrt{2 \times 6} + \sqrt{2 \times 3}\right) - \left(\sqrt{7 \times 6} + \sqrt{7 \times 3}\right)$$
$$= \left(\sqrt{12} + \sqrt{6}\right) - \left(\sqrt{42} + \sqrt{21}\right)$$
$$= \sqrt{12} + \sqrt{6} - \sqrt{42} - \sqrt{21}$$

$$= \sqrt{4 \times 3} + \sqrt{6} - \sqrt{42} - \sqrt{21}$$
$$= 2\sqrt{3} + \sqrt{6} - \sqrt{42} - \sqrt{21}$$
$$\therefore \left(\sqrt{2} - \sqrt{7}\right)\left(\sqrt{6} + \sqrt{3}\right) = 2\sqrt{3} + \sqrt{6} - \sqrt{42} - \sqrt{21}$$

d) $\left(\sqrt{5} + \sqrt{2}\right)\left(\sqrt{2} - \sqrt{3}\right)$
Solution

$$\left(\sqrt{5} + \sqrt{2}\right)\left(\sqrt{2} - \sqrt{3}\right) = \left[\sqrt{5}\left(\sqrt{2}\right)\right] - \left[\sqrt{5}\left(\sqrt{3}\right)\right] + \left[\sqrt{2}\left(\sqrt{2}\right)\right] - \left[\sqrt{2}\left(\sqrt{3}\right)\right]$$
$$= \sqrt{5 \times 2} - \sqrt{5 \times 3} + \sqrt{2 \times 2} - \sqrt{3 \times 2}$$
$$= \sqrt{10} - \sqrt{15} + 2 - \sqrt{6}$$
$$= 2 + \sqrt{10} - \sqrt{6} - \sqrt{15}$$
$$\therefore \left(\sqrt{5} + \sqrt{2}\right)\left(\sqrt{2} - \sqrt{3}\right) = 2 + \sqrt{10} - \sqrt{6} - \sqrt{15}$$

Another set of examples to try.

Example 12

By opening the brackets, evaluate each of the following surds. Present the answers in their simplest forms without brackets.

a) $\left(\sqrt{7} + 2\right)^2$ **b)** $\left(3\sqrt{6} - 4\sqrt{5}\right)^2$ **c)** $\sqrt{3} - \sqrt{2}\left(\sqrt{6} - \sqrt{24}\right)$ **d)** $\left(4\sqrt{3} - 2\right)\left(5\sqrt{10} + \sqrt{3}\right)$

What did you get? Find the solution below to double-check your answer.

Solution to Example 12

a) $\left(\sqrt{7} + 2\right)^2$
Solution

$$\left(\sqrt{7} + 2\right)^2 = \left(\sqrt{7} + 2\right)\left(\sqrt{7} + 2\right)$$
$$= \sqrt{7}\left(\sqrt{7} + 2\right) + 2\left(\sqrt{7} + 2\right)$$
$$= \sqrt{7}\left(\sqrt{7}\right) + 2\left(\sqrt{7}\right) + 2\left(\sqrt{7}\right) + 2\,(2)$$
$$= 7 + 4\left(\sqrt{7}\right) + 4 = 11 + 4\sqrt{7}$$
$$\therefore \left(\sqrt{7} + 2\right)^2 = 11 + 4\sqrt{7}$$

b) $\left(3\sqrt{6}-4\sqrt{5}\right)^2$

Solution

$$\left(3\sqrt{6}-4\sqrt{5}\right)^2 = \left(3\sqrt{6}-4\sqrt{5}\right)\left(3\sqrt{6}-4\sqrt{5}\right)$$

$$= \left[3\sqrt{6}\left(3\sqrt{6}-4\sqrt{5}\right)\right] - \left[4\sqrt{5}\left(3\sqrt{6}-4\sqrt{5}\right)\right]$$

$$= \left[3\sqrt{6}\left(3\sqrt{6}\right)\right] + \left[3\sqrt{6}\left(-4\sqrt{5}\right)\right] - \left[4\sqrt{5}\left(3\sqrt{6}\right)\right] - \left[4\sqrt{5}\left(-4\sqrt{5}\right)\right]$$

$$= \left[3^2 \times \left(\sqrt{6}\right)^2\right] - \left[(4\times3)\left(\sqrt{6}\times\sqrt{5}\right)\right] - \left[(4\times3)\left(\sqrt{5}\times\sqrt{6}\right)\right] + \left[4^2 \times \left(\sqrt{5}\right)^2\right]$$

$$= [9\times6] - \left[12\left(\sqrt{5\times6}\right)\right] - \left[12\left(\sqrt{5\times6}\right)\right] + [16\times5]$$

$$= 54 - 12\left(\sqrt{30}\right) - 12\left(\sqrt{30}\right) + 80$$

$$= 54 + 80 - 24\sqrt{30}$$

$$= 134 - 24\sqrt{30}$$

$$\therefore \left(3\sqrt{6}-4\sqrt{5}\right)^2 = \mathbf{134 - 24\sqrt{30}}$$

---

c) $\sqrt{3} - \sqrt{2}\left(\sqrt{6}-\sqrt{24}\right)$

Solution

$$\sqrt{3} - \sqrt{2}\left(\sqrt{6}-\sqrt{24}\right) = \left[\sqrt{3}\right] - \left[\sqrt{2}\left(\sqrt{6}\right)\right] - \left[\sqrt{2}\left(-\sqrt{24}\right)\right]$$

$$= \sqrt{3} - \sqrt{2\times6} + \sqrt{2\times24} = \sqrt{3} - \sqrt{12} + \sqrt{48}$$

$$= \sqrt{3} - \sqrt{4\times3} + \sqrt{16\times3} = \sqrt{3} - 2\sqrt{3} + 4\sqrt{3} = 3\sqrt{3}$$

$$\therefore \sqrt{3} - \sqrt{2}\left(\sqrt{6}-\sqrt{24}\right) = \mathbf{3\sqrt{3}}$$

---

d) $\left(4\sqrt{3}-2\right)\left(5\sqrt{10}+\sqrt{3}\right)$

Solution

$$\left(4\sqrt{3}-2\right)\left(5\sqrt{10}+\sqrt{3}\right) = \left[4\sqrt{3}\left(5\sqrt{10}\right)\right] + \left[4\sqrt{3}\left(\sqrt{3}\right)\right] - \left[2\left(5\sqrt{10}\right)\right] - \left[2\left(\sqrt{3}\right)\right]$$

$$= \left[(4\times5)\sqrt{3\times10}\right] + \left[4\left(\sqrt{3}\right)^2\right] - \left[(2\times5)\sqrt{10}\right] - \left[2\sqrt{3}\right]$$

$$= \left(20\sqrt{30}\right) + (4\times3) - \left(10\sqrt{10}\right) - \left(2\sqrt{3}\right)$$

$$= 20\sqrt{30} + 12 - 10\sqrt{10} - 2\sqrt{3}$$

$$= 12 + 20\sqrt{30} - 10\sqrt{10} - 2\sqrt{3}$$

$$\therefore \left(4\sqrt{3}-2\right)\left(5\sqrt{10}+\sqrt{3}\right) = \mathbf{12 + 20\sqrt{30} - 10\sqrt{10} - 2\sqrt{3}}$$

One final set of examples to try.

Example 13

Determine the cube of each of the following singular surds.

a) $\left(2\sqrt[3]{3}\right)$ b) $\left(3\sqrt[3]{13}\right)$ c) $\left(5\sqrt[3]{4}\right)$ d) $\left(10\sqrt[3]{5}\right)$

What did you get? Find the solution below to double-check your answer.

Solution to Example 13

HINT

This is a multiplication of a surd by itself repeatedly, just like what you would do with $5^3 = 5 \times 5 \times 5$.

a) $\left(2\sqrt[3]{3}\right)$
Solution

$$\left(2\sqrt[3]{3}\right)^3 = \left(2 \times \sqrt[3]{3}\right)^3$$
$$= 2^3 \times \left(\sqrt[3]{3}\right)^3 = 8 \times 3 = 24$$
$$\therefore \left(2\sqrt[3]{3}\right)^3 = 24$$

b) $\left(3\sqrt[3]{13}\right)$
Solution

$$\left(3\sqrt[3]{13}\right)^3 = \left(3 \times \sqrt[3]{13}\right)^3$$
$$= 3^3 \times \left(\sqrt[3]{13}\right)^3 = 27 \times 13 = 351$$
$$\therefore \left(3\sqrt[3]{13}\right)^3 = 351$$

c) $\left(5\sqrt[3]{4}\right)$
Solution

$$\left(5\sqrt[3]{4}\right)^3 = \left(5 \times \sqrt[3]{4}\right)^3$$
$$= 5^3 \times \left(\sqrt[3]{4}\right)^3 = 125 \times 4 = 500$$
$$\therefore \left(5\sqrt[3]{4}\right)^3 = 500$$

d) $\left(10\sqrt[3]{5}\right)$
Solution

$$\left(10\sqrt[3]{5}\right)^3 = \left(10 \times \sqrt[3]{5}\right)^3$$

$$= 10^3 \times \left(\sqrt[3]{5}\right)^3 = 1000 \times 5 = 5000$$

$$\therefore \left(10\sqrt[3]{5}\right)^3 = 5000$$

8.5.3 Division

Division of surds requires more specialised skills than those needed for other arithmetic operations. To consider this, we need to discuss a couple of concepts, namely: conjugate surds and rationalisation.

8.5.3.1 Conjugate Surds

Two binomial surds, compound or mixed, are said to be conjugates of each other if they only differ in the sign between their terms. Table 8.2 provides four possible cases of conjugates.

When conjugate surds are multiplied together, their product is a non-surd. In other words, their product is a rational number. This is a key attribute of conjugates; note this down. A binomial surd implies to a two-term surd.

Let's try some examples to illustrate this.

Example 14

State the conjugate of the following surds.

a) $13 + \sqrt{6}$ b) $\sqrt{13}$ c) $-3\sqrt{2} - \sqrt{5}$ d) $\sqrt{7} - 5$ e) $\sqrt[3]{2} + \sqrt[3]{5}$

TABLE 8.2
Conjugate Surds Illustrated

	Surd	Conjugate
a)	$a + \sqrt{b}$	$a - \sqrt{b}$
b)	$\sqrt{a} - \sqrt{b}$	$\sqrt{a} + \sqrt{b}$
c)	$a + b\sqrt{c}$	$a - b\sqrt{c}$
d)	$a\sqrt{b} + c\sqrt{d}$	$a\sqrt{b} - c\sqrt{d}$

What did you get? Find the solution below to double-check your answer.

Solution to Example 14

a) $13 + \sqrt{6}$
Solution
The conjugate of $13 + \sqrt{6}$ is

$$\mathbf{13 - \sqrt{6}}$$

b) $\sqrt{13}$
Solution
The conjugate of $\sqrt{13}$ is

$$\mathbf{\sqrt{13} \quad AND \quad -\sqrt{13}}$$

NOTE

- For conjugates, we generally intend this in relation to compound or mixed surds. If we however go by the fact that the product of conjugate surds is a rational number, then we can arrive at the above two options. For the same reason, the conjugate of $-\sqrt{13}$ are what we obtained above.

- In summary, the conjugate of a single surd is either its negation or itself. Also, their product is a rational number.

c) $-3\sqrt{2} - \sqrt{5}$
Solution
The conjugate of $-3\sqrt{2} - \sqrt{5}$ is

$$\mathbf{-3\sqrt{2} + \sqrt{5}}$$

NOTE

- Alternatively, the conjugate is $\mathbf{3\sqrt{2} - \sqrt{5}}$. This is because $-3\sqrt{2} - \sqrt{5}$ is the same as $-\sqrt{5} - 3\sqrt{2}$. If we change the sign between the two irrational numbers, we will have $-\sqrt{5} + 3\sqrt{2}$, which is the same as $3\sqrt{2} - \sqrt{5}$.
- However, $3\sqrt{2} + \sqrt{5}$ is not a conjugate of $-3\sqrt{2} - \sqrt{5}$.

d) $\sqrt{7} - 5$
Solution
The conjugate of $\sqrt{7} - 5$ is

$$\mathbf{\sqrt{7} + 5}$$

NOTE

- Alternatively, the conjugate is $-\sqrt{7} - 5$, but $-\sqrt{7} + 5$ is not its conjugate for the same reason stated above.
- In general, we only change the sign of one of the two terms but not both.

e) $\sqrt[3]{2} + \sqrt[3]{5}$

Solution

Given $\sqrt[3]{2} + \sqrt[3]{5}$, we do not have this in our list.

$$\therefore Not\ applicable$$

$\sqrt[3]{2} + \sqrt[3]{5}$ and $\sqrt[3]{2} - \sqrt[3]{5}$ are not conjugates. Whilst they differ in sign, their product does not produce a rational number. Here we consider only quadratic compound surds.

Now that we've discussed conjugates of surds, let's try examples relating to multiplication of conjugates.

Example 15

Simplify the following:

a) $\sqrt{13} \times \sqrt{13}$ **b)** $\sqrt{6} \times -\sqrt{6}$ **c)** $\left(1 - \sqrt{7}\right)\left(1 + \sqrt{7}\right)$

d) $\left(15 + \sqrt{5}\right)\left(15 - \sqrt{5}\right)$ **e)** $\left(\sqrt{2} - \sqrt{7}\right)\left(\sqrt{2} + \sqrt{7}\right)$ **f)** $\left(8\sqrt{3} - 2\sqrt{15}\right)\left(8\sqrt{3} + 2\sqrt{15}\right)$

What did you get? Find the solution below to double-check your answer.

Solution to Example 15

HINT

In this example we will be working with conjugates, and it will be helpful to remember that

$$\left(\sqrt{a} + \sqrt{b}\right)\left(\sqrt{a} - \sqrt{b}\right)$$
$$= \left(\sqrt{a}\right)^2 - \left(\sqrt{b}\right)^2$$

or to be more general,

$$\left(a\sqrt{b} + c\sqrt{d}\right)\left(a\sqrt{b} - c\sqrt{d}\right)$$
$$= \left(a\sqrt{b}\right)^2 - \left(c\sqrt{d}\right)^2$$

This is an application of the '**difference of two squares**' (Section 3.5.2).

a) $\sqrt{13} \times \sqrt{13}$

Solution

$$\sqrt{13} \times \sqrt{13} = \sqrt{13 \times 13}$$
$$= \sqrt{169} = 13$$
$$\therefore \sqrt{13} \times \sqrt{13} = 13$$

b) $\sqrt{6} \times -\sqrt{6}$

Solution

$$\sqrt{6} \times -\sqrt{6} = -\sqrt{6 \times 6}$$
$$= -\sqrt{36} = -6$$
$$\therefore \sqrt{6} \times -\sqrt{6} = -6$$

c) $\left(1 - \sqrt{7}\right)\left(1 + \sqrt{7}\right)$

Solution

$$\left(1 - \sqrt{7}\right)\left(1 + \sqrt{7}\right) = 1^2 - \left(\sqrt{7}\right)^2$$
$$= 1 - 7 = -6$$
$$\therefore \left(1 - \sqrt{7}\right)\left(1 + \sqrt{7}\right) = -6$$

d) $\left(15 + \sqrt{5}\right)\left(15 - \sqrt{5}\right)$

Solution

$$\left(15 + \sqrt{5}\right)\left(15 - \sqrt{5}\right) = 15\left(15 - \sqrt{5}\right) + \sqrt{5}\left(15 - \sqrt{5}\right)$$
$$= \left(15 \times 15 - 15 \times \sqrt{5}\right) + \left(15 \times \sqrt{5} - \sqrt{5} \times \sqrt{5}\right)$$
$$= \left(225 - 15\sqrt{5}\right) + \left(15\sqrt{5} - 5\right)$$
$$= 225 - 5 - 15\sqrt{5} + 15\sqrt{5}$$
$$= 220$$
$$\therefore \left(15 + \sqrt{5}\right)\left(15 - \sqrt{5}\right) = 220$$

e) $\left(\sqrt{2} - \sqrt{7}\right)\left(\sqrt{2} + \sqrt{7}\right)$

Solution

$$\left(\sqrt{2} - \sqrt{7}\right)\left(\sqrt{2} + \sqrt{7}\right) = \sqrt{2}\left(\sqrt{2} + \sqrt{7}\right) - \sqrt{7}\left(\sqrt{2} + \sqrt{7}\right)$$
$$= \left(\sqrt{2} \times \sqrt{2} + \sqrt{2} \times \sqrt{7}\right) - \left(\sqrt{2} \times \sqrt{7} + \sqrt{7} \times \sqrt{7}\right)$$
$$= \left(2 + \sqrt{14}\right) - \left(\sqrt{14} + 7\right)$$
$$= 2 - 7 + \sqrt{14} - \sqrt{14}$$
$$= -5$$
$$\therefore \left(\sqrt{2} - \sqrt{7}\right)\left(\sqrt{2} + \sqrt{7}\right) = -5$$

f) $\left(8\sqrt{3}-2\sqrt{15}\right)\left(8\sqrt{3}+2\sqrt{15}\right)$

Solution

$$\left(8\sqrt{3}-2\sqrt{15}\right)\left(8\sqrt{3}+2\sqrt{15}\right)=\left(8\sqrt{3}\right)^2-\left(2\sqrt{15}\right)^2$$
$$=64\,(3)-4\,(15)$$
$$=192-60=132$$
$$\therefore\ \left(8\sqrt{3}-2\sqrt{15}\right)\left(8\sqrt{3}+2\sqrt{15}\right)=\mathbf{132}$$

From Example 15(a) and 15(b), it can be observed that if a surd is multiplied by itself or its negative, the result is a rational number. In general,

$$\sqrt{a}\times\sqrt{a}=\sqrt{a\times a}=\left(\sqrt{a}\right)^2=a$$
$$\sqrt{a}\times-\sqrt{a}=-\sqrt{a\times a}=\left(\sqrt{a}\right)^2=-a$$

While this is not always treated as conjugates, it is very useful in dealing with division.

From Example 15(c) to 15(f), the following general rule can be applied.

$$\left(a+\sqrt{b}\right)\left(a-\sqrt{b}\right)=a^2-b$$
$$\left(\sqrt{a}+\sqrt{b}\right)\left(\sqrt{a}-\sqrt{b}\right)=a-b$$

This is very similar to the rule of difference of two squares.

$$(x+y)(x-y)=x^2-y^2$$

8.5.3.2 Rationalising the Denominators

Rationalisation is a method of simplifying a fraction having a surd either as its denominator or as both the denominator and numerator such that it can be re-written without a surd in its denominator. The surd in the denominator can either be a singular, mixed, or compound surd (Table 8.3).

As shown above, a surd can be changed into a rational number by multiplying it with itself or its conjugate. For this reason, this process is often referred to as '**rationalising the denominator**'. In other

TABLE 8.3

Rationalising the Denominator Illustrated

Case 1	Case 2	Case 3
$\dfrac{1}{\sqrt{b}}=\dfrac{1}{\sqrt{b}}\times\dfrac{\sqrt{b}}{\sqrt{b}}$ $=\dfrac{\sqrt{b}}{b}$	$\dfrac{1}{a+\sqrt{b}}=\dfrac{1}{a+\sqrt{b}}\times\dfrac{a-\sqrt{b}}{a-\sqrt{b}}$ $=\dfrac{a-\sqrt{b}}{a^2-\left(\sqrt{b}\right)^2}$ $=\dfrac{a-\sqrt{b}}{a^2-b}$	$\dfrac{1}{\sqrt{a}+\sqrt{b}}=\dfrac{1}{\sqrt{a}+\sqrt{b}}\times\dfrac{\sqrt{a}-\sqrt{b}}{\sqrt{a}-\sqrt{b}}$ $=\dfrac{\sqrt{a}-\sqrt{b}}{\left(\sqrt{a}\right)^2-\left(\sqrt{b}\right)^2}$ $=\dfrac{\sqrt{a}-\sqrt{b}}{a-b}$

words, to simplify a fractional surd, which has a surd as its denominator, we multiply its numerator and denominator by the conjugate of the denominator. Rationalising the denominator often takes one of the three forms in Table 8.3.

Let's quickly summarise the process of rationalising the denominator:

Step 1: Determine the conjugate of the denominator.
Step 2: Multiply the numerator and denominator by the conjugate identified in step 1.
Step 3: In the denominator, multiply the original surd with its conjugate to obtain a rational number. For the numerator, multiply the original expression with the identified conjugate in step 1.
Step 4: Simplify the expression using relevant rule(s).

That's the whole gist. Let's try some examples.

Example 16

By rationalising the denominator, show that the following can be written in the form $a\sqrt{b}$, where a and b are integers.

a) $\frac{3}{\sqrt{15}}$ b) $\frac{2\sqrt{3}}{\sqrt{2}}$ c) $\frac{5}{6\sqrt{2}}$ d) $\frac{4\sqrt{15}}{\sqrt{6}}$

What did you get? Find the solution below to double-check your answer.

Solution to Example 16

a) $\frac{3}{\sqrt{15}}$
Solution

$$\frac{3}{\sqrt{15}} = \frac{3}{\sqrt{15}} \times \frac{\sqrt{15}}{\sqrt{15}}$$
$$= \frac{3\sqrt{15}}{15} = \frac{\sqrt{15}}{5}$$
$$\therefore \frac{3}{\sqrt{15}} = \frac{1}{5}\sqrt{15}$$

b) $\frac{2\sqrt{3}}{\sqrt{2}}$
Solution

$$\frac{2\sqrt{3}}{\sqrt{2}} = \frac{2\sqrt{3}}{\sqrt{2}} \times \frac{\sqrt{2}}{\sqrt{2}}$$
$$= \frac{2\sqrt{6}}{2} = \sqrt{6}$$
$$\therefore \frac{2\sqrt{3}}{\sqrt{2}} = \sqrt{6}$$

c) $\dfrac{5}{6\sqrt{2}}$

Solution

$$\frac{5}{6\sqrt{2}} = \left[\frac{5}{6\sqrt{2}}\right] \times \left[\frac{\sqrt{2}}{\sqrt{2}}\right]$$

$$= \frac{5\sqrt{2}}{6\left(\sqrt{2}\right)^2} = \frac{5\sqrt{2}}{6 \times 2} = \frac{5}{12}\sqrt{2}$$

$$\therefore \frac{5}{6\sqrt{2}} = \frac{5}{12}\sqrt{2}$$

NOTE

In this and similar cases, do not use everything in the denominator to rationalise. Instead, take only the surd part and leave out the coefficient while rationalising, since the latter is already a rational number.

d) $\dfrac{4\sqrt{15}}{\sqrt{6}}$

Solution

$$\frac{4\sqrt{15}}{\sqrt{6}} = 4\sqrt{\frac{15}{6}} = 4\sqrt{\frac{5}{2}}$$

$$= \frac{4\sqrt{5}}{\sqrt{2}} = \left[\frac{4\sqrt{5}}{\sqrt{2}}\right] \times \left[\frac{\sqrt{2}}{\sqrt{2}}\right]$$

$$= \frac{4\sqrt{5 \times 2}}{\left(\sqrt{2}\right)^2} = \frac{4\sqrt{10}}{2} = 2\sqrt{10}$$

$$\therefore \frac{4\sqrt{15}}{\sqrt{6}} = 2\sqrt{10}$$

NOTE

In this type of question, it is convenient to apply the rule shown below and then rationalise the denominator. This is because rationalising without this will lead to a large radicand, which may be cumbersome to simplify.

$$\frac{\sqrt{a}}{\sqrt{b}} = \sqrt{\frac{a}{b}}$$

Another set of examples to try.

Example 17

Simplify the following by rationalising the denominator. Present your answers in simplified form.

a) $\dfrac{2}{1-\sqrt{5}}$ **b)** $\dfrac{\sqrt{3}+\sqrt{2}}{\sqrt{3}}$ **c)** $\dfrac{5-\sqrt{3}}{\sqrt{2}-\sqrt{3}}$ **d)** $\dfrac{1}{5\sqrt{7}+4\sqrt{11}}$ **e)** $\dfrac{3+\sqrt{5}}{3-\sqrt{5}}$

What did you get? Find the solution below to double-check your answer.

a) $\frac{2}{1-\sqrt{5}}$

Solution

$$\frac{2}{1-\sqrt{5}} = \frac{2}{1-\sqrt{5}} \times \frac{1+\sqrt{5}}{1+\sqrt{5}}$$

$$= \frac{2\left(1+\sqrt{5}\right)}{\left(1-\sqrt{5}\right)\left(1+\sqrt{5}\right)}$$

$$= \frac{2\left(1+\sqrt{5}\right)}{1^2-\left(\sqrt{5}\right)^2} = \frac{2\left(1+\sqrt{5}\right)}{1-5}$$

$$= \frac{2\left(1+\sqrt{5}\right)}{-4} = \frac{1+\sqrt{5}}{-2}$$

$$= -\frac{1}{2}\left(1+\sqrt{5}\right)$$

$$\therefore \frac{2}{1-\sqrt{5}} = -\frac{1}{2}\left(1+\sqrt{5}\right)$$

NOTE

The final answer can also be written as $-\frac{1}{2} - \frac{1}{2}\sqrt{5}$.

---

b) $\frac{\sqrt{3}+\sqrt{2}}{\sqrt{3}}$

Solution

$$\frac{\sqrt{3}+\sqrt{2}}{\sqrt{3}} = \left[\frac{\sqrt{3}+\sqrt{2}}{\sqrt{3}}\right] \times \left[\frac{\sqrt{3}}{\sqrt{3}}\right]$$

$$= \frac{\sqrt{3}\left(\sqrt{3}+\sqrt{2}\right)}{\left(\sqrt{3}\right)^2} = \frac{\left(\sqrt{3}\right)^2 + \left(\sqrt{3}\times\sqrt{2}\right)}{3}$$

$$= \frac{3+\sqrt{3\times2}}{3} = \frac{3+\sqrt{6}}{3}$$

$$\therefore \frac{\sqrt{3}+\sqrt{2}}{\sqrt{3}} = \frac{3+\sqrt{6}}{3}$$

NOTE

The final answer can also be written as $\frac{1}{3}\left(3+\sqrt{6}\right)$ **OR** $1 + \frac{1}{3}\sqrt{6}$.

c) $\dfrac{5-\sqrt{3}}{\sqrt{2}-\sqrt{3}}$

Solution

$$\frac{5-\sqrt{3}}{\sqrt{2}-\sqrt{3}} = \frac{5-\sqrt{3}}{\sqrt{2}-\sqrt{3}} \times \frac{\sqrt{2}+\sqrt{3}}{\sqrt{2}+\sqrt{3}}$$

$$= \frac{\left(5-\sqrt{3}\right)\left(\sqrt{2}+\sqrt{3}\right)}{\left(\sqrt{2}-\sqrt{3}\right)\left(\sqrt{2}+\sqrt{3}\right)}$$

$$= \frac{5\left(\sqrt{2}+\sqrt{3}\right)-\sqrt{3}\left(\sqrt{2}+\sqrt{3}\right)}{\left(\sqrt{2}\right)^{2}-\left(\sqrt{3}\right)^{2}}$$

$$= \frac{\left(5\sqrt{2}+5\sqrt{3}\right)-\left(\sqrt{6}+3\right)}{2-3}$$

$$= \frac{5\sqrt{2}+5\sqrt{3}-\sqrt{6}-3}{-1}$$

$$= -\left(5\sqrt{2}+5\sqrt{3}-\sqrt{6}-3\right)$$

$$\therefore \frac{5-\sqrt{3}}{\sqrt{2}-\sqrt{3}} = -\left(5\sqrt{2}+5\sqrt{3}-\sqrt{6}-3\right)$$

NOTE

The final answer can also be written as $3 + \sqrt{6} - 5\sqrt{2} - 5\sqrt{3}$.

---

d) $\dfrac{1}{5\sqrt{7}+4\sqrt{11}}$

Solution

$$\frac{1}{5\sqrt{7}+4\sqrt{11}} = \left[\frac{1}{5\sqrt{7}+4\sqrt{11}}\right] \times \left[\frac{5\sqrt{7}-4\sqrt{11}}{5\sqrt{7}-4\sqrt{11}}\right]$$

$$= \frac{5\sqrt{7}-4\sqrt{11}}{\left(5\sqrt{7}+4\sqrt{11}\right)\left(5\sqrt{7}-4\sqrt{11}\right)}$$

$$= \frac{5\sqrt{7}-4\sqrt{11}}{\left(5\sqrt{7}\right)^{2}-\left(4\sqrt{11}\right)^{2}} = \frac{5\sqrt{7}-4\sqrt{11}}{175-176}$$

$$= \frac{5\sqrt{7}-4\sqrt{11}}{-1} = 4\sqrt{11}-5\sqrt{7}$$

$$\therefore \frac{1}{5\sqrt{7}+4\sqrt{11}} = 4\sqrt{11}-5\sqrt{7}$$

e) $\frac{3+\sqrt{5}}{3-\sqrt{5}}$

Solution

$$\frac{3+\sqrt{5}}{3-\sqrt{5}} = \left[\frac{3+\sqrt{5}}{3-\sqrt{5}}\right] \times \left[\frac{3+\sqrt{5}}{3+\sqrt{5}}\right] = \frac{\left(3+\sqrt{5}\right)^2}{\left(3-\sqrt{5}\right)\left(3+\sqrt{5}\right)}$$

$$= \frac{3^2 + \left(\sqrt{5}\right)^2 + 2(3)\left(\sqrt{5}\right)}{3^2 - \left(\sqrt{5}\right)^2} = \frac{9+5+6\sqrt{5}}{9-5}$$

$$= \frac{14+6\sqrt{5}}{4} = \frac{1}{2}\left(7+3\sqrt{5}\right)$$

$$\therefore \frac{3+\sqrt{5}}{3-\sqrt{5}} = \frac{1}{2}\left(7+3\sqrt{5}\right)$$

Another example to try.

Example 18

Show that $\frac{\sqrt{2}+5\sqrt{18}}{\sqrt{2}-5\sqrt{18}}$ is a rational number and can be written as $\frac{a}{b}$, where a and b are integers.

What did you get? Find the solution below to double-check your answer.

Solution to Example 18

Solution

$$\frac{\sqrt{2}+5\sqrt{18}}{\sqrt{2}-5\sqrt{18}} = \left[\frac{\sqrt{2}+5\sqrt{18}}{\sqrt{2}-5\sqrt{18}}\right] \times \left[\frac{\sqrt{2}+5\sqrt{18}}{\sqrt{2}+5\sqrt{18}}\right]$$

$$= \frac{\left(\sqrt{2}+5\sqrt{18}\right)\left(\sqrt{2}+5\sqrt{18}\right)}{\left(\sqrt{2}-5\sqrt{18}\right)\left(\sqrt{2}+5\sqrt{18}\right)}$$

$$= \frac{\left(\sqrt{2}\right)^2 + \left(5\sqrt{18}\right)^2 + 2\left(\sqrt{2}\right)\left(5\sqrt{18}\right)}{\left(\sqrt{2}\right)^2 - \left(5\sqrt{18}\right)^2}$$

$$= \frac{2 + \left(5^2 \times 18\right) + 2(5)\left(\sqrt{2 \times 18}\right)}{2 - \left(5^2 \times 18\right)}$$

$$= \frac{2 + \left(25 \times 18\right) + 2(5)\left(\sqrt{2 \times 18}\right)}{2 - 450}$$

$$= \frac{2 + 450 + 10\sqrt{36}}{2 - 450}$$

$$= \frac{452 + 10\,(6)}{-448} = \frac{512}{-448} = -\frac{8}{7}$$

$$\therefore \frac{\sqrt{2} + 5\sqrt{18}}{\sqrt{2} - 5\sqrt{18}} = -\frac{8}{7}$$

It is apparent from the last two examples (17e and 18) that dividing a surd by its conjugate can either result in a rational or an irrational number. This is in contrary to when they are multiplied together, which always results in a rational number.

Let's try another set of examples that require simplification and rationalisation.

Example 19

Simplify each of the following:

a) $\frac{10}{\sqrt{2}} + \sqrt{8}$ b) $\frac{1}{\sqrt{2}} - \frac{1}{\sqrt{3}}$ c) $\frac{\sqrt{5}}{2} + \frac{1}{\sqrt{3}}$ d) $\frac{5}{\sqrt{2}} - \frac{\sqrt{3}}{\sqrt{6}}$

What did you get? Find the solution below to double-check your answer.

Solution to Example 19

HINT

In this example, we can either simplify the fractions and then rationalise the denominator or vice versa. It appears that we may need to adopt either of the two approaches as deemed appropriate.

a) $\frac{10}{\sqrt{2}} + \sqrt{8}$
Solution

$$\frac{10}{\sqrt{2}} + \sqrt{8} = \frac{10 + \sqrt{2} \times \sqrt{8}}{\sqrt{2}} = \frac{10 + \sqrt{16}}{\sqrt{2}}$$

$$= \frac{10 + 4}{\sqrt{2}} = \frac{14}{\sqrt{2}} = \left[\frac{14}{\sqrt{2}}\right] \times \left[\frac{\sqrt{2}}{\sqrt{2}}\right]$$

$$= \frac{14\sqrt{2}}{\left(\sqrt{2}\right)^2} = \frac{14\sqrt{2}}{2} = 7\sqrt{2}$$

$$\therefore \frac{10}{\sqrt{2}} + \sqrt{8} = 7\sqrt{2}$$

ALTERNATIVE METHOD

$$\frac{10}{\sqrt{2}} + \sqrt{8} = \left[\frac{10}{\sqrt{2}} \times \frac{\sqrt{2}}{\sqrt{2}}\right] + \sqrt{4 \times 2}$$

$$= \left[\frac{10\sqrt{2}}{\left(\sqrt{2}\right)^2}\right] + 2\sqrt{2} = \frac{10\sqrt{2}}{2} + 2\sqrt{2}$$

$$= 5\sqrt{2} + 2\sqrt{2} = 7\sqrt{2}$$

$$\therefore \frac{10}{\sqrt{2}} + \sqrt{8} = 7\sqrt{2}$$

b) $\frac{1}{\sqrt{2}} - \frac{1}{\sqrt{3}}$

Solution

$$\frac{1}{\sqrt{2}} - \frac{1}{\sqrt{3}} = \frac{\sqrt{3} - \sqrt{2}}{\sqrt{2 \times 3}} = \frac{\sqrt{3} - \sqrt{2}}{\sqrt{6}}$$

$$= \left[\frac{\left(\sqrt{3} - \sqrt{2}\right)}{\sqrt{6}}\right] \times \left[\frac{\sqrt{6}}{\sqrt{6}}\right] = \frac{\sqrt{6}\left(\sqrt{3} - \sqrt{2}\right)}{\left(\sqrt{6}\right)^2}$$

$$= \frac{\sqrt{3 \times 6} - \sqrt{2 \times 6}}{6} = \frac{\sqrt{18} - \sqrt{12}}{6}$$

$$= \frac{\sqrt{9 \times 2} - \sqrt{4 \times 3}}{6} = \frac{\left(3 \times \sqrt{2}\right) - \left(2 \times \sqrt{3}\right)}{6}$$

$$= \frac{1}{6}\left(3\sqrt{2} - 2\sqrt{3}\right)$$

$$\therefore \frac{1}{\sqrt{2}} - \frac{1}{\sqrt{3}} = \frac{1}{6}\left(3\sqrt{2} - 2\sqrt{3}\right)$$

ALTERNATIVE METHOD

$$\frac{1}{\sqrt{2}} - \frac{1}{\sqrt{3}} = \frac{1}{\sqrt{2}} \times \frac{\sqrt{2}}{\sqrt{2}} - \frac{1}{\sqrt{3}} \times \frac{\sqrt{3}}{\sqrt{3}}$$

$$= \frac{\sqrt{2}}{2} - \frac{\sqrt{3}}{3} = \frac{3\sqrt{2} - 2\sqrt{3}}{6}$$

$$\therefore \frac{1}{\sqrt{2}} - \frac{1}{\sqrt{3}} = \frac{1}{6}\left(3\sqrt{2} - 2\sqrt{3}\right)$$

c) $\dfrac{\sqrt{5}}{2} + \dfrac{1}{\sqrt{3}}$

Solution

$$\frac{\sqrt{5}}{2} + \frac{1}{\sqrt{3}} = \frac{\left(\sqrt{5}\times\sqrt{3}\right)+2}{2\sqrt{3}}$$

$$= \frac{\sqrt{15}+2}{2\sqrt{3}} = \frac{\sqrt{15}+2}{2\sqrt{3}}\times\frac{\sqrt{3}}{\sqrt{3}}$$

$$= \frac{\sqrt{3}\left(\sqrt{15}+2\right)}{2\sqrt{3}\times\sqrt{3}} = \frac{\sqrt{45}+2\sqrt{3}}{2\times3}$$

$$= \frac{\sqrt{9\times5}+2\sqrt{3}}{6} = \frac{3\sqrt{5}+2\sqrt{3}}{6}$$

$$\therefore \frac{\sqrt{5}}{2} + \frac{1}{\sqrt{3}} = \frac{3\sqrt{5}+2\sqrt{3}}{6}$$

ALTERNATIVE METHOD

$$\frac{\sqrt{5}}{2} + \frac{1}{\sqrt{3}} = \frac{\sqrt{5}}{2} + \frac{1}{\sqrt{3}}\times\frac{\sqrt{3}}{\sqrt{3}}$$

$$= \frac{\sqrt{5}}{2} + \frac{\sqrt{3}}{3} = \frac{3\sqrt{5}+2\sqrt{3}}{6}$$

$$\therefore \frac{\sqrt{5}}{2} + \frac{1}{\sqrt{3}} = \frac{1}{6}(3\sqrt{5}+2\sqrt{3})$$

d) $\dfrac{5}{\sqrt{2}} - \dfrac{\sqrt{3}}{\sqrt{6}}$

Solution

$$\frac{5}{\sqrt{2}} - \frac{\sqrt{3}}{\sqrt{6}} = \left[\frac{5}{\sqrt{2}}\times\frac{\sqrt{2}}{\sqrt{2}}\right] - \left[\frac{\sqrt{3}}{\sqrt{6}}\times\frac{\sqrt{6}}{\sqrt{6}}\right]$$

$$= \frac{5\sqrt{2}}{\left(\sqrt{2}\right)^2} - \frac{\sqrt{3\times6}}{\left(\sqrt{6}\right)^2} = \frac{5\sqrt{2}}{2} - \frac{\sqrt{18}}{6}$$

$$= \frac{5\sqrt{2}}{2} - \frac{\sqrt{9\times2}}{6} = \frac{5\sqrt{2}}{2} - \frac{3\sqrt{2}}{6}$$

$$= \frac{15\sqrt{2}-3\sqrt{2}}{6} = \frac{12\sqrt{2}}{6} = 2\sqrt{2}$$

$$\therefore \frac{5}{\sqrt{2}} - \frac{\sqrt{3}}{\sqrt{6}} = 2\sqrt{2}$$

ALTERNATIVE METHOD

$$\frac{5}{\sqrt{2}} - \frac{\sqrt{3}}{\sqrt{6}} = \frac{5}{\sqrt{2}} - \sqrt{\frac{3}{6}} = \frac{5}{\sqrt{2}} - \sqrt{\frac{1}{2}}$$

$$= \frac{5}{\sqrt{2}} - \frac{1}{\sqrt{2}} = \frac{5-1}{\sqrt{2}}$$

$$= \frac{4}{\sqrt{2}} = \frac{4}{\sqrt{2}} \times \frac{\sqrt{2}}{\sqrt{2}} = \frac{4\sqrt{2}}{\left(\sqrt{2}\right)^2}$$

$$= \frac{4\sqrt{2}}{2} = 2\sqrt{2}$$

$$\therefore \frac{5}{\sqrt{2}} - \frac{\sqrt{3}}{\sqrt{6}} = 2\sqrt{2}$$

Let's try this one too.

Example 20

Express $\frac{3\sqrt{2}+5\sqrt{6}}{3\sqrt{2}-5\sqrt{6}}$ in the form $a + b\sqrt{c}$, where a, b, and c are rational numbers. State the values of a, b, and c.

What did you get? Find the solution below to double-check your answer.

Solution to Example 20

$$\frac{3\sqrt{2}+5\sqrt{6}}{3\sqrt{2}-5\sqrt{6}} = \left[\frac{3\sqrt{2}+5\sqrt{6}}{3\sqrt{2}-5\sqrt{6}}\right] \times \left[\frac{3\sqrt{2}+5\sqrt{6}}{3\sqrt{2}+5\sqrt{6}}\right]$$

$$= \frac{\left(3\sqrt{2}+5\sqrt{6}\right)\left(3\sqrt{2}+5\sqrt{6}\right)}{\left(3\sqrt{2}-5\sqrt{6}\right)\left(3\sqrt{2}+5\sqrt{6}\right)}$$

$$= \frac{\left[\left(3\sqrt{2}\right)^2\right] + \left[\left(5\sqrt{6}\right)^2\right] + \left[2\left(3\sqrt{2}\right)\left(5\sqrt{6}\right)\right]}{\left(3\sqrt{2}\right)^2 - \left(5\sqrt{6}\right)^2}$$

$$= \frac{[9 \times 2] + [25 \times 6] + \left[2\left(3 \times 5\right)\left(\sqrt{2 \times 6}\right)\right]}{(9 \times 2) - (25 \times 6)}$$

$$= \frac{18 + 150 + 30\sqrt{12}}{18 - 150} = \frac{168 + 60\sqrt{3}}{-132}$$

$$\therefore \frac{3\sqrt{2}+5\sqrt{6}}{3\sqrt{2}-5\sqrt{6}} = -\frac{168}{132} - \frac{60\sqrt{3}}{132} = -\frac{14}{11} - \frac{5}{11}\sqrt{3}$$

Thus

$$a = -\frac{14}{11} \qquad b = -\frac{5}{11} \qquad \textbf{AND} \qquad c = \sqrt{3}$$

Final example to try.

Example 21

Given that $\beta = 8 + 3\sqrt{7}$, determine $\beta + \frac{1}{\beta}$ in the form $a + b\sqrt{c}$ where a, b, and c are rational numbers. State the values of a, b, and c.

What did you get? Find the solution below to double-check your answer.

Solution to Example 21

Given that

$$\beta = 8 + 3\sqrt{7}$$

then

$$\frac{1}{\beta} = \frac{1}{8 + 3\sqrt{7}}$$

$$= \left[\frac{1}{8 + 3\sqrt{7}}\right] \times \left[\frac{8 - 3\sqrt{7}}{8 - 3\sqrt{7}}\right]$$

$$= \frac{8 - 3\sqrt{7}}{\left(8 + 3\sqrt{7}\right)\left(8 - 3\sqrt{7}\right)} = \frac{8 - 3\sqrt{7}}{(8)^2 - \left(3\sqrt{7}\right)^2}$$

$$= \frac{8 - 3\sqrt{7}}{64 - 63} = 8 - 3\sqrt{7}$$

Hence

$$\beta + \frac{1}{\beta} = \left(8 + 3\sqrt{7}\right) + \left(8 - 3\sqrt{7}\right)$$

$$= 8 + 8 = 16$$

$$\therefore \beta + \frac{1}{\beta} = 16$$

From this, $a = 16$, $b = 0$, and $c = 0$.

8.6 APPLICATION

Surds are generally used to provide answers in the exact form, i.e without approximation. This can be very handy in geometry and trigonometry for the exact values of special angles, as an example. We will look at a few cases to illustrate this.

Example 22

Determine the length of the line joining A $(-5, -2)$ and B $(-3, 4)$ in a Cartesian plane. Present the answer in surd form.

What did you get? Check the solution below.

Solution to Example 22

The length of AB is given by

$$\overline{AB} = \sqrt{(x_2 - x_1)^2 + (y_2 - y_1)^2}$$

where (x_1, y_1) and (x_2, y_2) are the two points A and B respectively, connecting the line AB. For this case

$$(x_1, \ y_1) = (-5, -2) \text{ and } (x_2, y_2) = (-3, \ 4)$$

Therefore

$$\overline{AB} = \sqrt{(-3 + 5)^2 + (4 + 2)^2} = \sqrt{(2)^2 + (6)^2}$$
$$= \sqrt{4 + 36} = \sqrt{40}$$
$$= \sqrt{4 \times 10} = \sqrt{4} \times \sqrt{10}$$
$$= 2 \times \sqrt{10} = 2\sqrt{10}$$
$$\therefore \overline{AB} = 2\sqrt{10} \text{ unit}$$

Let's try another example.

Example 23

Given that $\tan 45° = 1$ and $\tan 60° = \sqrt{3}$, find the value of $\tan 15°$. Present the answer in simplified surd form.

What did you get? Find the solution below to double-check your answer.

Solution to Example 23

Let's use the addition formula

$$\tan (A \pm B) = \frac{\tan A \pm \tan B}{1 \mp \tan A\tan B}$$

We have

$$\tan 15° = \tan (60° - 45°)$$

That is $A = 60°$ and $B = 45°$. Hence, we will use

$$\tan (A - B) = \frac{\tan A - \tan B}{1 + \tan A\tan B}$$

which implies that

$$\tan 15° = \tan (60° - 45°)$$
$$= \frac{\tan 60° - \tan 45°}{1 + \tan 60°\tan 45°}$$
$$= \frac{\sqrt{3} - 1}{1 + \left(\sqrt{3} \times 1\right)} = \frac{\sqrt{3} - 1}{\sqrt{3} + 1}$$
$$= \left[\frac{\sqrt{3} - 1}{\sqrt{3} + 1}\right] \times \left[\frac{\sqrt{3} - 1}{\sqrt{3} - 1}\right] = \frac{\left(\sqrt{3} - 1\right)^2}{\left(\sqrt{3} + 1\right)\left(\sqrt{3} - 1\right)}$$
$$= \frac{\left(\sqrt{3}\right)^2 - 2\left(\sqrt{3}\right)(1) + 1^2}{\left(\sqrt{3}\right)^2 - 1^2} = \frac{3 - 2\sqrt{3} + 1}{3 - 1}$$
$$= \frac{4 - 2\sqrt{3}}{2} = \frac{2\left(2 - \sqrt{3}\right)}{2} = 2 - \sqrt{3}$$
$$\therefore \tan 15° = 2 - \sqrt{3}$$

Example 24

Determine the square roots of $11 + 2\sqrt{28}$.

What did you get? Find the solution below to double-check your answer.

Solution to Example 24

Let the square root of $11 + 2\sqrt{28}$ be $\sqrt{a} + \sqrt{b}$ for which a, $b \in \mathbb{R}$. Therefore

$$11 + 2\sqrt{28} = \left(\sqrt{a} + \sqrt{b}\right)^2$$
$$= a + b + 2\sqrt{ab}$$

Comparing the two sides of the above equation, we have:

$$a + b = 11 \qquad ----- \text{(i)}$$

and

$$2\sqrt{ab} = 2\sqrt{28}$$

Divide both sides by 2

$$\sqrt{ab} = \sqrt{28}$$

Square both sides

$$ab = 28 \qquad ----- \text{(ii)}$$

We've now formed simultaneous equations and we will use substitution method to solve for a and b. From (i)

$$b = 11 - a \qquad ----- \text{(iii)}$$

Substitute equation (iii) in equation (ii), we have:

$$a(11 - a) = 28$$
$$11a - a^2 = 28$$
$$a^2 - 11a + 28 = 0$$
$$(a - 4)(a - 7) = 0$$

Therefore, either

$$a - 4 = 0$$
$$a = 4$$

or

$$a - 7 = 0$$
$$a = 7$$

Now we need to find the corresponding value of b, thus when $a = 4$, from (iii)

$$b = 11 - 4 = 7$$

and when $a = 7$, from (iii)

$$b = 11 - 7 = 4$$

Taking the square root of $11 + 2\sqrt{27}$ to be $\sqrt{a} + \sqrt{b}$ for which $a, b \in \mathbb{R}$ implies that

$$\sqrt{a} + \sqrt{b} = \sqrt{7} + \sqrt{4} = \sqrt{7} + 2$$

and

$$\sqrt{a} + \sqrt{b} = \sqrt{4} + \sqrt{7} = 2 + \sqrt{7}$$

These are the same.

$$\therefore \sqrt{\left(11 + 2\sqrt{28}\right)} = 2 + \sqrt{7}$$

Another example to try.

Example 25

In a triangle ABC, $\overline{AB} = \overline{BC} = \sqrt{3} - 1$ cm and $\angle ACB = 30°$. Without using a calculator, calculate the length of AC and present the answer in exact form. Note that $\sin 30° = \cos 60° = \frac{1}{2}$ and $\sin 60° = \cos 30° = \frac{\sqrt{3}}{2}$.

What did you get? Find the solution below to double-check your answer.

Solution to Example 25

Solution
Using cosine rule, we have

$$b^2 = a^2 + c^2 - 2ac\cos B$$

where

$$\overline{AB} = c = \sqrt{3} - 1 \text{ cm}$$
$$\overline{BC} = a = \sqrt{3} - 1 \text{ cm}$$
$$\overline{AC} = b$$

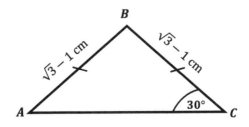

FIGURE 8.1 Solution to Example 25.

Also,

$$\angle BAC = \angle ACB = 30°$$
$$\because \overline{AB} = \overline{BC}$$
$$\therefore \widehat{B} = 180° - (\angle BAC + \angle ACB)$$
$$= 180° - (30° + 30°)$$
$$= 120°$$

We can now find the length AC as:

$$b^2 = \left(\sqrt{3} - 1\right)^2 + \left(\sqrt{3} - 1\right)^2 - 2\left(\sqrt{3} - 1\right)\left(\sqrt{3} - 1\right)\cos 120°$$
$$= 2\left(\sqrt{3} - 1\right)^2 - 2\left(\sqrt{3} - 1\right)^2 (-\cos 60°)$$
$$= 2\left[\left(\sqrt{3}\right)^2 + (-1)^2 - 2\sqrt{3}\right] - 2\left(\left(\sqrt{3}\right)^2 + (-1)^2 - 2\sqrt{3}\right)\left(-\frac{1}{2}\right)$$
$$= 2\left[3 + 1 - 2\sqrt{3}\right] + \left(3 + 1 - 2\sqrt{3}\right)$$
$$= 2\left[4 - 2\sqrt{3}\right] + \left(4 - 2\sqrt{3}\right)$$
$$= 3\left[4 - 2\sqrt{3}\right]$$
$$= 12 - 6\sqrt{3}$$
$$\therefore b = \sqrt{\left(12 - 6\sqrt{3}\right)}$$

Now it's time to determine the square roots of $12 - 6\sqrt{3}$. We will use the same approach as shown in Example 24.

Let the square root of $\mathbf{12 - 6\sqrt{3}}$ be $\sqrt{x} - \sqrt{y}$ for which $x, y \in \mathbb{R}$. Therefore,

$$12 - 6\sqrt{3} = \left(\sqrt{x} - \sqrt{y}\right)^2$$
$$= x + y - 2\sqrt{xy}$$

Comparing the two sides of the above equation, we have

$$x + y = 12 \qquad ----- \text{(i)}$$

and

$$-2\sqrt{xy} = -6\sqrt{3}$$

Divide both sides by -2

$$\sqrt{xy} = 3\sqrt{3} = \sqrt{27}$$

Square both sides

$$xy = 27 \qquad ----- \text{(ii)}$$

We've now formed simultaneous equations and we will use the substitution method to solve for x and y.

From (i)

$$y = 12 - x \qquad ----- \textbf{(iii)}$$

Substitute equation (iii) in equation (ii),

$$x^2 - 12x + 27 = 0$$
$$(x - 9)(x - 3) = 0$$

Therefore, either

$$x - 9 = 0$$
$$x = 9$$

or

$$x - 3 = 0$$
$$x = 3$$

We now need to find the corresponding value of y, thus when $x = 9$, from (iii)

$$y = 12 - 9 = 3$$

and when $x = 3$, from (iii)

$$y = 12 - 3 = 9$$

Hence, the square root of $12 - 6\sqrt{3}$ are

$$\sqrt{x} - \sqrt{y} = \sqrt{9} - \sqrt{3}$$
$$= 3 - \sqrt{3}$$

and

$$\sqrt{a} + \sqrt{b} = \sqrt{3} - \sqrt{9}$$
$$= \sqrt{3} - 3$$
$$\therefore \sqrt{\left(12 - 6\sqrt{3}\right)} = \pm\left(3 - \sqrt{3}\right)$$

Since the length AC can only be a positive value, the only answer here is

$$\overline{AC} = 3 - \sqrt{3} \text{ cm}$$

This is because $\sqrt{3} - 3$ is negative.

One final example to try.

Example 26

If $\sin \theta = \frac{2}{3}$, without using a calculator, determine the value of $\tan \theta$. Present the answer in simplified surd form.

What did you get? Find the solution below to double-check your answer.

Solution to Example 26

Solution
Given that

$$\sin \theta = \frac{2}{3}$$

we know that

$$\sin^2\theta + \cos^2\theta = 1$$

which implies that

$$\cos^2\theta = 1 - \sin^2\theta$$

$$\cos \theta = \sqrt{1 - \sin^2\theta}$$

substitute for $\sin \theta$, we therefore have:

$$\cos \theta = \sqrt{1 - \left(\frac{2}{3}\right)^2} = \sqrt{1 - \frac{4}{9}}$$

$$= \sqrt{\frac{9-4}{9}} = \sqrt{\frac{5}{9}}$$

$$= \frac{\sqrt{5}}{\sqrt{9}} = \frac{\sqrt{5}}{3}$$

But

$$\tan \theta = \frac{\sin \theta}{\cos \theta}$$

Now substitute for sine and cosine of θ

$$\tan \theta = \frac{2/3}{\sqrt{5}/3} = \left[\frac{2}{3}\right] \times \left[\frac{3}{\sqrt{5}}\right] = \frac{2}{\sqrt{5}}$$

$$= \left[\frac{2}{\sqrt{5}}\right] \times \left[\frac{\sqrt{5}}{\sqrt{5}}\right] = \frac{2\sqrt{5}}{\left(\sqrt{5}\right)^2} = \frac{2}{5}\sqrt{5}$$

$$\therefore \tan \theta = \frac{2}{5}\sqrt{5}$$

8.7 CHAPTER SUMMARY

1) A surd is a number or an expression of the form $\pm k\sqrt[n]{x}$, which cannot be expressed exactly as a fraction of two integers.

- n can be any integer except 0 (zero) and 1 (one). However, the most common case is $n = 2$ as in $\sqrt[2]{x}$ (or $\sqrt{x}$). This is an example of **quadratic surds** or **quadratic irrationals**. Similarly, $\sqrt[3]{x}$ is an example of **cubic surds**.
- x can be any positive rational number.
- The constant k is any real number excluding 0 (zero); it can be regarded as a coefficient of a surd, just like 3 is the coefficient of the term $3x$. This shows that a surd can either be positive or negative and has a value between $-\infty$ and ∞.

2) The general expression for a mixed surd is:

$$\boxed{a \pm b\sqrt{c}}$$

3) A **compound surd** is made up of two irrational parts and takes the general form:

$$\boxed{\sqrt[n]{a} \pm \sqrt[n]{b}}$$

Mixed surd is sometimes considered synonymous to compound surd and they are therefore used interchangeably.

4) Two or more surds can be multiplied together as:

$$\boxed{k_1\sqrt{a} \times k_2\sqrt{b} = k_1 k_2 \sqrt{a \times b}}$$

If $k_1 = k_2 = 1$, the above can be reduced to:

$$\boxed{\sqrt{a} \times \sqrt{b} = \sqrt{a \times b}}$$

5) The rules of addition and subtraction of surds:

$$\boxed{m\sqrt{a} + n\sqrt{a} = (m + n)\sqrt{a}} \text{ AND } \boxed{m\sqrt{b} - n\sqrt{b} = (m - n)\sqrt{b}}$$

On the contrary, the following are not valid:

$$\boxed{\sqrt{a} + \sqrt{b} \neq \sqrt{a + b}} \text{ AND } \boxed{\sqrt{a} - \sqrt{b} \neq \sqrt{a - b}}$$

6) The rule of the division of surds:

$$\frac{\sqrt{a}}{\sqrt{b}} = \sqrt{\frac{a}{b}} \quad \text{OR} \quad \sqrt{a} \div \sqrt{b} = \sqrt{\frac{a}{b}}$$

7) Simplifying surds implies reducing the value of the radicands where possible. Reverse multiplication rule will be used for whole numbers whilst reverse division rule will be relevant for fractions.

8) Two surds, compound or mixed, are said to be conjugates of each other if they only differ in the sign between their terms.

9) When conjugate surds are multiplied together, their product is a non-surd. In other words, their product is a rational number.

10) Rationalisation is a method of simplifying a fraction having a surd either as its denominator or as both the denominator and numerator such that it can be re-written without a surd in its denominator. This is achieved by multiplying the numerator and denominator with the conjugate of the denominator; this process is therefore referred to as 'rationalising the denominator'.

11) Surds are generally used to provide answers in the exact form and has application in geometry and trigonometry.

****

8.8 FURTHER PRACTICE

To access complementary contents, including additional exercises, please go to www.dszak.com.

9 Polynomials

Learning Outcomes

Once you have studied the content of this chapter, you should be able to:

- Explain the terms dividend, divisor, quotient, and remainder
- Carry out addition, subtraction, multiplication, and division of polynomials
- Use long division method to carry out division of polynomials
- Use the remainder theorem to solve algebraic expressions
- Use the factor theorem to factorise algebraic expressions
- Factorise cubic and quartic expressions

9.1 INTRODUCTION

It may be interesting to learn that the algebraic equations that we covered in Chapters 6 and 7 belong to the family of polynomials, though they have only one or two valid values for their unknown variables. You can imagine what it will be if there are three or even more valid values for a variable and what the equations (henceforth called polynomials) will look like. This chapter will cover, among others, principles relevant to evaluating polynomials, determining zeros of polynomials, factor theorem, and remainder theorem.

9.2 WHAT IS A POLYNOMIAL?

A polynomial is an algebraic expression consisting of a finite sum of terms and a constant. Each unit is a product of (i) a coefficient, and (ii) a variable raised to a power of a positive integer, as illustrated in Figure 9.1.

The general format of a polynomial is:

$$\boxed{f(x) = a_m x^n + a_{m-1} x^{n-1} + a_{m-2} x^{n-2} + a_{m-3} x^{n-3} + a_{m-4} x^{n-4} + \cdots + a_0} \qquad (9.1)$$

where $a_m, a_{m-1}, a_{m-2}, a_{m-3}, \ldots, a_0$ are real numbers, $a_m \neq 0$, and n is a positive integer. In the above formula, the variable is x, which is common, but any letter (or physical quantity) can be used instead.

Monomial (e.g., x^7), binomial (e.g., $x^2 + y$), and trinomial (e.g., $y^2 + y + 1$) refer to one-term, two-term, and three-term expressions respectively. In general, these three technical terms are used in a general sense, and as such, the power can be negative and each term can have one or more variables.

You may have noticed that we use $f(x)$ in the above general algebraic expression. This means that the value of the RHS expression depends on the value attached to the independent variable x and as

DOI:10.1201/9781003027928-9

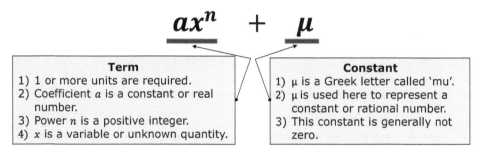

$$ax^n \; + \; \mu$$

Term	Constant
1) 1 or more units are required.	1) μ is a Greek letter called 'mu'.
2) Coefficient a is a constant or real number.	2) μ is used here to represent a constant or rational number.
3) Power n is a positive integer.	3) This constant is generally not zero.
4) x is a variable or unknown quantity.	

FIGURE 9.1 Constituents of a polynomial illustrated.

such, it's a function of x. f standing for function. Other letters such as g, h, p, etc., may also be used instead of f just like other letters can replace the variable x. This was introduced in Chapter 6.

9.3 DEGREE OF A POLYNOMIAL

The degree of a polynomial is the highest power (i.e., n) of the variable in the expression. Thus, if:

- $n = 1$, then we have $ax + b$. This is a polynomial of degree 1 and it is specially called a **linear expression**.

- $n = 2$, then we have $ax^2 + bx + c$. This is a polynomial of degree 2 and it is specially called a **quadratic expression**.

- $n = 3$, then we have $ax^3 + bx^2 + cx + d$. This is a polynomial of degree 3 and it is specially called a **cubic expression**.

- $n = 4$, then we have $ax^4 + bx^3 + cx^2 + dx + e$. This is a polynomial of degree 4 and it is specially called a **quartic expression**.

- $n = 5$, then we have $ax^5 + bx^4 + cx^3 + dx^2 + ex + f$. This is a polynomial of degree 5 and it is specially called a **quintic expression**.

where $a, b, c, d, e,$ and f are real numbers.

It is possible to say that a constant or number, such as 7, $-2e$, π, $\sqrt{5}$, etc., is a polynomial with a degree of zero, i.e., $n = 0$. However, this is not generally the intended meaning of a polynomial or polynomial function.

You might have noticed that a polynomial of degree n will have a maximum number of $(n+1)$ terms in its simplied form. For example, a polynomial of degree 3 or cubic expression will be $2x^3 - x^2 + 7x - 11$. It can however have less than four terms, provided that the highest power of n is 3. Therefore, the following are also cubic expressions:

a) Three terms: $x^3 - x^2 + 3$ and $4x^3 - 8x - 13$.

b) Two terms: $\frac{1}{2}x^3 - 1.1$ and $2 - x^3$.

You may also have noticed that we've thus far written the expressions in descending power of x, sometimes it can be presented in ascending power, i.e., $1 + 3x - 4x^2$, or no particular order, i.e., $7x + 1 - 6x^2 - 5x^3$. Orderliness is not a criterion, but it only makes it easier to read or evaluate.

Good start! Let's pause and try a few examples.

Example 1

State whether each of the following is a polynomial and state the degree of the polynomial (where applicable).

a) $x^3 - 4x + 23$

b) $x^3 - \dfrac{4}{x} - \dfrac{1}{7}$

c) $2(1 - 3x) - x^2$

d) $x^{2.5} + x^2 - x^{1.5} + 4$

e) $x^5 + 3x^4 - 4x^3 + 7x^2 - x + 9$

f) $\sqrt{x^2 - 5x + 6}$

What did you get? Find the solution below to double-check your answer.

Solution to Example 1

a) $x^3 - 4x + 23$
Solution
It is a polynomial of degree **3**.

b) $x^3 - \dfrac{4}{x} - \dfrac{1}{7}$
Solution
It is NOT a polynomial because one of the powers is a negative integer, i.e., $-\dfrac{4}{x} = -4x^{-1}$.

c) $2(1 - 3x) - x^2$
Solution
Let's open the brackets and simplify the expression as:

$$2(1 - 3x) - x^2 = 2 - 6x - x^2$$

From the above, we can say that the expression is a polynomial of degree **2**.

d) $x^{2.5} + x^2 - x^{1.5} + 4$
Solution
It is NOT a polynomial because at least one of the powers, i.e., $x^{2.5}$ and $-x^{1.5}$ is a fraction or a non-integer.

e) $x^5 + 3x^4 - 4x^3 + 7x^2 - x + 9$
Solution
It is a polynomial of degree **5**.

f) $\sqrt{x^2 - 5x + 6}$
Solution
It is NOT a polynomial. Although $(x^2 - 5x + 6)$ is a polynomial, the inclusion of the square root makes the whole expression a non-polynomial expression.

9.4 EQUALITY OF POLYNOMIALS

Two polynomials are equal if their corresponding coefficients and powers are equal. For example, if
$f(x) = a_4x^4 + a_3x^3 + a_2x^2 + a_1x + a_0$ and $g(x) = 5x^4 - 1.6x^3 + 8x^2 - x + 51$ are equal, then using
the equality principle, we can conclude that:

$$a_4 = 5, \quad a_3 = -1.6, \quad a_2 = 8, \quad a_1 = -1, \quad a_0 = 51$$

Simple concept, isn't it? Let's try some examples.

Example 2

Find the value of the unknown constant(s) for which the following pairs of functions are equal.

a) $f(x) = ax^2 + bx + c$ **and** $g(x) = x^2 - 7x - \frac{3}{4}$

b) $h(x) = x^{n+2} - x^{m-3} + cx^0$ **and** $p(x) = x^5 - x - 14$

c) $\phi(x) = ax^{k+3} + bx^2 - c$ **and** $\mu(x) = x^5 - 7x^2 + 51$

What did you get? Find the solution below to double-check your answer.

Solution to Example 2

a) $f(x) = ax^2 + bx + c$ and $g(x) = x^2 - 7x - \frac{3}{4}$

Solution

If $f(x)$ and $g(x)$ are equal, then comparing both functions we have:

$ax^2 \equiv x^2$	$bx \equiv -7x$	Also	$c \equiv -\frac{3}{4}$
$\therefore a = 1$	$\therefore b = -7$		$\therefore c = -\frac{3}{4}$

b) $h(x) = x^{n+2} - x^{m-3} + cx^0$ and $p(x) = x^5 - x - 14$

Solution

If $h(x)$ and $p(x)$ are equal, then comparing both functions we have:

$x^{n+2} \equiv x^5$	$-x^{m-3} \equiv -x$		$cx^0 \equiv -14$
$n + 2 = 5$	$m - 3 = 1$	Also	$\therefore c = -14$
$n = 5 - 2$	$m = 3 + 1$		
$\therefore n = 3$	$\therefore m = 4$		

c) $\phi(x) = ax^{k+3} + bx^2 - c$ and $\mu(x) = x^5 - 7x^2 + 51$

Solution

If $\phi(x)$ and $\mu(x)$ are equal, then comparing both functions we have:

$ax^{k+3} \equiv x^5$	$k + 3 = 5$	$bx^2 \equiv -7x^2$	$-c \equiv 51$
$\therefore a = 1$	$k = 5 - 3$	$\therefore b = -7$	Also $\therefore c = -51$
	$\therefore k = 2$		

9.5 EVALUATING POLYNOMIALS

A polynomial $p(x)$ is a function of a variable x, and the value of the function depends on the value assigned to the variable x. We may be asked to find the value of $p(x)$ when the variable x is equal to k. This is written as $p(k)$ and obtained by replacing x with k in the expression. With examples things become clearer, let's try some.

Example 3

Given that $\phi(x) = ax^3 - ax - 4$ and $g(x) = x^2 - 7x - 1$, evaluate the following:

a) $\phi(-1)$ **b)** $g(1)$ **c)** $\phi\left(\frac{1}{2}\right)$ **d)** $g(-3)$ **e)** $\phi(0) + g(0)$ **f)** $g(-2) - g(2)$

What did you get? Find the solution below to double-check your answer.

Solution to Example 3

a) $\phi(-1)$
Solution
For $x = -1$, we have:

$$\phi(x) = ax^3 - ax - 4$$
$$\phi(-1) = a(-1)^3 - a(-1) - 4$$
$$= a(-1) + a - 4$$
$$= -a + a - 4$$
$$\therefore \phi(-1) = -4$$

b) $g(1)$
Solution
For $x = 1$, we have:

$$g(x) = x^2 - 7x - 1$$
$$g(1) = (1)^2 - 7(1) - 1$$
$$= 1 - 7 - 1$$
$$\therefore g(1) = -7$$

c) $\phi\left(\frac{1}{2}\right)$
Solution
For $x = \frac{1}{2}$, we have:

$$\phi(x) = ax^3 - ax - 4$$
$$\phi\left(\frac{1}{2}\right) = a\left(\frac{1}{2}\right)^3 - a\left(\frac{1}{2}\right) - 4$$
$$= a\left(\frac{1}{8}\right) - \frac{1}{2}a - 4$$

$$\frac{a}{8} - \frac{a}{2} - 4 = \frac{a}{8} - \frac{4a}{8} - \frac{32}{8}$$

$$= \frac{a - 4a - 32}{8}$$

$$= \frac{-3a - 32}{8}$$

$$\therefore \phi\left(\frac{1}{2}\right) = -\frac{(3a + 32)}{8}$$

---

d) $g(-3)$

Solution

For $x = -3$, we have:

$$g(x) = x^2 - 7x - 1$$

$$g(-3) = (-3)^2 - 7(-3) - 1$$

$$= 9 + 21 - 1$$

$$\therefore g(-3) = 29$$

---

e) $\phi(0) + g(0)$

Solution

For $\phi(0)$ at $x = 0$, we have:

$$\phi(x) = ax^3 - ax - 4$$

$$\phi(0) = a(0)^3 - a(0) - 4$$

$$= 0 + 0 - 4$$

$$\therefore \phi(0) = -4$$

For $g(0)$ at $x = 0$, we have:

$$g(x) = x^2 - 7x - 1$$

$$g(0) = (0)^2 - 7(0) - 1$$

$$= 0 - 0 - 1$$

$$\therefore g(0) = -1$$

Hence

$$\phi(0) + g(0) = -4 - 1$$

$$\therefore \phi(0) + g(0) = -5$$

---

f) $g(-2) - g(2)$

Solution

For $x = -2$, we have:

$$g(x) = x^2 - 7x - 1$$

$$g(-2) = (-2)^2 - 7(-2) - 1$$

$$= 4 + 14 - 1$$

$$\therefore g(-2) = 17$$

For $x = 2$, we have:

$$g(x) = x^2 - 7x - 1$$
$$g(2) = (2)^2 - 7(2) - 1$$
$$= 4 - 14 - 1$$
$$\therefore g(2) = -11$$

Hence

$$g(-2) - g(2) = 17 - (-11)$$
$$\therefore g(-2) - g(2) = 28$$

9.6 FUNDAMENTAL OPERATIONS OF POLYNOMIALS

Basic arithmetic operators (addition, subtraction, multiplication, and division) can be applied to polynomials. The division process follows a different procedure compared to the other three, and it will be covered in this chapter along with the others. Let's treat them one after the other, starting with the easy operations.

9.6.1 ADDITION AND SUBTRACTION

Two or more polynomials can be added together (or one is subtracted from another) irrespective of whether they have the same degree. To carry out these operations, add or subtract the coefficients of the like terms. By 'like terms' we mean the terms that have the same power.

If a term is missing in a function, it can be introduced with zero as its coefficient. For example, two terms are missing in a polynomial $5x^3 - 1$. These can be written as $0x^2$ and $0x$. As a result, the polynomial becomes $5x^3 + 0x^2 + 0x - 1$. These missing terms need not be written before carrying out addition and subtraction, but they do help when one is new to working with polynomials.

That's everything we can say about these two operations. Let's try some examples.

Example 4

Given that $g(x) = 3 - x - 5x^2$, $h(x) = 7x^3 - 5x^2 + 3x - 1$, $\phi(x) = x^4 - 1$, and $\mu(x) = x - 31$, evaluate the following:

a) $g(x) + h(x)$ **b)** $\mu(x) - g(x)$ **c)** $\mu(x) + \phi(x)$ **d)** $\phi(x) - h(x)$

What did you get? Find the solution below to double-check your answer.

Solution to Example 4

HINT

Although not particularly required, let's first write all the functions in standard and full form.

$$g(x) = 3 - x - 5x^2 = -5x^2 - x + 3$$
$$h(x) = 7x^3 - 5x^2 + 3x - 1$$
$$\phi(x) = x^4 - 1 = x^4 + 0x^3 + 0x^2 + 0x - 1$$
$$\mu(x) = x - 31$$

Here we go.

a) $g(x) + h(x)$
Solution

$$\begin{aligned} g(x) + h(x) &= \left(-5x^2 - x + 3\right) + \left(7x^3 - 5x^2 + 3x - 1\right) \\ &= -5x^2 - x + 3 + 7x^3 - 5x^2 + 3x - 1 \\ &= \mathbf{7x^3 - 10x^2 + 2x + 2} \end{aligned}$$

b) $\mu(x) - g(x)$
Solution

$$\begin{aligned} \mu(x) - g(x) &= (x - 31) - \left(-5x^2 - x + 3\right) \\ &= x - 31 + 5x^2 + x - 3 \\ &= \mathbf{5x^2 + 2x - 34} \end{aligned}$$

c) $\mu(x) + \phi(x)$
Solution

$$\begin{aligned} \mu(x) + \phi(x) &= (x - 31) + \left(x^4 + 0x^3 + 0x^2 + 0x - 1\right) \\ &= x - 31 + x^4 + 0x^3 + 0x^2 + 0x - 1 \\ &= \mathbf{x^4 + x - 32} \end{aligned}$$

d) $\phi(x) - h(x)$
Solution

$$\begin{aligned} \phi(x) - h(x) &= \left(x^4 + 0x^3 + 0x^2 + 0x - 1\right) - \left(7x^3 - 5x^2 + 3x - 1\right) \\ &= x^4 + 0x^3 + 0x^2 + 0x - 1 - 7x^3 + 5x^2 - 3x + 1 \\ &= \mathbf{x^4 - 7x^3 + 5x^2 - 3x} \end{aligned}$$

9.6.2 MULTIPLICATION

Multiplication involving polynomials can be of two types:

Case 1 Multiplication by a scalar or number

This is when a polynomial $p(x)$ is multiplied by a real number or constant k. The new polynomial becomes $kp(x)$. For example, given $p(x) = x^4 - 5x^3 + 3x^2 + 2x - 7$, we can multiply the polynomial by 3 and -2 as follows:

For $k = 3$, we have:

$$kp(x) = 3p(x) = 3\left(x^4 - 5x^3 + 3x^2 + 2x - 7\right)$$
$$= 3x^4 - 15x^3 + 9x^2 + 6x - 21$$

For $k = -2$, we have:

$$kp(x) = -2p(x) = -2\left(x^4 - 5x^3 + 3x^2 + 2x - 7\right)$$
$$= -2x^4 + 10x^3 - 6x^2 - 4x + 14$$

The above process is like opening brackets. Let's try some examples of this operation.

Example 5

Given that $g_1(x) = x - 5$ and $g_2(x) = x^2 - 2x - 3$, evaluate the following:

a) $3g_1(x)$　　　　b) $-1.5g_2(x)$　　　　c) $2g_1(x) - 5g_1(x)$　　　　d) $7g_1(x) - 2g_2(x)$

What did you get? Find the solution below to double-check your answer.

Solution to Example 5

a) $3g_1(x)$
Solution
Here we go:

$$3g_1(x) = 3[g_1(x)]$$
$$= 3[x - 5]$$
$$= 3x - 15$$
$$\therefore 3g_1(x) = 3x - 15$$

b) $-1.5g_2(x)$
Solution
Here we go:

$$-1.5g_2(x) = -1.5[g_2(x)]$$
$$= -1.5[x^2 - 2x - 3]$$
$$= -1.5x^2 + 3x + 4.5$$
$$\therefore -1.5g_2(x) = 4.5 + 3x - 1.5x^2$$

c) $2g_1(x) - 5g_1(x)$
Solution
Here we go:

$$
\begin{aligned}
2g_1(x) - 5g_1(x) &= 2\left[g_1(x)\right] - 5\left[g_1(x)\right] \\
&= 2\left[x - 5\right] - 5\left[x - 5\right] \\
&= 2x - 10 - 5x + 25 \\
&= -3x + 15 \\
\therefore 2g_1(x) - 5g_1(x) &= \mathbf{15 - 3x}
\end{aligned}
$$

OR

$$\therefore 2g_1(x) - 5g_1(x) = 3(5 - x)$$

OR

$$\therefore 2g_1(x) - 5g_1(x) = -3g_1(x)$$

---

d) $7g_1(x) - 2g_2(x)$
Solution
Here we go:

$$
\begin{aligned}
7g_1(x) - 2g_2(x) &= 7\left[g_1(x)\right] - 2\left[g_2(x)\right] \\
&= 7\left[x - 5\right] - 2\left[x^2 - 2x - 3\right] \\
&= 7x - 35 - 2x^2 + 4x + 6 \\
&= -2x^2 + 11x - 29 \\
\therefore 7g_1(x) - 2g_2(x) &= \mathbf{-2x^2 + 11x - 29}
\end{aligned}
$$

Case 2 Multiplication with another polynomial

Multiplication of two (or more) polynomials has been partly covered in Chapter 3, where we used the opening brackets method, as this is exactly what multiplication of two polynomials is all about. What is important to mention here before we proceed with examples is that when a polynomial of degree m is multiplied by another of degree n, the resulting polynomial will be of a degree $(m + n)$. It is a principle that is not negotiable.

Let's try some examples to illustrate this.

Example 6

Given that $P_1(t) = t + 4$, $P_2(t) = t^2 - 3$ and $P_3(t) = t^3 - t + 2$, evaluate the following and state the degree of the resulting polynomial.

a) $P_1(t) \times P_2(t)$ **b)** $P_1(t) \times P_3(t)$ **c)** $P_2(t) \times P_3(t)$

What did you get? Find the solution below to double-check your answer.

Solution to Example 6

HINT

You will notice that the degree of the result is the sum of the degree of the multiplier and multiplicand.

a) $P_1(t) \times P_2(t)$
Solution
Here we go:

$$\begin{aligned} P_1(t) \times P_2(t) &= (t+4)\left(t^2 - 3\right) \\ &= t\left(t^2 - 3\right) + 4\left(t^2 - 3\right) \\ &= t^3 - 3t + 4t^2 - 12 \\ \therefore P_1(t) \times P_2(t) &= t^3 + 4t^2 - 3t - 12 \end{aligned}$$

The degree of the resulting polynomial is **3**.

ALTERNATIVE METHOD

				t	$+$	4
				t^2	$-$	3
			$-$	$3t$	$-$	12
t^3	$+$	$4t^2$				
t^3	$+$	$4t^2$	$-$	$3t$	$-$	12

This is a long multiplication and can become cumbersome, especially when some terms are missing in the functions.

b) $P_1(t) \times P_3(t)$
Solution

$$\begin{aligned} P_1(t) \times P_3(t) &= (t+4)\left(t^3 - t + 2\right) \\ &= t\left(t^3 - t + 2\right) + 4\left(t^3 - t + 2\right) \\ &= t^4 - t^2 + 2t + 4t^3 - 4t + 8 \\ &= t^4 + 4t^3 - t^2 + 2t - 4t + 8 \\ \therefore P_1(t) \times P_3(t) &= t^4 + 4t^3 - t^2 - 2t + 8 \end{aligned}$$

The degree of the resulting polynomial is **4**.

c) $P_2(t) \times P_3(t)$

Solution

Here we go:

$$\begin{aligned} P_2(t) \times P_3(t) &= \left(t^2 - 3\right)\left(t^3 - t + 2\right) \\ &= t^2\left(t^3 - t + 2\right) - 3\left(t^3 - t + 2\right) \\ &= t^5 - t^3 + 2t^2 - 3t^3 + 3t - 6 \\ &= t^5 - t^3 - 3t^3 + 2t^2 + 3t - 6 \\ \therefore P_2(t) \times P_3(t) &= \boldsymbol{t^5 - 4t^3 + 2t^2 + 3t - 6} \end{aligned}$$

The degree of the resulting polynomial is **5**.

9.6.3 DIVISION

Division is always a tricky operator and usually requires a special method; you might have noticed this while going through the previous chapters. The approach to use for division will be based on our understanding and procedure of working with division of numbers. Let's use 5 and 13 to illustrate this important concept.

In the early days of studying mathematics, we understand that 5 divided by 13 (or $\frac{5}{13}$) is impossible because the numerator (5) is less than the denominator (13). However, the operation of $\frac{13}{5}$ is possible but with a '**left-over**', technically called **remainder**. This is because there are two lots of 5 in 13 with 3 as a remainder. We can write this operation as: $13 = (5 \times 2) + 3$, which is fully described in Figure 9.2.

We can therefore conclude that:

$$\boxed{\textbf{Dividend} = (\textbf{Divisor}) \times (\textbf{Quotient}) + \textbf{Remainder}} \qquad (9.2)$$

The above description is applicable to the division of a polynomial by another polynomial. Note down the new jargons, as we will be referring to them constantly henceforth.

If the degree of the polynomials $g(x)$ and $f(x)$ are $\boldsymbol{m}$ and $\boldsymbol{n}$ respectively, then $f(x)$ can be divided by $g(x)$, i.e., $\frac{f(x)}{g(x)}$, only if $n \geq m$. This is an important criterion for dividing polynomials, and it matches the explanation on the division of numbers we outlined above.

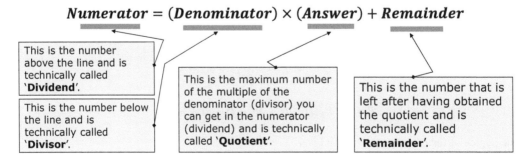

FIGURE 9.2 Division of polynomials' constituents explained.

To illustrate, if $f(x)$ is a polynomial of degree 3 such that $f(x) = 7x^3 + 5x^2 - 2x + 11$, then we can divide $f(x)$ by any of the following polynomials:

a) $x^3 - 3x - 1$: This is because it is of a degree which is ≤ 3 (i.e., 3 in this case). When this process is completed, the result will be a number excluding zero.

b) $x^2 - 2x + 3$: This is because it is of a degree which is ≤ 3 (i.e., 2 in this case). When this process is completed, the result will be a polynomial of degree 1 with or without a remainder.

c) $x + 5$: This is because it is of a degree which is ≤ 3 (i.e., 1 in this case). When this process is completed, the result will be a polynomial of degree 2 with or without a remainder.

Hence, we can state that:

$$\frac{f(x)}{g(x)} = \text{Quotient} + \frac{\text{Remainder}}{g(x)}$$

If the quotient and the remainder are given as $Q(x)$ and R respectively, then

$$\boxed{\frac{f(x)}{g(x)} = Q(x) + \frac{R}{g(x)}} \tag{9.3}$$

OR

$$\boxed{f(x) = g(x) \times Q(x) + R} \tag{9.4}$$

We've now covered the basics of division involving polynomials. We will make this clear(er) using an example with a polynomial of degree 2 (i.e., $5x^2 - 3x - 10$) as our dividend and 1 as the divisor (i.e., $x - 2$). In other words, we want to carry out the following division:

$$\frac{5x^2 - 3x - 10}{x - 2}$$

The following steps will be taken:

Step 1: Using the long division method, we will write it out as shown below. Notice that there are three sections labelled 1, 2, and 3.

(1)
Quotient

(2)
Divisor | **Dividend (3)**

So, we have:

$$x - 2 \,\big|\, 5x^2 - 3x - 10$$

Step 2: Divide the first term of the dividend $5x^2$ by the first term of the divisor x. In other words, $\frac{5x^2}{x} = 5x$. Write the answer in the quotient section.

So, we have:

$$
\begin{array}{r}
5x \\
x - 2 \enclose{longdiv}{5x^2 - 3x - 10}
\end{array}
$$

Step 3: Multiply each term of the divisor with the answer obtained in step 2 above, that is, (i) $(x) \times (5x) = \mathbf{5x^2}$ and (ii) $(-2) \times (5x) = \mathbf{-10x}$. Write the results below the dividend in that order as:

$$
\begin{array}{r}
5x \\
x - 2 \enclose{longdiv}{5x^2 - 3x - 10} \\
5x^2 - 10x
\end{array}
$$

Step 4: Now subtract the second row from the first row in the dividend section as:

$$
\begin{array}{r}
5x \\
x - 2 \enclose{longdiv}{5x^2 - 3x - 10} \\
5x^2 - 10x \\
\hline
7x
\end{array}
$$

Step 5: You can see that we have used the first two terms of the (original) dividend in this process. We now need to drop the last term as shown below.

$$
\begin{array}{r}
5x \\
x - 2 \enclose{longdiv}{5x^2 - 3x - 10} \\
5x^2 - 10x \\
\hline
7x - 10
\end{array}
$$

This (i.e., $7x - 10$) is our 'new' dividend and we need to repeat steps 2 to 5 until the degree of any 'new dividend' is less than that of the divisor. As this is not the case yet, we need to repeat steps 2 to 5. Essentially, we need to go through the loop of **Divide** $\rightarrow$ **Multiply** $\rightarrow$ **Subtract**.

Step 6: Divide the first term of the dividend $7x$ by the first term of the divisor x. In other words, $\frac{7x}{x} = 7$. Write the answer in the quotient section as its second term.

So, we have:

$$
\begin{array}{r}
5x + 7 \\
x - 2 \enclose{longdiv}{5x^2 - 3x - 10} \\
5x^2 - 10x \\
\hline
7x - 10
\end{array}
$$

Step 7: Multiply each term of the divisor with the answer obtained in step 6 above, that is, (i) $(x) \times (7) = \mathbf{7x}$ and (ii) $(-2) \times (7) = \mathbf{-14}$. Write the answer below the dividend in this order.

$$
\begin{array}{r}
5x + 7 \\
\hline
x - 2 \quad\big|\quad 5x^2 - 3x - 10 \\
5x^2 - 10x \\
\hline
7x - 10 \\
7x - 14
\end{array}
$$

Step 8: Now subtract the fourth row from the third row in the dividend section as:

$$
\begin{array}{r}
5x + 7 \\
\hline
x - 2 \quad\big|\quad 5x^2 - 3x - 10 \\
5x^2 - 10x \\
\hline
7x - 10 \\
7x - 14 \\
\hline
4
\end{array}
$$

The 'new dividend' (i.e., 4) is a number, which implies that the degree of the divisor is more than that of the 'new dividend'. Note that 4 is our remainder. This is the end of our long division.

Since we know that:

$$\boxed{\textbf{Dividend} = (\textbf{Divisor}) \times (\textbf{Quotient}) + \textbf{Remainder}} \qquad (9.5)$$

Let us write our example in this format as:

$$5x^2 - 3x - 10 = (x - 2)(5x + 7) + 4$$

If we expand and simplify the RHS, we will end up getting the LHS. Hence, the LHS is equal to the RHS.

That's all about dividing polynomials. It's a long one but we got there. Before we attempt examples on this, we should mention that if the divisor is a monomial (a single-term expression), in this case then we simply divide each term of the dividend by the divisor. For example, to divide $5x^5 - 3x^3 + 2x^2 - 1$ by x^2, we have

$$\frac{5x^5 - 3x^3 + 2x^2 - 1}{x^2} = \frac{5x^5}{x^2} + \frac{-3x^3}{x^2} + \frac{2x^2}{x^2} + \frac{-1}{x^2}$$

$$= 5x^3 - 3x + 2 - \frac{1}{x^2}$$

Let's try some examples.

Example 7

Determine the quotient and remainder when each of the following is divided by $x^2 - 3$.

a) $2x^2 + x - 3$ **b)** $4x^3 - 7x - 8$ **c)** $x^4 + 2x^3 - x^2 + 2x - 7$

What did you get? Find the solution below to double-check your answer.

Solution to Example 7

HINT

In this case, we will try to write the polynomial in its complete form to make workings clearer, but we do not have to.

a) $2x^2 + x - 3$
Solution

$$
\begin{array}{r}
2 \\
x^2 - 3 \enclose{longdiv}{2x^2 + x - 3} \\
2x^2 \quad - 6 \\
\hline
x + 3
\end{array}
$$

Therefore,

$$\textit{Quotient} : \mathbf{2} \quad \textit{Remainder} : \mathbf{x+3}$$

b) $4x^3 - 7x - 8$
Solution

$$
\begin{array}{r}
4x \\
x^2 - 3 \enclose{longdiv}{4x^3 + 0x^2 - 7x - 8} \\
4x^3 \quad - 12x \\
\hline
5x - 8
\end{array}
$$

Therefore,

$$\textit{Quotient} : \mathbf{4x} \quad \textit{Remainder} : \mathbf{5x-8}$$

c) $x^4 + 2x^3 - x^2 + 2x - 7$

Solution

$$
\begin{array}{r}
x^2 + 2x + 2 \\
x^2 - 3 \overline{\smash{\big)}\ x^4 + 2x^3 - x^2 + 2x - 7}
\end{array}
$$

$$
\begin{array}{rrrrrr}
 & x^2 & + & 2x & + & 2 \\
x^2 - 3\ \big)\ \ x^4 & + & 2x^3 & - & x^2 & + & 2x & - & 7 \\
x^4 & & & - & 3x^2 \\
\hline
 & & 2x^3 & + & 2x^2 & + & 2x \\
 & & 2x^3 & & & - & 6x \\
\hline
 & & & & 2x^2 & + & 8x & - & 7 \\
 & & & & 2x^2 & & & - & 6 \\
\hline
 & & & & & & 8x & - & 1
\end{array}
$$

Therefore,

Quotient : $x^2 + 2x + 2$ Remainder : $8x - 1$

9.7 ZEROS OF POLYNOMIALS

The zeros of a polynomial are the solutions obtained when the polynomial is equated to zero. In other words, the values of the variable when substituted into the polynomial will result in zero. If a polynomial is given as $p(x)$, then solving $p(x) = 0$ will yield its zeros (or roots). We've looked at zeros of polynomials of degree one (linear equation) and two (quadratic equation) in Chapters 6 and 7, though we did not specifically call them zeros, rather we termed them roots.

It is also essential to mention that the degree of a polynomial is equal to the number of solutions expected from a polynomial equation. For example, a quartic equation will give four solutions and a cubic will give three solutions. However, sometimes not all the solutions are real numbers, as previously noted in Chapters 6 and 7.

9.8 REMAINDER THEOREM

We've used the long division method to divide one polynomial with another. It can be noted that this process, though long, provides us with both a quotient and a remainder, and enables us to get the dividend back should we need to. However, if we need the remainder or want to check whether there is a remainder when the division is carried out, there is a short way to do this. This is the basis of the **remainder theorem**, and it states that:

- If a polynomial $f(x)$ is divided by $x - a$, the remainder is $f(a)$.

- If a polynomial $f(x)$ is divided by $ax + b$, the remainder is $f(-\frac{b}{a})$.

This two-step process below can be used to find the remainder when a polynomial is divided by another:

Step 1: Take the divisor, equate it to zero and solve for the variable. Let us say this is $x = \alpha$.
Step 2: Evaluate the dividend at $x = \alpha$. In other words, find $f(\alpha)$. The result of this is the remainder.

Let's try some examples.

Example 8

Find the remainder when $7x^3 - 5x^2 + 3x - 1$ is divided by $x + 2$.

    a) Use the long division method.

    b) Use the remainder theorem method.

    c) Comment on the answers obtained in (a) and (b) above.

What did you get? Find the solution below to double-check your answer.

Solution to Example 8

a) Long division method
Solution

$$
\begin{array}{r}
7x^2 - 19x + 41 \\
x+2 \enclose{longdiv}{7x^3 \quad - \quad 5x^2 \quad + \quad 3x \quad - \quad 1} \\
\underline{7x^3 \quad + \quad 14x^2} \\
-19x^2 \quad + \quad 3x \\
\underline{-19x^2 \quad - \quad 38x} \\
41x \quad - \quad 1 \\
\underline{41x \quad + \quad 82} \\
- \quad 83
\end{array}
$$

$$\therefore Remainder : -83$$

b) Remainder theorem
Solution
Let us equate the divisor to zero as:

$$x + 2 = 0$$
$$x = -2$$

Next, evaluate the dividend when $x = -2$ as:

$$f(x) = 7x^3 - 5x^2 + 3x - 1$$
$$f(-2) = 7(-2)^3 - 5(-2)^2 + 3(-2) - 1$$
$$= 7(-8) - 5(4) - 6 - 1$$
$$= -56 - 20 - 6 - 1$$
$$\therefore f(-2) = -83$$

Hence,

$$Remainder : -83$$

c) Comment
Solution

- The remainder in both (a) and (b) is the same.
- $f(-2) = -83 \neq 0$ means that $x + 2$ is not a factor of $f(x)$.
- Although the long division method in (a) provides additional information, the latter however is quicker and easier.

Example 9

Find the remainder when $3x^4 - x^2 + 7x - 12$ is divided by $x^2 - 5$.

a) Use the long division method.

b) Use the remainder theorem method.

c) Comment on the answers obtained in (a) and (b) above.

What did you get? Find the solution below to double-check your answer.

Solution to Example 9

a) Long division method
Solution

$$
\begin{array}{r}
3x^2 + 14 \\
x^2 - 5 \overline{\smash{\big)}\ 3x^4 + 0x^3 - x^2 + 7x - 12} \\
\underline{3x^4 + 0x^3 - 15x^2} \\
14x^2 + 7x - 12 \\
\underline{14x^2 + 0x - 70} \\
7x + 58
\end{array}
$$

$$\therefore Remainder : 7x + 58$$

b) Remainder theorem
Solution
Let us equate the divisor to zero as:

$$x^2 - 5 = 0$$
$$x^2 = 5$$

Next, evaluate the dividend when $x^2 = 5$. Here, we will be using $f(x^2 = 5)$ to imply that we will only replace x^2 with 5 and make no changes for occurrence(s) of x. Thus:

$$f(x) = 3x^4 - x^2 + 7x - 12$$
$$= 3(x^2)^2 - (x^2) + 7x - 12$$
$$f(x^2 = 5) = 3(5)^2 - (5) + 7x - 12$$
$$= 75 - 5 + 7x - 12$$
$$= 7x + 75 - 5 - 12$$
$$\therefore f(5) = 7x + 58$$

Hence,

$$\textbf{\textit{Remainder}} : \textbf{7x + 58}$$

---

c) Comment
Solution

- The remainder in both (a) and (b) is the same.
- Although the long division method in (a) provides additional information, which is missing in the remainder method, the latter however is quicker and easier.
- Also, the remainder approach here requires using x^2 instead of x.

Example 10

Find the remainder when $x^3 - 6x^2 + 5$ is divided by $x^2 - 2$.

   a) Use the long division method.
   b) Use the remainder theorem method.
   c) Comment on the answers obtained in (a) and (b) above.

What did you get? Find the solution below to double-check your answer.

Solution to Example 10

a) Long division method
Solution
Here we go:

$$x - 6$$

$$x^2 - 2 \,\big|\; x^3 \quad - \quad 6x^2 \quad + \quad 0x \quad + \quad 5$$
$$x^3 \quad + \quad 0x^2 \quad - \quad 2x$$

$$- \quad 6x^2 \quad + \quad 2x \quad + \quad 5$$
$$- \quad 6x^2 \quad + \quad 0x \quad + \quad 12$$

$$2x \quad - \quad 7$$

$$\therefore Remainder : 2x - 7$$

b) Remainder theorem
Solution
Let us equate the divisor to zero as:

$$x^2 - 2 = 0$$
$$x^2 = 2$$

Next, evaluate the dividend when $x^2 = 2$:

$$f(x) = x^3 - 6x^2 + 5$$
$$= (x^2)x - 6(x^2) + 5$$
$$f(x^2 = 2) = (2)x - 6(2) + 5$$
$$= 2x - 12 + 5$$
$$\therefore f(x^2 = 2) = 2x - 7$$

Hence,

$$Remainder : 2x - 7$$

c) Comment
Solution

- The remainder in both (a) and (b) is the same.
- Although the long division method in (a) provides additional information, which is missing in the remainder method, the latter is quicker and easier.
- Also, the remainder approach here requires using x^2 instead of x. This approach often works.

Example 11

Find the remainder when $x^4 + 6$ is divided by $x - 3$.

a) Use the long division method.

b) Use the remainder theorem method.

c) Comment on the answers obtained in (a) and (b) above.

What did you get? Find the solution below to double-check your answer.

Solution to Example 11

a) Long division method
Solution
Here we go:

$$
\begin{array}{r}
x^3 + 3x^2 + 9x + 27 \\
\hline
x-3 \,\big|\, x^4 + 0x^3 - 0x^2 + 0x + 6 \\
x^4 - 3x^3 \\
\hline
3x^3 - 0x^2 \\
3x^3 - 9x^2 \\
\hline
9x^2 + 0x \\
9x^2 - 27x \\
\hline
27x + 6 \\
27x - 81 \\
\hline
87
\end{array}
$$

$$\therefore Remainder : 87$$

---

b) Remainder theorem
Solution
Let us equate the divisor to zero as:

$$x - 3 = 0$$
$$x = 3$$

Next, evaluate the dividend when $x = 3$:

$$f(x) = x^4 + 6$$
$$f(3) = (3)^4 + 6$$
$$= 81 + 6$$
$$\therefore f(3) = 87$$

Hence,

$$Remainder : 87$$

---

c) Comment
Solution

- The remainder in both (a) and (b) is the same.

- Although the long division method in (a) provides additional information, missing in the remainder method, the latter however is quicker and easier.

The examples we've just looked at prove the applicability of the remainder theorem. Let's now try another example different from the ones above.

Example 12

The remainder when $3x^3 + kx^2 - 4x + 11$ is divided by $x - 2$ is -1. Determine the value of k.

What did you get? Find the solution below to double-check your answer.

Solution to Example 12

Solution

Let us equate the divisor to zero as:

$$x - 2 = 0$$
$$x = 2$$

Next, evaluate the dividend when $x = 2$:

$$f(x) = 3x^3 + kx^2 - 4x + 11$$
$$f(2) = 3(2)^3 + k(2)^2 - 4(2) + 11$$
$$= 3(8) + k(4) - 8 + 11$$
$$= 24 + 4k - 8 + 11$$
$$f(2) = 4k + 27$$

But $f(2) = -1$, therefore:

$$-1 = 4k + 27$$
$$4k = -28$$
$$\therefore k = -7$$

One last example to try on the remainder theorem.

Example 13

The remainder when $f(x) = x^3 + px^2 - 7x - p$ is divided by $x + 1$ is the negative of the remainder when $f(x)$ is divided by $x + 4$. Determine the value of p.

What did you get? Find the solution below to double-check your answer.

Solution to Example 13

Let us equate the first divisor to zero as:

$$x + 1 = 0$$
$$x = -1$$

Next, evaluate the dividend $f(x)$ at $x = -1$:

$$f(x) = x^3 + px^2 - 7x - p$$
$$f(-1) = (-1)^3 + p(-1)^2 - 7(-1) - p$$
$$= -1 + p + 7 - p = 6$$
$$f(-1) = 6$$

Also, let us equate the second divisor to zero as:

$$x + 4 = 0$$
$$x = -4$$

Now evaluate the dividend $f(x)$ when $x = -4$:

$$f(x) = x^3 + px^2 - 7x - p$$
$$f(-4) = (-4)^3 + p(-4)^2 - 7(-4) - p$$
$$= -64 + 16p + 28 - p = 15p - 36$$
$$f(-4) = 15p - 36$$

But $-f(-4) = f(-1)$, therefore:

$$-(15p - 36) = 6$$
$$15p - 36 = -6$$
$$15p = 36 - 6$$
$$15p = 30$$
$$p = \frac{30}{15}$$
$$\therefore p = 2$$

9.9 FACTOR THEOREM

This is a special case of the remainder theorem which states that:

- If a polynomial $f(x)$ is divided by $x - a$ and the remainder is zero (i.e., $f(a) = 0$), then $x - a$ is a factor of $f(x)$.

OR

- If $x - a$ is a factor of the polynomial $f(x)$, then $f(a) = 0$. Alternatively, if $f(a) = 0$, then $x - a$ is a factor of $f(x)$.

OR

- If $ax - b$ is a factor of the polynomial $f(x)$, then $f\left(\frac{b}{a}\right) = 0$. Alternatively, if $f\left(\frac{b}{a}\right) = 0$, then $ax - b$ is a factor of $f(x)$.

OR

- If $f(x)$ is a polynomial such that $f(a) = 0$, then $x - a$ is a factor of the polynomial.

Great and easy! Since this theorem is a special case of the remainder theorem, all (or most of) the key points mentioned earlier apply too.

Good time to try some examples.

Example 14

Show that $2x - 5$ is a factor of $2x^2 - 7x + 5$.

What did you get? Find the solution below to double-check your answer.

Solution to Example 14

Let us equate the divisor to zero as:

$$2x - 5 = 0$$
$$x = \frac{5}{2}$$

Next, evaluate the dividend when $x = \frac{5}{2}$.

$$f(x) = 2x^2 - 7x + 5$$
$$f\left(\frac{5}{2}\right) = 2\left(\frac{5}{2}\right)^2 - 7\left(\frac{5}{2}\right) + 5$$
$$= 2\left(\frac{25}{4}\right) - 7\left(\frac{5}{2}\right) + 5$$
$$= 12.5 - 17.5 + 5$$
$$\therefore f\left(\frac{5}{2}\right) = 0$$

Hence, $2x - 5$ is a factor of $2x^2 - 7x + 5$.

Example 15

Determine the value of β and γ if $(x - 2)$ and $(x + 4)$ are factors of the polynomial $f(x) = \beta x^3 + \gamma x^2 - 22x - 8$. Write $f(x)$ completely with the values of the unknowns found.

What did you get? Find the solution below to double-check your answer.

Solution to Example 15

Let us equate the first factor to zero as:

$$x - 2 = 0$$
$$x = 2$$

Next, evaluate the dividend when $x = 2$.

$$f(x) = \beta x^3 + \gamma x^2 - 22x - 8$$
$$f(2) = \beta(2)^3 + \gamma(2)^2 - 22(2) - 8$$
$$= 8\beta + 4\gamma - 44 - 8$$
$$= 8\beta + 4\gamma - 52$$

Because $(x - 2)$ is a factor of $f(x)$, we have:

$$8\beta + 4\gamma - 52 = 0$$
$$8\beta + 4\gamma = 52$$
$$\therefore 2\beta + \gamma = 13 - - - - - -(i)$$

Let us equate the second factor to zero as:

$$x + 4 = 0$$
$$x = -4$$

Now evaluate the dividend when $x = -4$.

$$f(x) = \beta x^3 + \gamma x^2 - 22x - 8$$
$$f(2) = \beta(-4)^3 + \gamma(-4)^2 - 22(-4) - 8$$
$$= -64\beta + 16\gamma + 88 - 8$$
$$= -64\beta + 16\gamma + 80$$

Because $(x + 4)$ is a factor of $f(x)$, we have:

$$-64\beta + 16\gamma + 80 = 0$$
$$-64\beta + 16\gamma = -80$$
$$\therefore -4\beta + \gamma = -5 - - - - - -(ii)$$

Equations (i) and (ii) need to be solved simultaneously as follows:

- Equation (i) minus equation (ii)

$$6\beta = 18$$
$$\beta = \frac{18}{6}$$
$$\therefore \beta = 3$$

- From equation (i), we have:

$$2\beta + \gamma = 13$$
$$\gamma = 13 - 2\beta$$
$$= 13 - 2\,(3)$$
$$= 13 - 6$$
$$\therefore \gamma = 7$$

Therefore,

$$f(x) = 3x^3 + 7x^2 - 22x - 8$$

Example 16

Determine the value of k if $2x + 1$ is a factor of $kx^3 - 3x^2 + 2kx + 3$.

What did you get? Find the solution below to double-check your answer.

Solution to Example 16

Let us equate the divisor to zero as:

$$2x + 1 = 0$$
$$x = -\frac{1}{2}$$

Now evaluate the dividend when $x = -\frac{1}{2}$ as:

$$f(x) = kx^3 - 3x^2 + 2kx + 3$$
$$f\left(-\frac{1}{2}\right) = k\left(-\frac{1}{2}\right)^3 - 3\left(-\frac{1}{2}\right)^2 + 2k\left(-\frac{1}{2}\right) + 3$$
$$= k\left(-\frac{1}{8}\right) - 3\left(\frac{1}{4}\right) - k + 3$$
$$= -\frac{k}{8} - \frac{3}{4} - k + 3$$
$$= \frac{9}{4} - \frac{9k}{8}$$

But $f\left(-\frac{1}{2}\right) = 0$, therefore:

$$\frac{9}{4} - \frac{9k}{8} = 0$$
$$\frac{9}{4} = \frac{9k}{8}$$
$$\frac{k}{8} = \frac{1}{4}$$
$$4k = 8$$
$$\therefore k = 2$$

We can therefore say that $2x + 1$ is a factor of $2x^3 - 3x^2 + 4x + 3$.

Example 17

Show that $2x + 1$ is a factor of $f(x) = 2x^3 - 5x^2 + x + 2$. Hence, factorise $f(x)$ completely.

What did you get? Find the solution below to double-check your answer.

Solution to Example 17

Let us equate the divisor to zero as:

$$2x + 1 = 0$$
$$x = -\frac{1}{2}$$

Now evaluate the dividend when $x = -\frac{1}{2}$:

$$f(x) = 2x^3 - 5x^2 + x + 2$$
$$f\left(-\frac{1}{2}\right) = 2\left(-\frac{1}{2}\right)^3 - 5\left(-\frac{1}{2}\right)^2 - \frac{1}{2} + 2$$
$$= 2\left(-\frac{1}{8}\right) - 5\left(\frac{1}{4}\right) - \frac{1}{2} + 2$$
$$= -\frac{1}{4} - \frac{5}{4} - \frac{1}{2} + 2$$
$$\therefore f\left(-\frac{1}{2}\right) = 0$$

Hence, $2x + 1$ is a factor of $2x^3 - 5x^2 + x + 2$.
To factorise, we need to carry out a long division method as:

$$
\begin{array}{r}
x^2 - 3x + 2
\end{array}
$$

Therefore, we can write that

$$2x^3 - 5x^2 + x + 2 = (2x + 1)(x^2 - 3x + 2)$$

It is easy to factorise the quotient $x^2 - 3x + 2$ as:

$$x^2 - 3x + 2 = (x - 1)(x - 2)$$

Thus, the complete factorisation of $f(x)$ is

$$
\begin{aligned}
f(x) &= 2x^3 - 5x^2 + x + 2 \\
&= (2x + 1)(x^2 - 3x + 2) \\
&= \mathbf{(2x + 1)(x - 1)(x - 2)}
\end{aligned}
$$

One final example.

Example 18

Calculate the zeros of the polynomial $f(x) = 5x^3 + 8x^2 - 19x + 6$.

What did you get? Find the solution below to double-check your answer.

Solution to Example 18

This is a polynomial of degree three and we expect three factors and 3 zeros (distinct or repeated). Let us write this as:

$$f(x) = (x \pm \alpha)(x \pm \beta)(x \pm \gamma)$$

If the RHS of the above is expanded, we have $\pm\alpha\beta\gamma$ as the constant term, which implies that $\pm\alpha\beta\gamma = 6$. This means that α, β, and γ are factors of 6. We can try $x \pm 1$, $x \pm 2$, $x \pm 3$, and $x \pm 6$.

- Let us start with $x + 1$, thus:

$$x + 1 = 0$$
$$x = -1$$

We therefore have:

$$f(x) = 5x^3 + 8x^2 - 19x + 6$$
$$f(-1) = 5(-1)^3 + 8(-1)^2 - 19(-1) + 6$$
$$= 5(-1) + 8(1) + 19 + 6$$
$$= -5 + 8 + 19 + 6$$
$$\therefore f(-1) = 28$$

Hence, $x + 1$ is NOT a factor of $f(x)$.

- Let us try $x - 1$, thus:

$$x - 1 = 0$$
$$x = 1$$

We therefore have:

$$f(x) = 5x^3 + 8x^2 - 19x + 6$$
$$f(-1) = 5(1)^3 + 8(1)^2 - 19(1) + 6$$
$$= 5(1) + 8(1) - 19 + 6$$
$$= 5 + 8 - 19 + 6$$
$$\therefore f(1) = 0$$

Hence, $x - 1$ is a factor of $f(x)$. We can continue to try others, but it is better to factorise using the long division method as:

$$
\begin{array}{r}
5x^2 + 13x - 6 \\
x - 1 \enclose{longdiv}{5x^3 + 8x^2 - 19x + 6} \\
\underline{5x^3 - 5x^2} \\
13x^2 - 19x \\
\underline{13x^2 - 13x} \\
-6x + 6 \\
\underline{-6x + 6} \\
\cdot
\end{array}
$$

Therefore, we can write that

$$5x^3 + 8x^2 - 19x + 6 = (x - 1)(5x^2 + 13x - 6)$$

It is easy to factorise the quotient $5x^2 + 13x - 6$ as:

$$5x^2 + 13x - 6 = (x + 3)(5x - 2)$$

Thus, the complete factorisation of $f(x)$ is:

$$f(x) = 5x^3 + 8x^2 - 19x + 6$$
$$= (x - 1)(5x^2 + 13x - 6)$$
$$= (x - 1)(x + 3)(5x - 2)$$

To find the zeros of the polynomial we need to equate $f(x)$ to zero. Thus,

$$5x^3 + 8x^2 - 19x + 6 = 0$$
$$(x - 1)(x + 3)(5x - 2) = 0$$

$(x - 1) = 0$	$(x + 3) = 0$	$(5x - 2) = 0$
$x - 1 = 0$ **OR**	$x + 3 = 0$ **OR**	$5x - 2 = 0$
$\therefore x = 1$	$\therefore x = -3$	$5x = 2$
		$\therefore x = \dfrac{2}{5}$

Hence, the zeros of $f(x)$ are:

$$x = 1, \quad x = -3, \quad x = \frac{2}{5}$$

9.10 GRAPHS OF POLYNOMIALS

The graph of a polynomial is always continuous without any gap or discontinuity, but the shape is determined primarily by the degree of the polynomial. The focus here is to look at the shapes and make notes on each, as sketching/plotting have been covered for polynomials of degrees 1 and 2 (see Chapters 4–7).

We will limit our brief discussion to polynomials of degree one to four, i.e., linear to quartic. Let's look at them one after the other.

Degree 1 when $n = 1$

It has a general expression of $f(x) = ax + b$ and produces a straight-line shape of either / or \. There is only one zero (Figure 9.3) as the lines cross the x-axis once.

Degree 2 when $n = 2$

It has a general expression of $f(x) = ax^2 + bx + c$ and it is a parabola of either ∪ or ∩. We expect a maximum of two zeros as shown in Figure 9.4, where the curve crosses the x-axis. The zeros can be real, complex, or a combination, and they can have distinct or repeated values.

Degree 3 when $n = 3$

It has a general expression of $f(x) = ax^3 + bx^2 + cx + d$. It appears like a double laterally merged parabola of either ∪∩ or ∪∩. We expect a maximum of three zeros as shown in Figure 9.5(a), but it's possible that the curve crosses only once as shown in Figure 9.5(b). The zeros can be real, complex, or a combination, and they can have distinct or repeated values.

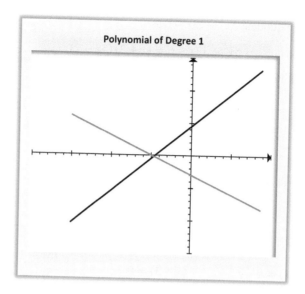

FIGURE 9.3 A polynomial of degree 1.

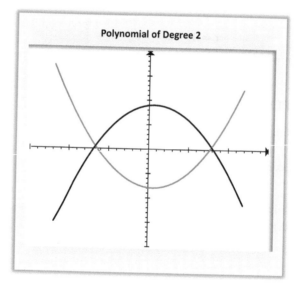

FIGURE 9.4 A polynomial of degree 2.

Degree 4 when $n = 4$

It has a general expression of $f(x) = ax^4 + bx^3 + cx^2 + dx + e$. The shape depends on the value of the coefficients a, b, c, d, and e. We expect a maximum of four as shown in Figure 9.6(a), but it's possible that the curve crosses the x-axis less than this number of times as shown in Figure 9.6(b). The zeros can be real, complex, or a combination, and they can have distinct or repeated values.

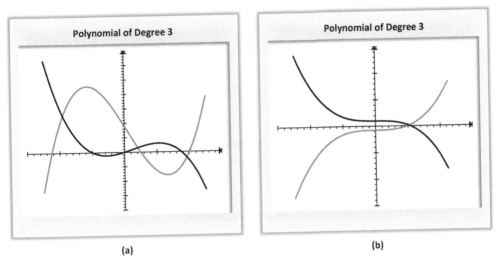

FIGURE 9.5 A polynomial of degree 3: (a) three distinct roots, and (b) one real root.

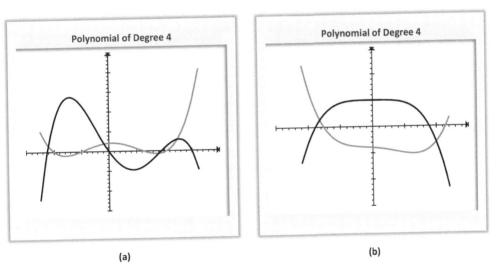

FIGURE 9.6 A polynomial of degree 4: (a) four distinct roots, and (b) two real roots.

9.11 CHAPTER SUMMARY

1) A polynomial is an algebraic expression that is a summation of units and a constant. Each unit is a product of:

- a coefficient, and
- a variable raised to the power of a positive integer.

2) The general format of a polynomial is:

$$f(x) = a_m x^n + a_{m-1}x^{n-1} + a_{m-2}x^{n-2} + a_{m-3}x^{n-3} + a_{m=4}x^{n-4} + \cdots + a_0$$

where $a_m, a_{m-1}, a_{m-2}, a_{m-3}, \ldots a_0$ are real numbers, $a_m \neq 0$, and n is a positive integer.

3) The degree of a polynomial is the highest power (i.e., n) of the variable in the expression. Thus, if:

- $n = 1$, then we have $ax + b$. This is a polynomial of degree 1 and it is practically called a **linear expression**.
- $n = 2$, then we have $ax^2 + bx + c$. This is a polynomial of degree 2 and it is specially called a **quadratic expression**.
- $n = 3$, then we have $ax^3 + bx^2 + cx + d$. This is a polynomial of degree 3 and it is specially called a **cubic expression**.
- $n = 3$, then we have $ax^4 + bx^3 + cx^2 + dx + e$. This is a polynomial of degree 4 and it is specially called a **quartic expression**.

4) Two polynomials are equal if their corresponding coefficients and powers are equal. For example, if $f(x) = a_4x^4 + a_3x^3 + a_2x^2 + a_1x + a_0$ and $g(x) = 5x^4 - 1.6x^3 + 8x^2 - x + 51$, then, using the equality principle, we can conclude that:

$$a_4 = 5, \quad a_3 = -1.6, \quad a_2 = 8, \quad a_1 = -1, \quad a_0 = 51$$

5) A polynomial $p(x)$ is a function of a variable x, and the value of the function depends on the value assigned to the variable x. We may be asked to find the value of $p(x)$ when the variable x is equal to k. This is written as $p(k)$ and obtained by replacing x with k in the expression.

6) The basic arithmetic operators (addition, subtraction, multiplication, and division) can be applied to polynomials. The division process takes a different procedure from the other three.

7) We can state that:

$$\boxed{\textbf{Dividend} = (\textbf{Divisor}) \times (\textbf{Quotient}) + \textbf{Remainder}}$$

If the quotient and the remainder are given as $Q(x)$ and R respectively, then

$$\boxed{\frac{f(x)}{g(x)} = Q(x) + \frac{R}{g(x)}}$$

OR

$$\boxed{f(x) = g(x) \times Q(x) + R}$$

8) The zeros of a polynomial are the solutions obtained when the polynomial is equated to zero. If a polynomial is given as $p(x)$, then solving $p(x) = 0$ will yield its zeros (or roots).

9) The **remainder theorem** states that:

- If a polynomial $f(x)$ is divided by $x - a$, the remainder is $f(a)$.
- If a polynomial $f(x)$ is divided by $ax + b$, the remainder is $f(-\frac{b}{a})$.

10) The **factor theorem** is a special case of the remainder theorem, which states that:

- If a polynomial $f(x)$ is divided by $x - a$ and the remainder is zero (i.e., $f(a) = 0$), then $x - a$ is a factor of $f(x)$.

ument。

OR

- If $x - a$ is a factor of the polynomial $f(x)$, then $f(a) = 0$.

OR

- If $f(x)$ is a polynomial such that $f(a) = 0$, then $x - a$ is a factor of the polynomial.

11) The shape of the graph of a polynomial is determined by the degree of the polynomial.

- **When $n = 1$**: It has a general expression of $ax + b$ and produces a straight-line shape of either / or \. There is only one zero.
- **When $n = 2$**: It has a general expression of $ax^2 + bx + c$ and it is a parabola of either ∪ or ∩. There are two real zeros.
- **When $n = 3$**: It has a general expression of $ax^3 + bx^2 + cx + d$. It is as though a double laterally merged parabola of either ∪∩ or ∩∪. There are three real zeros.
- **When $n = 4$**: It has a general expression of $ax^4 + bx^3 + cx^2 + dx + e$. The shape depends on the value of the coefficients $a, b, c, d,$ and e. There are four real zeros.

****

9.12 FURTHER PRACTICE

To access complementary contents, including additional exercises, please go to www.dszak.com.

10 Inequalities

Learning Outcomes

Once you have studied the content of this chapter, you should be able to:

- Explain inequalities and their notations
- Discuss the fundamental rules of inequalities
- Solve problems involving linear inequalities
- Solve problems involving quadratic inequalities
- Solve simultaneous inequalities

10.1 INTRODUCTION

An inequality (plural inequalities) is more evident in our day-to-day affairs now than ever. Like its sister, i.e., equality, an inequality also plays a crucial role in mathematics and will be discussed here in this chapter, covering linear, quadratic, and simultaneous inequalities.

10.2 NOTATIONS FOR INEQUALITIES

Recall that the equality of two expressions is shown using the symbol $=$. However, there are four different symbols to show inequalities, and these are illustrated in Table 10.1.

The $<$ and $>$ symbols are a pair and so are the $\leq$ and $\geq$ symbols. However, $<$ and $\geq$ are complementary just like $>$ complements $\leq$. What it means is that in any given set of numbers, if an inequality sign is used to enlist some members, then its complementary must be used to enlist the remaining members of the same set.

To illustrate, let's consider a set of numbers x such that $x: -5, -4, -3, -2, -1, 0, 1, 2, 3, 4, 5$. This list is a set of numbers given in the range of -5 and 5 inclusive and it is represented as $-5 \leq x \leq 5$. Notice that only the combination of the complementary would produce the entire set, as illustrated in Table 10.2 for when -2 is chosen as the reference number.

Notice that we can produce all the numbers when we use:

i) $<$ and $\geq$, i.e., $(x < -2) + (x \geq -2)$, and

ii) $>$ and $\leq$, i.e., $(x > -2) + (x \leq -2)$

DOI:10.1201/9781003027928-10

TABLE 10.1
Notations for Inequalities Explained

Symbol	Name	What It Does/Shows	Example
<	Less than	• The left-hand side expression (or number) is less than the right-hand side expression (or number). • We use → to represent it on a number line. Notice the direction of the arrow and the hollow circle or unfilled circular part, denoted as ○.	• $2 < 3$ • $5x - 2 < 0$
>	Greater than	• The left-hand side expression (or number) is greater than the right-hand side expression. • We use ← to represent it on a number line.	• $10 > 3$ • $b - 1 > \frac{1}{4}$
≤	Less than or equal to	• The left-hand side expression (or number) is less than **OR** equal to the right-hand side expression (or number). • It is possible to think about this expression as a combination of < and =. Alternatively, it is like when we use 'inclusive'. For example, our holiday is from Monday to Saturday inclusive, which means Monday and Saturday are included in the holiday. • We use → to represent it on a number line. Notice the direction of the arrow and the solid circle or filled circular part, denoted as •.	• $y + 1 \leq y^2 - 4y$
≥	Greater than or equal to	• The left-hand side expression (or number) is greater than **OR** equal to the right-hand side expression (or number). • It is possible to think about this expression as a combination of > and =. • We use ← to represent it on a number line.	• $ab - a \leq a^2 + 4$

TABLE 10.2

Complementary Inequalities Illustrated

	Inequality	Members
i)	$x < -2$	$-5, \ -4, \ -3$
	$x \geq -2$	$-2, \ -1, \ 0, \ 1, \ 2, \ 3, \ 4, \ 5$
ii)	$x > -2$	$-1, \ 0, \ 1, \ 2, \ 3, \ 4, \ 5$
	$x \leq -2$	$-5, \ -4, \ -3, \ -2,$

All numbers appear once and only once. This will be valid irrespective of the number chosen. For instance, if the reference number is 1, then the following pairs of complementary inequalities will produce all the numbers within the given range $-5 \leq x \leq 5$:

a) $(x < 1) + (x \geq 1)$
b) $(x > 1) + (x \leq 1)$

Let's try some examples.

Example 1

For $x \in \mathbb{Z}$, determine the three smallest integers in the range defined by the following inequalities.

a) $x > 3$ b) $x > -2$ c) $x \geq 11$ d) $x > \dfrac{5}{2}$ e) $30 \leq x$

What did you get? Find the solution below to double-check your answer.

Solution to Example 1

HINT

- It may be useful to list more than three elements of the set and then choose the smallest three.
- As we are only interested in the integers, we can ignore fractions in the set.

a) $x > 3$
Solution
$x > 3$ implies that x can be 4, 5, 6, 7, 8, 9 ... . The smallest three are:

$$4, 5, 6$$

b) $x > -2$
Solution
$x > -2$ implies that x can be $-1, 0, 1, 2, 3, 4 \ldots$ The smallest three are:

$$-1, 0, 1$$

c) $x \geq 11$
Solution
$x \geq 11$ implies that x can be $11, 12, 13, 14, 15 \ldots$ The smallest three are:

$$11, 12, 13$$

d) $x > \frac{5}{2}$
Solution
$x > \frac{5}{2}$ is the same as $x > 2.5$, which implies that x can be $3, 4, 5, 6, 7, 8 \ldots$ The smallest three are:

$$3, 4, 5$$

e) $30 \leq x$
Solution
$30 \leq x$ is the same as $x \geq 30$, which implies that x can be $30, 31, 32, 33, 34, 35 \ldots$ The smallest three are:

$$30, 31, 32$$

Another set of examples to try.

Example 2

For $m \in \mathbb{Z}$, determine the three largest integers in the range defined by the following inequalities.

a) $m < 5$ **b)** $m < -5$ **c)** $m \leq 0$ **d)** $m < \sqrt{2}$ **e)** $-\frac{75}{10} \geq m$

What did you get? Find the solution below to double-check your answer.

Solution to Example 2

HINT

- It may be useful to list more than three elements of the set and then choose the largest three.
- As we are only interested in the integers, we can ignore fractions in the set.

a) $m < 5$

Solution

$m < 5$ implies that m can be 4, 3, 2, 1, 1, -1 ... . The largest three are:

$$2,\ 3,\ 4$$

---

b) $m < -5$

Solution

$m < -5$ implies that m can be $-6,\ -7,\ -8,\ -9,\ -10$ ... . The largest three are:

$$-8,\ -7,\ -6$$

---

c) $m \leq 0$

Solution

$m \leq 0$ implies that m can be $0, -1, -2, -3, -4, -5$ ... . The largest three are:

$$-2,\ -1,\ 0$$

---

d) $m < \sqrt{2}$

Solution

$\sqrt{2}$ is irrational, but we know that it's 1.414 (3 d.p.). We can therefore say that $m < \sqrt{2}$ implies that m can be 1, 0, $-1, -2, -3, -4$ ... . The largest three are:

$$-1,\ 0,\ 1$$

---

e) $-\dfrac{75}{10} \geq m$

Solution

$-\dfrac{75}{10} \geq m$ is the same as $m \leq -\dfrac{75}{10}$ or $m \leq -7.5$. We can therefore say that $m \leq -7.5$ implies that m can be $-8,\ -9,\ -10,\ -11,\ -12, ...$. The largest three are:

$$-10, -9, -8$$

The above inequalities represent an **open range**, such that the list (or members of the set) continues infinitely to the left or right. For example, $x \geq 1$ is a list of all numbers from 1 inclusive to the right up to ∞ while $x < -5$ is a list of all numbers from -5 exclusive to the left up to $-\infty$.

On the other hand, we can use inequalities to represent a **closed range** such that the members of the set are finite. We do this by using $<$ or $\leq$ twice or their combination. For example, we can write $-2 < x < 10$ or $5 < x \leq 8$ or $-2 \leq x < 10$ or $-7 \leq x \leq -2$. What this means is that the variable (x in this case) is a number between the left and the right numbers; both are inclusive in the list when we use $\leq$ or both are exclusive when we use $<$ or a mix. To illustrate, $5 < x \leq 8$ represents a set of $\{6, 7, 8\}$. The curly brackets $\{\}$ are used to list a set of numbers.

In this case, the left number must be less than the right number. As a result, it is invalid to write $8 < x \leq 5$, as it is impossible for 8 to be less than x and at the same time x is less than or equal to 5. Occasionally, $>$ or $\geq$ is used to specify a close range, for this case the left number must be greater than. Thus, we can write $10 > x > -2$ instead of $-2 < x < 10$ to represent the same list.

Let's try some examples.

| Example 3 |

For $y \in \mathbb{Z}$, list the numbers in the range defined by the following inequalities.

a) $-3 < y < 3$ **b)** $4 < y \leq 7$ **c)** $-5 \leq y < 0$ **d)** $-14 \leq y \leq -9$ **e)** $6 \geq y \geq -2$

What did you get? Find the solution below to double-check your answer.

| Solution to Example 3 |

a) $-3 < y < 3$
Solution

$$\{-2, -1, 0, 1, 2\}$$

b) $4 < y \leq 7$
Solution

$$\{5, 6, 7\}$$

c) $-5 \leq y < 0$
Solution

$$\{-5, -4, -3, -2, -1\}$$

d) $-14 \leq y \leq -9$
Solution

$$\{-14, -13, -12, -11, -10, -9\}$$

e) $6 \geq y \geq -2$
Solution
We can re-write $6 \geq y \geq -2$ as $-2 \leq y \leq 6$

$$\{-2, -1, 0, 1, 2, 3, 4, 5, 6\}$$

10.3 FUNDAMENTAL RULES OF INEQUALITIES

To solve problems involving inequalities, we will use all the rules and tips that we have learnt and applied to equations. However, there are additional rules that are peculiar to inequalities, which are covered below.

TABLE 10.3

Rules of Addition and Subtraction Illustrated

Inequality	Addition	Subtraction
$a > b$	$a + c > b + c$	$a - c > b - c$
	For example, if $13 > 5$, then $13 + 4 > 5 + 4$ is also true. That is, $17 > 9$.	For example, if $13 > 5$, then $13 - 4 > 5 - 4$ is also true. That is, $9 > 1$.
$a < b$	$a + c < b + c$	$a - c < b - c$
	For example, if $7 < 21$, then $7 + 4 < 21 + 4$ is also true. That is, $11 < 25$.	For example, if $7 < 21$, then $7 - 4 < 21 - 4$ is also true. That is, $3 < 17$.
$a \geq b$	Given that $a \geq b$, it follows that $$a + c \geq b + c$$ is also valid.	Given that $a \geq b$, it follows that $$a - c \geq b - c$$ is also valid.
$a \leq b$	Given that $a \leq b$, it follows that $$a + c \leq b + c$$ is also valid.	Given that $a \leq b$, it follows that $$a - c \leq b - c$$ is also valid.

10.3.1 Addition and Subtraction

The inequality sign will not change by adding or subtracting a constant positive value c (i.e., $c > 0$) from both sides of the inequality. This is illustrated in Table 10.3 for $c = 4$ in the first two instances.

Note that adding a positive is the same as subtracting a negative.

10.3.2 Multiplication and Division

The inequality remains unchanged if both sides are multiplied by or divided by the same positive value c, i.e., $c > 0$. This is illustrated in Table 10.4 for $c = 3$ in the first two instances.

However, if the constant multiplier or divisor is negative (i.e., $c < 0$), then the inequality sign (sense or direction) is reversed. This is illustrated in Table 10.5 for $c = -2.5$ in the first two instances.

The logic behind the switch of direction is illustrated (Table 10.6) using two methods. Let's say we are given $-2x < 6$ and want to solve for x.

Although the answers are ordered differently, they are essentially the same. In other words, x is greater than -3 implies that -3 is less than x.

TABLE 10.4

Rules of Multiplication and Division Illustrated for Positive Factor

Inequality	Multiplication	Division
$a < b$	$ac < bc$	$\dfrac{a}{c} < \dfrac{b}{c}$
	For example, if $1 < 2$, then $1 \times 3 < 2 \times 3$ is also true. That is, $3 < 6$.	For example, if $4.2 < 6$, then $\dfrac{4.2}{3} < \dfrac{6}{3}$ is also true. That is, $1.4 < 2$.
$a > b$	$ac > bc$	$\dfrac{a}{c} > \dfrac{b}{c}$
	For example, if $5 > 2$, then $5 \times 3 > 2 \times 3$ is also true. That is, $15 > 6$.	For example, if $27 > 21$, then $\dfrac{27}{3} > \dfrac{21}{3}$ is also true. That is, $9 > 7$.
$a + b \leq 1$	Given that $a + b \leq 1$, it follows that $$(a + b)c \leq c$$ is also valid.	Given that $a + b \leq 1$, it follows that $$\dfrac{a + b}{c} \leq \dfrac{1}{c}$$ is also valid.
$a + b \geq p$	Given that $a + b \geq p$, it follows that $$(a + b)c \geq pc$$ is also valid for $c > 0$.	Given that $a + b \geq p$, it follows that $$\dfrac{(a + b)}{c} \geq \dfrac{p}{c}$$ is also valid for $c > 0$.

10.3.3 POWER

When both sides of the inequality are raised to the same positive value m (i.e., $m > 0$) the inequality remains the same. This is illustrated in Table 10.7 for $m = 2$. However, if the power is negative (i.e., $m < 0$) then the inequality sign is reversed. This is also illustrated below for $m = -2$.

The above conditions are valid for positive values of a, b, c, and d.

10.3.4 INVERSION

When the expressions are laterally swapped, the inequality sign is equally swapped. In other words, if the entire LHS expression (along with their respective signs) is moved to the RHS then the equality sign is reversed, i.e., '**greater than**' sign becomes '**less than**' sign, and vice versa. This is illustrated in Table 10.8.

TABLE 10.5

Rules of Multiplication and Division Illustrated for Negative Factor

Inequality	Multiplication	Division
$a > b$	$ac < bc$	$\dfrac{a}{c} < \dfrac{b}{c}$
	For example, if $7 > -4$, then $7 \times -2.5 < -4 \times -2.5$ is also true. That is, $-17.5 < 10$.	For example, if $9 > 4$, then $\dfrac{9}{-2.5} < \dfrac{4}{-2.5}$ is also true. That is, $-3.6 < -1.6$.
$a < b$	$ac > bc$	$\dfrac{a}{c} > \dfrac{b}{c}$
	For example, if $-4 < 1$, then $-4 \times -2.5 > 1 \times -2.5$ is also true. That is, $10 > -2.5$.	For example, if $15 < 25$, then $\dfrac{15}{-2.5} > \dfrac{25}{-2.5}$ is also true. That is, $-6 > -10$.
$a - b \le 1$	Given that $a - b \le 1$, it follows that $(a-b)c \ge c$ is also valid for $c < 0$.	Given that $a - b \le 1$, it follows that $\dfrac{a-b}{c} \ge \dfrac{1}{c}$ is also valid for $c < 0$.
$a - b \ge p$	Given that $a - b \ge p$, it follows that $(a-b)c \le pc$ is also valid for $c < 0$.	Given that $a - b \ge p$, it follows that $\dfrac{(a-b)}{c} \le \dfrac{p}{c}$ is also valid for $c < 0$.

TABLE 10.6

Alternative Method to Reversing Inequality Sign When Multiplying and Dividing with a Negative Factor Illustrated

Using the Rule	Alternative Method
Using the rule of division by a negative, we will divide both sides by -2 and flip the direction of the inequality as: $$-2x < 6$$ $$-\frac{2x}{-2} > \frac{6}{-2}$$ $$x > -3$$	Alternatively, we can move $-2x$ to the right and 6 to the left and divide by 2 as: $$-6 < 2x$$ $$-\frac{6}{2} < \frac{2x}{2}$$ $$-3 < x$$

TABLE 10.7
Rules of Power Illustrated

Inequality	Positive Power (for $m = 2$)	Negative Power (for $m = -2$)
$a < b$	$a^m < b^m$	$a^m > b^m$
	For example, if $6 < 7$, then $6^2 < 7^2$ is also true. That is, $36 < 49$.	For example, if $10 < 15$, then $10^{-2} > 15^{-2}$ is also true. That is, $\frac{1}{10^2} > \frac{1}{15^2}$.
$a > b$	$a^m > b^m$	$a^m < b^m$
	For example, if $10 > 5$, then $10^2 > 5^2$ is also true. That is, $100 > 25$.	For example, if $8 > 3$, then $8^{-2} < 3^{-2}$ is also true. That is, $\frac{1}{8^2} < \frac{1}{3^2}$.
$a + b \leq c$	Given that $a + b \leq c$, it follows that $$(a + b)^m \leq c^m$$ is also valid.	Given that $a + b \leq c$, it follows that $$(a + b)^{-m} \geq c^{-m}$$ is also valid.
$a + b \geq c + d$	Given that $a + b \geq c + d$, it follows that $$(a + b)^m \geq (c + d)^m$$ is also valid.	Given that $a + b \geq c + d$, it follows that $$(a + b)^{-m} \leq (c + d)^{-m}$$ is also valid.

TABLE 10.8
Rules of Inversion Illustrated

Inequality	Inversion
$a > b$	$b < a$
$m + n < 0$	$0 > m + n$
$s + c \leq t + d$	$t + d \geq s + c$
$u \geq v$	$v \leq u$

10.3.5 VARIABLE

An unknown variable, such as x, y, and a, should not be used to multiply or divide both sides of an inequality. This is because the numerical value of the variable can be positive or negative, and each has a different effect on an inequality sign as shown in the previous rules. For example, if we multiply both sides of $\frac{1}{x} < 5$ by x to have $1 < 5x$. This resulting inequality is true only and only if x is positive (i.e., $x > 0$) otherwise the multiplication would result in $1 > 5x$. Notice that the sense of the inequality has been reversed. As we would not know or have the information beforehand, it becomes invalid to use this approach. Furthermore, the variable can be zero and this may even have other effects, including asymptotes, on the inequality.

10.3.6 SQUARE FUNCTIONS

This is used to solve inequalities involving a square function of the form $[f(x)]^2 < k$ where $f(x)$ is a function of the variable x and $k \in \mathbb{Z}^+$. Table 10.9 summarises the rules relating to this for the four inequalities.

Notice that we have simplified the examples to give the final answers; the process of carrying out this will be further covered shortly in this chapter (Section 10.5.2.4). It is also important to mention that the rule can be used to solve quadratic inequalities, especially when we need to use the completing the square method because the quadratic expression cannot be factorised easily.

These are the fundamental rules of working with inequalities; we will be using them shortly to solve problems.

TABLE 10.9
Rules of Inversion Illustrated

Inequality	Solution	Example
$x^2 < a$	$-\sqrt{a} < x < \sqrt{a}$	$x^2 < 5$ implies $-\sqrt{5} < x < \sqrt{5}$
$x^2 > a$	$x > \sqrt{a}$ and $x < -\sqrt{a}$	$x^2 > 4$ implies $x > \sqrt{4}$ and $x < -\sqrt{4}$ This can be further simplified to $x > 2$ and $x < -2$
$x^2 \le b$	$-\sqrt{b} \le x \le \sqrt{b}$	$(x-3)^2 \le 10$ implies $-10 \le (x-3) \le 10$ This can be further simplified to $-10 \le x - 3 \le 10$ then $-7 \le x \le 13$
$x^2 \ge b$	$x \ge \sqrt{b}$ and $x \le -\sqrt{b}$	$(2x-5)^2 - 9 \ge 0$ this can be re-written as $(2x-5)^2 \ge 9$ This implies $2x - 5 \ge \sqrt{9}$ and $2x - 5 \le -\sqrt{9}$ This can be further simplified to $2x - 5 \ge 3$ and $2x - 5 \le -3$ then $x \ge 4$ and $x \le 1$

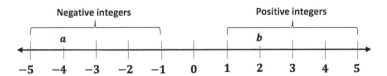

FIGURE 10.1 Number line with centre zero illustrated.

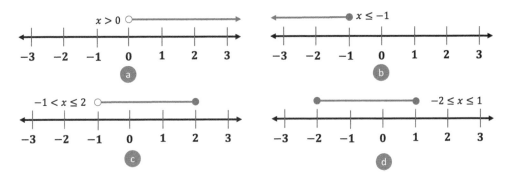

FIGURE 10.2 Inequalities shown on number lines: (a) x is less that zero, (b) x is less than or equal to -1, (c) x is a number between -1 (excluded) and 2 (included), and (d) x is a number between -2 (included) and 1 (included).

10.4 NUMBER LINE

Think about a number line as the x-axis in an x-y Cartesian plane, and it is generally used to visually show the members or elements of a particular inequality.

A number line is a horizontal line which is demarcated with zero as the centre or reference point. Positive integers are evenly placed to the right of the origin and corresponding negative integers are placed to the left of this origin, as shown in Figure 10.1.

Given two points a and b on the real number line such that a is placed to the left of b, we can deduce that:

$$a < b \quad \textbf{OR} \quad b > a$$

In other words, the point (or number) to the right is said to be **greater than** the one to its left. Conversely, the point (or number) to the left is said to be **less than** the one to its right. Figure 10.2 shows four examples of number lines, labelled (a) to (d).

It's now a good time to pause and try some examples.

Example 4

Show the following inequalities on a number line.

a) $x < -3$
d) $1 \le x < \dfrac{9}{2}$

b) $-2 \le x$
e) $-4 \le x \le 4$

c) $-1.5 < x < 3$
f) $1 > x > -5$

What did you get? Find the solution below to double-check your answer.

Solution to Example 4

a) $x < -3$
Solution (Figure 10.3)

---

b) $-2 \leq x$
Solution (Figure 10.4)
Note that $-2 \leq x$ is the same as $x \geq -2$. This is based on the inversion rule discussed above.

---

c) $-1.5 < x < 3$
Solution (Figure 10.5)

---

d) $1 \leq x < \frac{9}{2}$
Solution (Figure 10.6)
$1 \leq x < \frac{9}{2}$ is the same as $1 \leq x < 4.5$.

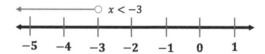

FIGURE 10.3 Solution to Example 4(a).

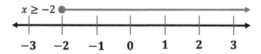

FIGURE 10.4 Solution to Example 4(b).

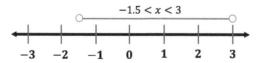

FIGURE 10.5 Solution to Example 4(c).

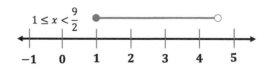

FIGURE 10.6 Solution to Example 4(d).

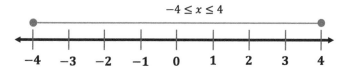

FIGURE 10.7 Solution to Example 4(e).

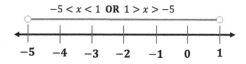

FIGURE 10.8 Solution to Example 4(f).

e) $-4 \leq x \leq 4$
Solution (Figure 10.7)

---

f) $1 > x > -5$
Solution
Note that $1 > x > -5$ is the same as $-5 < x < 1$ as earlier discussed. We can show this by splitting $1 > x > -5$ into two as:

    i) $1 > x$, which is the same as $x < 1$, and

    ii) $x > -5$, which is the same as $-5 < x$

The above two inequalities can then be recombined into $-5 < x < 1$. This is shown in the number line in Figure 10.8.

10.5 SOLVING PROBLEMS INVOLVING INEQUALITIES

Solving inequalities usually follows similar procedures as its equivalent equations. In other words, a linear inequality will be solved in a similar way as we would solve a linear equation; the same applies to a quadratic inequality. In summary, two key tips to take away are:

Note 1 Treat an inequality as its equivalent equation when using a positive constant. Do the same with a negative constant but reverse the sense of the inequality.
Note 2 Do not multiply or divide with a variable.

10.5.1 LINEAR INEQUALITY

A linear inequality in one variable can take any of the following general expressions:

$$ax + b < c \quad ax + b > c$$
$$ax + b \leq c \quad ax + b \geq c$$

where a, b, and c are real numbers.

It should be noted that the RHS side can be an algebraic expression if the linear inequality is not in its simplest form, e.g., $3x + 4 < x - 5$. However, this can quickly be simplified and transformed to any of the four formats above. Whilst b and c can be zero, it is not possible that a is zero.

As previously mentioned, solving a linear inequality is very similar to that of a linear equation with the exception of the rules discussed above. Nevertheless, we expect just one solution for the unknown variable as it is a linear equation.

Let's try some examples.

Example 5

Solve the following inequalities.

a) $3x - 5 < 0$ **b)** $7 - 2x > 1$ **c)** $12x - 4 \le 8x - 7$

d) $\frac{1}{3}x - 4 \ge \frac{1}{2} - x$ **e)** $\frac{x-11}{5} < \frac{2-x}{3}$ **f)** $3(1 - x) < 4\left(\frac{x}{2} - 3\right)$

What did you get? Find the solution below to double-check your answer.

Solution to Example 5

a) $3x - 5 < 0$
Solution

$$3x - 5 < 0$$
$$3x < 5$$
$$\therefore x < \frac{5}{3}$$

---

b) $7 - 2x > 1$
Solution

$$7 - 2x > 1$$
$$-2x > 1 - 7$$
$$-2x > -6$$
$$-\frac{2x}{-2} < -\frac{6}{-2}$$

Notice that the sense of the inequality changes as we divide with a negative number.

$$\therefore x < 3$$

ALTERNATIVE METHOD

$$7 - 2x > 1$$
$$7 - 1 > 2x$$
$$6 > 2x$$
$$3 > x \text{ OR } x < 3$$
$$\therefore x < 3$$

---

c) $12x - 4 \leq 8x - 7$

Solution

$$12x - 4 \leq 8x - 7$$
$$12x - 8x \leq 4 - 7$$
$$4x \leq -3$$
$$\therefore x \leq -\frac{3}{4}$$

---

d) $\frac{1}{3}x - 4 \geq \frac{1}{2} - x$

Solution

$$\frac{1}{3}x - 4 \geq \frac{1}{2} - x$$
$$\frac{1}{3}x + x \geq 4 + \frac{1}{2}$$
$$\frac{4}{3}x \geq \frac{9}{2}$$
$$x \geq \frac{9}{2} \times \frac{3}{4}$$
$$\therefore x \geq \frac{27}{8}$$

---

e) $\frac{x-11}{5} < \frac{2-x}{3}$

Solution

$$\frac{x - 11}{5} < \frac{2 - x}{3}$$
$$3(x - 11) < 5(2 - x)$$
$$3x - 33 < 10 - 5x$$
$$3x + 5x < 10 + 33$$
$$8x < 43$$
$$\therefore x < \frac{43}{8}$$

f) $3(1-x) < 4\left(\frac{x}{2} - 3\right)$

Solution

$$3(1-x) < 4\left(\frac{x}{2} - 3\right)$$
$$3 - 3x < 2x - 12$$
$$-3x - 2x < -3 - 12$$
$$-5x < -15$$
$$5x > 15$$
$$x > \frac{15}{5}$$
$$\therefore x > 3$$

Let's try another example.

Example 6

Determine the set of values x which satisfy $-4 \le 2x + 1 \le 5$. Represent the solutions on a number line.

What did you get? Find the solution below to double-check your answer.

Solution to Example 6

Solution

Method 1

For this approach, we need to make the middle expression $(2x + 1)$ as x only. Let's go.

$$-4 \le 2x + 1 \le 5$$

Subtract 1 from each section as:

$$-4 - 1 \le 2x + 1 - 1 \le 5 - 1$$
$$-5 \le 2x \le 4$$

Divide each section by 2 as:

$$-\frac{5}{2} \le \frac{2x}{2} \le \frac{4}{2}$$
$$-2.5 \le x \le 2$$

This solution is shown on a number line in Figure 10.9.

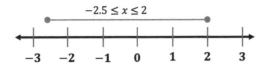

FIGURE 10.9 Solution to Example 6 – Method 1.

Method 2

In this method, we will split the inequality into two as:

$$-4 \le 2x + 1 \text{ and } 2x + 1 \le 5$$

Let's do this one at a time

- $-4 \le 2x + 1$

$$-4 \le 2x + 1$$
$$-4 - 1 \le 2x$$
$$-5 \le 2x$$
$$-2.5 \le x$$

- $2x + 1 \le 5$

$$2x + 1 \le 5$$
$$2x \le 5 - 1$$
$$2x \le 4$$
$$x \le 2$$

The two solutions are:

$$x \ge -2.5 \text{ AND } x \le 2$$

These are shown in Figure 10.10; the region of overlap for the two solutions is:

$$-2.5 \le x \le 2$$

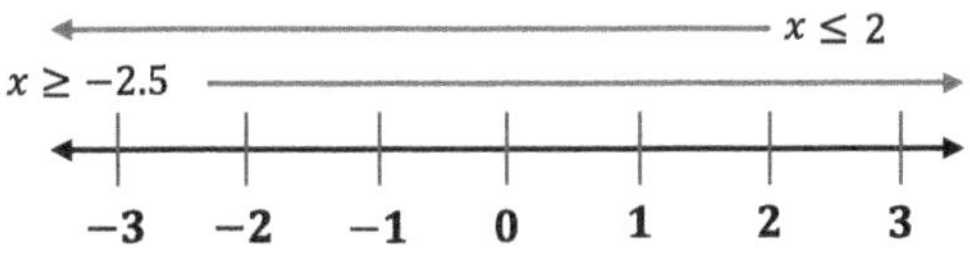

$$x \ge -2.5 \qquad\qquad x \le 2$$

$$\begin{array}{ccccccc} & -3 & -2 & -1 & 0 & 1 & 2 & 3 \end{array}$$

FIGURE 10.10 Solution to Example 6 – Method 2.

10.5.2 Quadratic Inequality

A quadratic inequality in one variable can take any of the following general expressions:

$$\begin{array}{ll} 1)\ ax^2 + bx + c < 0 & 2)\ ax^2 + bx + c > 0 \\ 3)\ ax^2 + bx + c \leq 0 & 4)\ ax^2 + bx + c \geq 0 \end{array}$$

Solving a quadratic inequality is like solving a quadratic equation with a few additional steps that will be covered shortly. Four methods (i.e., graphical method, author's method, square functions, and intervals and tables method) will be used here to solve any of the above cases.

10.5.2.1 Graphical Method

We will use the following three-step approach to solve a quadratic inequality graphically:

Step 1: Arrange the inequality in any of the above four formats, i.e., with the expression $ax^2 + bx + c$ entirely on one side (usually the LHS) and 0 on the other side (usually the RHS).

Step 2: Solve $ax^2 + bx + c = 0$ and sketch its graph (see Chapters 6 and 7).

Step 3: Use the inequality sign ($<$, $>$, $\leq$, or $\geq$) to determine the solution. There are two regions of interest:

     a) **Above the x-axis**: This is made up of $>$ and $\geq$. Greater than ($>$) is everything above the x-axis while greater than or equal to ($\geq$) is everything above including the x-axis.

     b) **Below the x-axis**: This is made up of $<$ and $\leq$. Less than ($<$) is everything below the x-axis while less than or equal to ($\leq$) is everything below including the x-axis.

Generally, the area that satisfies the inequality on the graph is shaded. The problem will determine which region is to be shaded, i.e., the region which satisfies the inequality. Note that in another convention, the shaded region is that which does not satisfy the inequality.

Good! Let's try a couple of examples.

Example 7

Solve the following inequalities.

a) $x^2 - 3x < 2 - 2x$ **b)** $5x + 4 \leq 3x\,(x + 2)$

What did you get? Find the solution below to double-check your answer.

Solution to Example 7

a) $x^2 - 3x < 2 - 2x$

Solution

Step 1: Re-arrange

$$x^2 - 3x < 2 - 2x$$
$$x^2 - 3x + 2x - 2 < 0$$
$$x^2 - x - 2 < 0$$

Step 2: Solve and sketch the quadratic function

The quadratic expression is $x^2 - x - 2$, hence we solve

$$x^2 - x - 2 = 0$$
$$(x - 2)(x + 1) = 0$$

Therefore

$$x - 2 = 0 \quad \textbf{OR} \quad x + 1 = 0$$
$$\therefore x = 2 \qquad\qquad \therefore x = -1$$

Thus, we have:

- **x-intercept**: $(-1, 0), (2, 0)$.
- **y-intercept**: $(0, -2)$.
- **Shape**: U-like shape as $a > 0$.

The graph of $f(x)$ is given in Figure 10.11.

Step 3: Determine the solution.

Looking at the inequality $x^2 - x - 2 < 0$, our solution is below zero or the x-axis, as this is where $y < 0$ or y is negative. We can confirm this by substituting a value between the roots of the equation, i.e., -1 and 2. Let's choose $x = 1$ and substitute as

$$f(x) = x^2 - x - 2$$
$$f(1) = x^2 - x - 2$$
$$= 1^2 - 1 - 2$$
$$= -2 < 0$$

We've shown the region that satisfies $x^2 - x - 2 < 0$ with a paler colour in Figure 10.12. Notice the use of hollow circle at $x = -1$ and $x = 2$ to show that the points are **excluded** in the solution of this inequality. In other words, the x-axis is not included in the solution. This is generally the case when we use the '**less than**' sign $<$ or '**greater than**' sign $>$.

We will therefore write this solution as:

$$-1 < x < 2$$

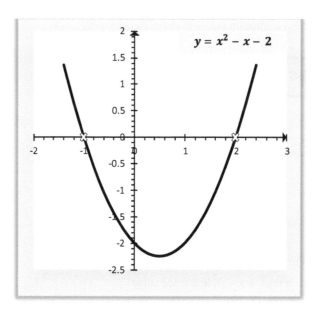

FIGURE 10.11 Solution to Example 7(a) – Part I.

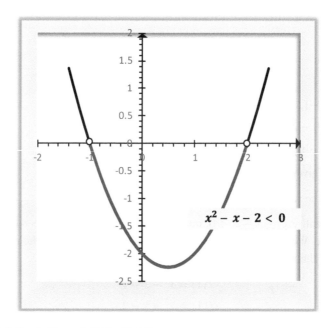

FIGURE 10.12 Solution to Example 7(a) – Part II.

b) $5x + 4 \le 3x(x + 2)$

Solution

Step 1: Re-arrange

$$5x + 4 \le 3x(x + 2)$$
$$5x + 4 \le 3x^2 + 6x$$

$$-3x^2 - 6x + 5x + 4 \leq 0$$
$$-3x^2 - x + 4 \leq 0$$
$$3x^2 + x - 4 \geq 0$$

Step 2: Solve and sketch the quadratic function.

The quadratic expression is $3x^2 + x - 4$, hence we solve

$$3x^2 + x - 4 = 0$$
$$(3x + 4)(x - 1) = 0$$

Therefore

$$3x + 4 = 0 \qquad \mathbf{OR} \qquad x - 1 = 0$$
$$\therefore x = -\frac{4}{3} \qquad\qquad \therefore x = 1$$

Thus, we have:

- **x-intercept**: $\left(-\frac{4}{3}, 0\right), (1, 0)$.
- **y-intercept**: $(0, -4)$.
- **Shape**: U-like shape as $a > 0$.

Step 3: Determine the solution.

Looking at the inequality $3x^2 + x - 4 \geq 0$, our solution is everything above and including the x-axis. Again, we can confirm this by substituting a value that is less than $-\frac{4}{3}$ or greater than 1. Let's choose $x = -2$ and substitute as

$$f(x) = 3x^2 + x - 4$$
$$f(-2) = 3(-2)^2 + (-2) - 4$$
$$= 3(4) - 2 - 4$$
$$= 6 > 0$$

Let's try a value that is greater than 1; we will go for $x = 3$ and substitute as

$$f(x) = 3x^2 + x - 4$$
$$f(3) = 3(3)^2 + (3) - 4$$
$$= 3(9) + 3 - 4$$
$$= 26 > 0$$

We've shown the region that satisfies $3x^2 + x - 4 \geq 0$ with a paler colour in Figure 10.14. Notice the use of filled circle ● at $x = -\frac{4}{3}$ and $x = 1$ to show that the points are **included** in the solution of this inequality. This is generally the case when we use the '**less than or equal to**' sign $\leq$ or '**greater than or equal to**' sign $\geq$.

We will therefore write this solution as:

$$x \leq -\frac{4}{3} \text{ and } x \geq 1$$

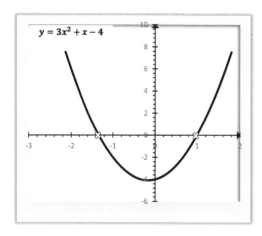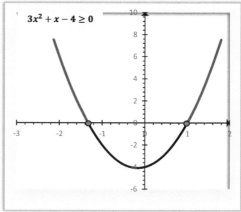

FIGURE 10.13   Solution to Example 7(b) – Method 1, Part I.

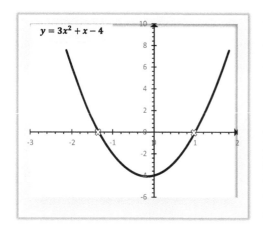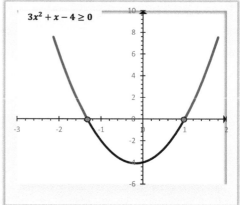

FIGURE 10.14   Solution to Example 7(b) – Method 1, Part II.

ALTERNATIVE METHOD

Step 1: Re-arrange

$$5x + 4 \leq 3x(x+2)$$
$$5x + 4 \leq 3x^2 + 6x$$
$$-3x^2 - 6x + 5x + 4 \leq 0$$
$$-3x^2 - x + 4 \leq 0$$

Notice that, in contrast to the first method, we did not multiply both sides by -1 in this alternative method. Let's see how this pans out.

Step 2: Solve and sketch the quadratic function

The quadratic expression is $-3x^2 - x + 4$ or $4 - x - 3x^2$, hence we solve

$$4 - x - 3x^2 = 0$$
$$(4 + 3x)(1 - x) = 0$$

Therefore

$$4 + 3x = 0 \qquad \qquad 1 - x = 0$$
$$\therefore x = -\frac{4}{3} \qquad \textbf{OR} \qquad \therefore x = 1$$

Thus, we have:

- **x-intercept**: $\left(-\frac{4}{3}, 0\right), (1, 0)$.

- **y-intercept**: $(0, 4)$.

- **Shape**: Inverted U-like shape as $a < 0$.

Step 3: Determine the solution (Figure 10.15).

Looking at the inequality $4 - x - 3x^2 \leq 0$, our solution is below and including x-axis, where x is either less than or equal to $-\frac{4}{3}$ or greater than or equal to 1. This is shown with a paler colour (Figure 10.16).

We will therefore write this solution as:

$$x \leq -\frac{4}{3} \text{ and } x \geq 1$$

NOTE

Notice that the shaded areas are not the same in both cases, but the resulting inequalities are the same.

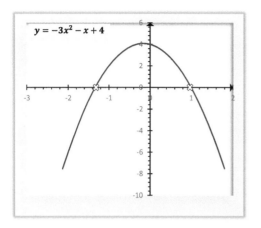
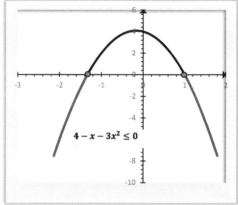

FIGURE 10.15   Solution to Example 7(b) – Method 2, Part I.

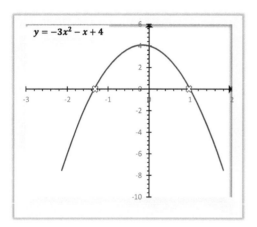
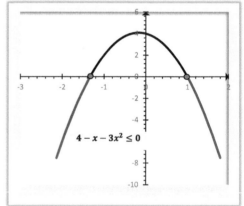

FIGURE 10.16   Solution to Example 7(a) – Method 2, Part II.

10.5.2.2 Square Functions Method

We discussed this rule in Section 10.3.6. We would like to use it here to solve quadratic inequalities, with a focus on those that cannot be factorised easily. However, the method can be applied to any quadratic inequality provided that the discriminant is not less than zero.

Let's try an example.

Example 8

Determine the range of values x which satisfy $3x^2 + 6x - 1 \geq 0$.

What did you get? Find the solution below to double-check your answer.

Solution to Example 8

Solution

Given $3x^2 + 6x - 1 \geq 0$, we need to arrange it in the square function as

$$3x^2 + 6x - 1 \geq 0$$

$$3x^2 + 6x \geq 1$$

$$x^2 + 2x \geq \frac{1}{3}$$

Let's complete the square of $x^2 + 2x$, we have

$$(x + 1)^2 - 1^2 \geq \frac{1}{3}$$

$$(x + 1)^2 - 1 \geq \frac{1}{3}$$

We now need to simplify as

$$(x + 1)^2 \geq 1 + \frac{1}{3}$$

$$(x + 1)^2 \geq \frac{4}{3}$$

This is now in the form $x^2 \geq b$, so we have

$$x + 1 \leq -\sqrt{\frac{4}{3}} \qquad\qquad x + 1 \geq \sqrt{\frac{4}{3}}$$

$$x + 1 \leq -\frac{2}{\sqrt{3}} \qquad \textbf{AND} \qquad x + 1 \geq \frac{2}{\sqrt{3}}$$

$$x \leq -\frac{2}{\sqrt{3}} - 1 \qquad\qquad x \geq \frac{2}{\sqrt{3}} - 1$$

$$\therefore x \leq -\frac{3 + 2\sqrt{3}}{3} \qquad\qquad \therefore x \geq \frac{-3 + 2\sqrt{3}}{3}$$

10.5.2.3 Author's Method

This method relies on the premise that a quadratic equation has two distinct real roots α and β such that $\alpha < \beta$. We therefore expect the solutions of the quadratic equation to have three clear regions as:

a) $x < \alpha$

b) $\alpha < x < \beta$

c) $x > \beta$

From the above, we can form just two solutions that will fit any inequality. The two are:

1) $x < \alpha$ and $x > \beta$

2) $\alpha < x < \beta$

α and β are sometimes referred to as critical values, as the intended solutions are separated by these values. The exceptions to this are $\leq$ and $\geq$ when the critical values are also included in the solution.

Like the graphical method, we will use a three-step approach to solve a quadratic inequality using the author's method.

Step 1: Arrange the inequality, with the expression $ax^2 + bx + c$ entirely on one side (usually the left-hand side) and 0 on the other side (usually the right-hand side).

Step 2: Solve and sketch $ax^2 + bx + c = 0$ to obtain the roots of the equation.

Step 3: Determine the solution by choosing from the two options in Table 10.10.

The summary of Table 10.10 is that option 1 should be used if a and y are of the same sign otherwise option 2 should be used. y in this case is the left-hand expression, i.e., when each of the four expressions in Figure 10.17 are placed on the LHS. We've only used $<$ and $>$, the other two inequalities, $\leq$ and $\geq$, can also be used.

That's all about the author's method, let's try a couple of examples.

Example 9

Solve the following inequalities.

a) $3 > 27x^2$ **b)** $2x^2 + 5x - 3 \geq 0$

TABLE 10.10
Author's Method Illustrated

Option	Solution	Condition	Example
1)	$x < \alpha$ and $x > \beta$	• $a > 0$ and $y > 0$	• $x^2 - 2x - 3 > 0$
		• $a < 0$ and $y < 0$	• $5 - 3x - 2x^2 < 0$
2)	$\alpha < x < \beta$	• $a > 0$ and $y < 0$	• $x^2 - 2x - 3 < 0$
		• $a < 0$ and $y > 0$	• $5 - 3x - 2x^2 > 0$

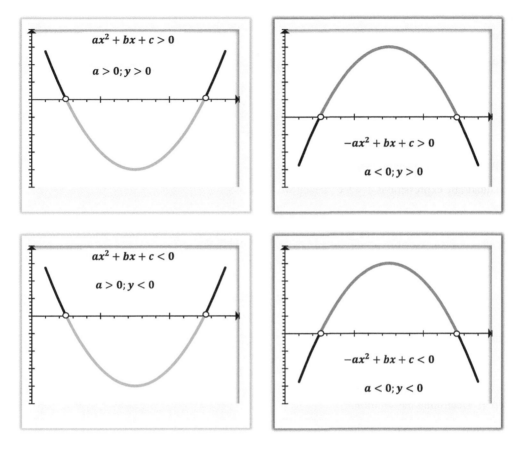

FIGURE 10.17 Author's method illustrated for: (a) $ax^2 + bx + c > 0$, (b) $ax^2 + bx + c < 0$, (c) $-ax^2 + bx + c > 0$, and (d) $-ax^2 + bx + c < 0$.

What did you get? Find the solution below to double-check your answer.

Solution to Example 9

HINT

- Step **2** is the main part of the solution. Use any suitable method to accomplish this.
- When a and y are both less than **0** (or negative) or both greater than **0** (or positive), the solution is within the range $x < \alpha$ and $x > \beta$. If one is less than zero (or negative) and the other is greater than **0** (or positive), then the solution is $\alpha < x < \beta$.

a) $3 > 27x^2$

Solution

Step 1: Re-arrange

$$3 > 27x^2$$
$$1 > 9x^2$$
$$1 - 9x^2 > 0$$

Step 2: Solve the quadratic equation.

The quadratic expression is $1 - 9x^2$, hence we solve as:

$$1 - 9x^2 = 0$$
$$(1 + 3x)(1 - 3x) = 0$$

Therefore

$$1 + 3x = 0 \qquad \mathbf{OR} \qquad 1 - 3x = 0$$
$$\therefore x = -\frac{1}{3} \qquad\qquad \therefore x = \frac{1}{3}$$

Thus, we have:

$$\alpha = -\frac{1}{3} \text{ and } \beta = \frac{1}{3}$$

Step 3: Determine the solution.

From above, $a < 0$ and $y > 0$, hence, the solution is:

$$-\frac{1}{3} < x < \frac{1}{3}$$

ALTERNATIVE METHOD

Alternatively, we could have re-arranged as:

$$3 > 27x^2$$
$$1 > 9x^2$$
$$0 > 9x^2 - 1$$

But the above is the same as:

$$9x^2 - 1 < 0$$

Steps 2 and 3 will be as in the first approach above, the solution is:

$$-\frac{1}{3} < x < \frac{1}{3}$$

b) $2x^2 + 5x - 3 \geq 0$

Solution

Step 1: Re-arrange

$$2x^2 + 5x - 3 \geq 0$$

No further arrangement is required.

Step 2: Solve the quadratic equation.

The quadratic expression is $2x^2 + 5x - 3$, hence we solve as:

$$2x^2 + 5x - 3 = 0$$

$\begin{Vmatrix} 1 & 3 \\ 2 & -1 \end{Vmatrix}$ See Chapter 6 for how to use this method to factorise quadratic expression especially when $a \neq 1$.

$$(x + 3)(2x - 1) = 0$$

Therefore

$$
\begin{array}{ccc}
x + 3 = 0 & & 2x - 1 = 0 \\
\therefore x = -3 & \textbf{OR} & \therefore x = \frac{1}{2}
\end{array}
$$

Thus, we have:

$$\alpha = -3 \text{ and } \beta = \frac{1}{2}$$

Step 3: Determine the solution.

From above, $a > 0$ and $y > 0$, hence, our solutions are:

$$\boldsymbol{x \leq -3 \text{ and } x \geq \frac{1}{2}}$$

Another set of examples to try.

Example 10

Determine the value(s) of k for which the following quadratic equations have real roots.

a) $x^2 - 3x + k + 2 = 0$ **b)** $kx^2 + (k + 2)x - \frac{1}{4} = 0$

What did you get? Find the solution below to double-check your answer.

Solution to Example 10

a) $x^2 - 3x + k + 2 = 0$
Solution

$$x^2 - 3x + k + 2 = 0$$
$$x^2 - 3x + (k + 2) = 0$$

Therefore

$$a = 1, \; b = -3, \; c = k + 2$$

Thus, to have two real roots (distinct or repeated), we must have

$$b^2 - 4ac \geq 0$$

This implies that

$$(-3)^2 - 4(1)(k + 2) \geq 0$$
$$9 - 4(k + 2) \geq 0$$
$$9 - 4k - 8 \geq 0$$
$$1 - 4k \geq 0$$
$$1 \geq 4k$$
$$\frac{1}{4} \geq k$$
$$\therefore k \leq \frac{1}{4}$$

---

b) $kx^2 + (k + 2)x - \frac{1}{4} = 0$
Solution
For this case,

$$a = k, \; b = k + 2, \; c = -\frac{1}{4}$$

Thus, to have two real roots (distinct or repeated), we must have

$$b^2 - 4ac \geq 0$$

This implies that

$$(k + 2)^2 - 4(k)\left(-\frac{1}{4}\right) \geq 0$$
$$k^2 + 4k + 4 + k \geq 0$$
$$k^2 + 5k + 4 \geq 0$$

Since this is a quadratic inequality, let's follow one of the methods discussed above.

Step 1: Re-arrange

$$k^2 + 5k + 4 \geq 0$$

No further arrangement is required.

Step 2: Solve

The quadratic expression is $k^2 + 5k + 4$, hence we solve as

$$k^2 + 5k + 4 = 0$$
$$(k + 1)(k + 4) = 0$$

Therefore

$$k + 4 = 0 \quad \textbf{OR} \quad k + 1 = 0$$
$$\therefore k = -4 \qquad\qquad \therefore k = -1$$

Thus, we have:

$$\alpha = -4 \quad \textbf{AND} \quad \beta = -1$$

Step 3: Determine the solution.

From above, $a > 0$ and $y > 0$, hence

$$k \leq -4 \quad \textbf{AND} \quad k \geq -1$$

The result implies that the quadratic equation $kx^2 + (k + 2)x - \frac{1}{4} = 0$ will have real roots only when $k \leq -4$ and $k \geq -1$.

10.5.2.4 Using Intervals and Table

This is our last method on solving quadratic inequalities and will be used to solve $x^2 + bx + c < 0$, but equally applicable when $y > 0$, $y \leq 0$ and $y \geq 0$ and where $a \neq 1$.

Follow these steps to solve $x^2 + bx + c < 0$ using this method.

Step 1: Arrange the inequality with the expression $ax^2 + bx + c$ entirely on one side (usually the LHS) and 0 on the other side (usually the RHS).
Step 2: Factorise the LHS of $x^2 + bx + c = 0$ as $(x - \alpha)(x - \beta) = 0$.
Step 3: Determine the roots α and β, such that $\alpha < \beta$.
Step 4: Create Table 10.11 with relevant information as shown.
Step 5: Choose appropriate values for α and β within the three ranges $x < \alpha$, $\alpha < x < \beta$ and $x > \beta$. Substitute these in $(x - \alpha)$ and $(x - \beta)$. If the result is less than 0 insert minus '$-$' otherwise add plus '$+$'. The solution is shown in Table 10.12 for R1 and R2.
Step 6: Multiply R1 with R2 and indicate the result in R3 as shown in Table 10.13.
Step 7: Determine the solution.

- When $y < 0$ or $y \leq 0$ choose the range with a negative sign (or signs). In this case, $\alpha < x < \beta$.
- When $y > 0$ or $y \geq 0$ choose the range with a positive sign (or signs). In this case, $x < \alpha$ and $x > \beta$.

TABLE 10.11
Solving Quadratic Inequalities Using Intervals Illustrated – Part I

		$x < \alpha$	$\alpha < x < \beta$	$x > \beta$
R1	$(x - \alpha)$			
R2	$(x - \beta)$			
R3	$(x - \alpha)(x - \beta)$			

TABLE 10.12
Solving Quadratic Inequalities Using Intervals Illustrated – Part II

		$x < \alpha$	$\alpha < x < \beta$	$x > \beta$
R1	$(x - \alpha)$	−	−	+
R2	$(x - \beta)$	−	+	+
R3	$(x - \alpha)(x - \beta)$			

TABLE 10.13
Solving Quadratic Inequalities Using Intervals Illustrated – Part III

		$x < \alpha$	$\alpha < x < \beta$	$x > \beta$
R1	$(x - \alpha)$	−	−	+
R2	$(x - \beta)$	−	+	+
R3	$(x - \alpha)(x - \beta)$	+	−	+

NOTE

Columns 3 and 5 (counting from the left) will always have the same sign in $R3$ while column 4 will always have a sign opposite to that of 3 and 5 in the same row $R3$.

That's all we need for this method. Let's work through a couple of examples.

Example 11

Solve the following quadratic inequalities using the intervals and table method.

a) $x^2 - 2x - 15 < 0$ **b)** $6x^2 - 13x + 5 \geq 0$

What did you get? Find the solution below to double-check your answer.

HINT

Once we are familiar with this method, some of the steps are no longer necessary or can at least be combined.

a) $x^2 - 2x - 15 < 0$
Solution

Step 1: Re-arrange if necessary

$$x^2 - 2x - 15 < 0$$

Step 2: Solve the quadratic function.

The quadratic expression is $x^2 - 2x - 15$, hence we solve as:

$$x^2 - 2x - 15 = 0$$
$$(x + 3)(x - 5) = 0$$

Therefore

$$\begin{array}{ccc} x + 3 = 0 & & x - 5 = 0 \\ \therefore x = -3 & \textbf{OR} & \therefore x = 5 \end{array}$$

Thus, we have:

$$\alpha = -3 \quad \textbf{AND} \quad \beta = 5$$

Step 3: Create the table (Table 10.14).
Step 4: We choose -4, 1, and 6 as values in the range $x < -3$, $-3 < x < 5$, and $x > 5$ respectively.
Step 5: Now complete the table with minus or plus (Table 10.15).
Step 6: Since $y < 0$, then our solution is in the range shown in the third column, counting from the left.

$$\therefore -3 < x < 5$$

b) $6x^2 - 13x + 5 \geq 0$
Solution

Step 1: Re-arrange if necessary

$$6x^2 - 13x + 5 \geq 0$$

Step 2: Solve the quadratic function. The quadratic expression is $6x^2 - 13x + 5$, hence we solve as:

$$6x^2 - 13x + 5 = 0$$

$\begin{Vmatrix} 2 & -1 \\ 3 & -5 \end{Vmatrix}$ See Chapter 6 for how to use this method to factorise quadratic expression, especially when $a \neq 1$.

$$(2x - 1)(3x - 5) = 0$$

TABLE 10.14
Solution to Example 11(a) – Part I

	$x < -3$	$-3 < x < 5$	$x > 5$
$(x+3)$			
$(x-5)$			
$(x+3)(x-5)$			

TABLE 10.15
Solution to Example 11(a) – Part II

	$x < -3$	$-3 < x < 5$	$x > 5$
$(x+3)$	–	+	+
$(x-5)$	–	–	+
$(x+3)(x-5)$	+	–	+

Therefore

$$2x - 1 = 0 \qquad \qquad 3x - 5 = 0$$
$$\therefore x = \frac{1}{2} \quad \text{OR} \quad \therefore x = \frac{5}{3}$$

Thus

$$\alpha = \frac{1}{2} \quad \text{AND} \quad \beta = \frac{5}{3}$$

Step 3: Create the table (Table 10.16).
Step 4: We choose 0, 1, and 2 as values in the range $x < \frac{1}{2}$, $\frac{1}{2} < x < \frac{5}{3}$, and $x > \frac{5}{3}$, respectively.
Step 5: Now complete the table with minus or plus (Table 10.17).
Step 6: Since $y \geq 0$, then our solution is in the range shown in the second and fourth columns.

Thus

$$x \leq \frac{1}{2} \quad \text{AND} \quad x \geq \frac{5}{3}$$

10.5.3 Simultaneous Inequalities

In this section, we will be looking at solving two inequalities to determine the range of values that satisfy both. We will consider two cases, namely:

TABLE 10.16
Solution to Example 11(b) – Part I

	$x \leq \dfrac{1}{2}$	$\dfrac{1}{2} \leq x \leq \dfrac{5}{3}$	$x \geq \dfrac{5}{3}$
$(2x - 1)$			
$(3x - 5)$			
$(2x - 1)(3x - 5)$			

TABLE 10.17
Solution to Example 11(b) – Part II

	$x \leq \dfrac{1}{2}$	$\dfrac{1}{2} \leq x \leq \dfrac{5}{3}$	$x \geq \dfrac{5}{3}$
$(2x - 1)$	–	+	+
$(3x - 5)$	–	–	+
$(2x - 1)(3x - 5)$	+	–	+

a) Two linear inequalities

b) A linear inequality and a quadratic inequality

10.5.3.1 Two Linear Inequalities

This takes the general form

$$ax + b < c \quad \text{and} \quad dx + e < f$$

where a, b, c, d, e, and f are real numbers.

The inequality sign < in the above can be replaced with >, ≤, or ≥. For this case, we will solve the two inequalities independently to have two separate solutions. This is different from the procedure we use to solve simultaneous linear equations.

The solutions produced may also be exclusive, i.e., they do not overlap. This means that there are no values of the unknown variable which satisfy both inequalities. This is similar to when we say that simultaneous equations are not consistent.

On the other hand, when they overlap or are inclusive, it means that there are values where both inequalities are true together. In this case, we can combine them together as one solution in one of these two ways.

a) $x < r$ where x is the unknown variable and r is a real number. For example, if the two solutions are $x < 3$ and $x < 2$, we can combine this to become $x < 2$. This is what satisfies both together.

b) $r < x < s$ where x is the unknown variable and r and s are real numbers. For example, if the two solutions are $x < 5$ and $x > -2$, we can combine this to become $-2 < x < 5$. This is what satisfies both together.

We observe that in (a) the direction of the solutions is the same while in (b) it is the opposite.

That's it for now. Let's try some examples.

Example 12

Determine the range of values x which satisfy the following pair of inequalities. Represent the solutions on a number line.

a) $2x - 4 < x + 1$ **and** $3 - 5x < 2x - 4$ $\quad$ **b)** $\frac{x-4}{2} \geq \frac{12-3x}{3}$ **and** $\frac{5-2x}{4} < 3(x+1)$

c) $7 - \frac{x}{2} < 8 - x$ **and** $3(1+x) \leq 5x - 3$ $\quad$ **d)** $2\left(x - \frac{1}{10}\right) \leq \frac{4}{5} + x$ **and** $5 - 2x \leq \frac{1}{2}(7 - x)$

What did you get? Find the solution below to double-check your answer.

Solution to Example 12

a) $2x - 4 < x + 1$ **and** $3 - 5x < 2x - 4$
Solution
Let's do this one at a time:

• $2x - 4 < x + 1$

$$2x - 4 < x + 1$$
$$2x - x < 4 + 1$$
$$x < 5$$

• $3 - 5x < 2x - 4$

$$3 - 5x < 2x - 4$$
$$3 + 4 < 2x + 5x$$
$$7 < 7x$$
$$1 < x$$

The two solutions are

$$x < 5 \text{ and } x > 1$$

Let's represent these on a number line (Figure 10.18).

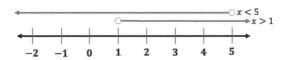

FIGURE 10.18 Solution to Example 12(a).

From the number line we can write the combined solution as:

$$1 < x < 5$$

---

b) $\frac{x-4}{2} \geq \frac{12-3x}{3}$ **and** $\frac{5-2x}{4} < 3(x+1)$

Solution

Let's do this one at a time

- $\frac{x-4}{2} \geq \frac{12-3x}{3}$

$$\frac{x-4}{2} \geq \frac{12-3x}{3}$$
$$3(x-4) \geq 2(12-3x)$$
$$3x - 12 \geq 24 - 6x$$
$$3x + 6x \geq 24 + 12$$
$$9x \geq 36$$
$$x \geq 4$$

- $\frac{5-2x}{4} < 3(x+1)$

$$\frac{5-2x}{4} < 3(x+1)$$
$$5 - 2x < 12(x+1)$$
$$5 - 2x < 12x + 12$$
$$5 - 12 < 12x + 2x$$
$$-7 < 14x$$
$$-\frac{1}{2} < x$$

The two solutions are

$$x \geq 4 \textbf{ AND } x > -\frac{1}{2}$$

Let's represent these on a number line (Figure 10.19). From the number line, we can write the combined solution as:

$$x \geq 4$$

FIGURE 10.19 Solution to Example 12(b).

c) $7 - \frac{x}{2} < 8 - x$ **and** $3(1 + x) \le 5x - 3$

Solution

Let's do this one at a time:

• $7 - \frac{x}{2} < 8 - x$

$$7 - \frac{x}{2} < 8 - x$$
$$x - \frac{x}{2} < 8 - 7$$
$$\frac{x}{2} < 1$$
$$\boldsymbol{x < 2}$$

• $3(1 + x) \le 5x - 3$

$$3(1 + x) \le 5x - 3$$
$$3 + 3x \le 5x - 3$$
$$3 + 3 \le 5x - 3x$$
$$6 \le 2x$$
$$3 \le x$$
$$\boldsymbol{x \ge 3}$$

The two solutions are

$$\boldsymbol{x < 2} \;\; \textbf{AND} \;\; \boldsymbol{x \ge 3}$$

Let's represent these on a number line (Figure 10.20). From the number line, we can see that no values of x for which both inequalities are true together. Therefore, there is no solution for these simultaneous inequalities.

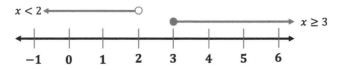

FIGURE 10.20 Solution to Example 12(c).

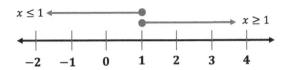

FIGURE 10.21   Solution to Example 12(d).

d) $2\left(x - \frac{1}{10}\right) \leq \frac{4}{5} + x$ **and** $5 - 2x \leq \frac{1}{2}(7 - x)$

Solution

Let's do this one at a time

• $2\left(x - \frac{1}{10}\right) \leq \frac{4}{5} + x$

$$2\left(x - \frac{1}{10}\right) \leq \frac{4}{5} + x$$

$$2x - \frac{1}{5} \leq \frac{4}{5} + x$$

$$2x - x \leq \frac{4}{5} + \frac{1}{5}$$

$$x \leq \frac{5}{5}$$

$$x \leq 1$$

• $5 - 2x \leq \frac{1}{2}(7 - x)$

$$5 - 2x \leq \frac{1}{2}(7 - x)$$

$$10 - 4x \leq 7 - x$$

$$10 - 7 \leq 4x - x$$

$$3 \leq 3x$$

$$1 \leq x$$

$$x \geq 1$$

The two solutions are

$$x \leq 1 \text{ AND } x \geq 1$$

Let's represent these on a number line (Figure 10.21). From the number line, we can see that the inequalities are true together when $x = 1$.

10.5.3.2 A Linear and a Quadratic Inequality

This case is not in any way new or special. All we need to do is to solve each of the inequalities separately and find the point of overlap of the solutions using any suitable method discussed earlier.

Let's try a couple of examples to illustrate this.

Determine the set of values x which satisfy the following pair of inequalities.

a) $x - 3 > 4x - 2$ **and** $7 - x \leq 2x - 5x^2$ **b)** $2\left(x - \frac{1}{6}\right) \leq 1 - \frac{2x}{3}$ **and** $x^2 - 5x + 4 < 0$

What did you get? Check the solution below.

a) $x - 3 > 4x - 2$ **and** $7 - x \leq 2x - 5x^2$
Solution

- **Linear**

$$x - 3 > 4x - 2$$
$$x - 4x > 3 - 2$$
$$-3x > 1$$
$$x < -\frac{1}{3}$$

- **Quadratic**

Step 1: Re-arrange

$$7 - x \leq 2x - 5x^2$$
$$5x^2 - x - 2x + 7 \leq 0$$
$$5x^2 - 3x + 7 \leq 0$$

Step 2: Solve the quadratic function.

The quadratic expression is $5x^2 - 3x + 7$, hence we solve as:

$$5x^2 - 3x + 7 = 0$$

But let's check if this has real roots. For this, we have:

$$a = 5,\ b = -3,\ c = 7$$

Therefore

$$b^2 - 4ac = (-3)^2 - 4\,(5)\,(7)$$
$$= 9 - 140 = -131 < 0$$

Thus, $5x^2 - 3x + 7 \le 0$ does not have real roots.

No solutions

What it means is that there are no values of x for which both inequalities are true together.

---

b) $2\left(x - \frac{1}{6}\right) \le 1 - \frac{2x}{3}$ **and** $x^2 - 5x + 4 < 0$

Solution

- **Linear**

$$2\left(x - \frac{1}{6}\right) \le 1 - \frac{2x}{3}$$
$$2x - \frac{2}{6} \le 1 - \frac{2x}{3}$$
$$2x - \frac{1}{3} \le 1 - \frac{2x}{3}$$
$$2x + \frac{2x}{3} \le \frac{1}{3} + 1$$

Multiply through by 3, we have:

$$6x + 2x \le 1 + 3$$
$$8x \le 4$$
$$x \le \frac{4}{8}$$
$$x \le 0.5$$

- **Quadratic**

Step 1: Re-arrange

$$x^2 - 5x + 4 < 0$$

No further arrangement is required.

Step 2: Solve the quadratic function

The quadratic expression is $x^2 - 5x + 4$, hence we solve as:

$$x^2 - 5x + 4 = 0$$
$$(x - 1)(x - 4) = 0$$

Therefore

$$x - 1 = 0$$
$$\therefore x = 1$$

OR

$$x - 4 = 0$$
$$\therefore x = 4$$

Thus, we have:

$$\alpha = 1 \qquad \text{AND} \qquad \beta = 4$$

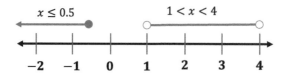

FIGURE 10.22 Solution to Example 13(b).

Step 3: Determine the solution

From above, $a > 0$ and $y < 0$, hence

$$1 < x < 4$$

The solutions do not overlap, hence there are no real values that satisfy the simultaneous inequalities.

10.6 MODULUS

The modulus (also known as absolute value) of a number k is its value without any sign (positive or negative) attached to it. In this case, -4 and $+4$ are the same or have the same absolute value of 4.

We use two vertical straight lines to denote modulus, so $|k|$ is read as '**modulus of** k' or '**absolute value of** k', as such $|4| = |-4| = 4$. In MATLAB, Excel, or similar software applications, '**abs**' is used. Thus, abs $(4) =$ abs $(-4) = 4$.

We can solve inequalities involving modulus using the rules in Table 10.17, where we consider x to be a variable and $a, b \in \mathbb{R}$. You may notice that the solutions look like what we have for square functions in Section 10.3.6.

Solutions using number lines to the examples given in Table 10.18 are shown in Figure 10.23. It is obvious from the figure that the reference point is zero or origin. This is because $|x|$ is the same as $|x + 0|$ and $|x| < 3$ represent all values that are less than 3 from zero.

TABLE 10.18
Rules of Inversion Illustrated

Inequality	Solution	Example
$\lvert x \rvert < a$	$-a < x < a$	$\lvert x \rvert < 3$ implies $-3 < x < 3$
$\lvert x \rvert > a$	$x > a$ and $x < -a$	$\lvert x \rvert > 4$ implies $x > 4$ and $x < -4$
$\lvert x \rvert \leq b$	$-b \leq x \leq b$	$\lvert x \rvert \leq 5$ implies $-5 \leq x \leq 5$
$\lvert x \rvert \geq b$	$x \geq b$ and $x \leq -b$	$\lvert x \rvert \geq 6$ implies $x \geq 6$ and $x \leq -6$

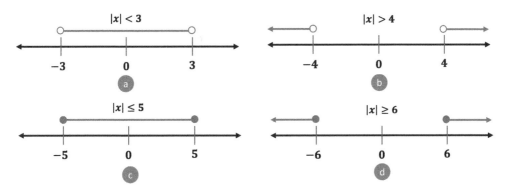

FIGURE 10.23 Inequalities involving modulus for the four inequality symbols illustrated: (a) $|x| < 3$, (b) $|x| > 4$, (c) $|x| \leq 5$, and (d) $|x| \geq 6$.

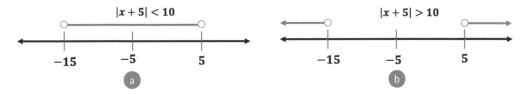

FIGURE 10.24 Inequalities involving modulus with reference other than the origin zero illustrated: (a) $|x + 5| < 10$ and (b) $|x + 5| > 10$.

On the other hand, $|x + 5| < 10$ implies that all values that are less than 10 are now measured from -5, which is $-15 < x < 5$ obtained by solving the compound inequality $-10 < x + 5 < 10$. Similarly, $|x + 5| > 10$ means that all values that are greater than 10 measured from -5, which implies is $x > 5$ and $x < -15$. These are shown in Figure 10.24.

That's all for modulus and let's finish this with a few examples.

Example 14

Solve the following inequalities.

a) $|3t - 5| < 2$ **b)** $|10 - 7t| \leq 5$ **c)** $|3t + 4| \geq 2$

What did you get? Find the solution below to double-check your answer.

Solution to Example 14

a) $|3t - 5| < 2$
Solution
Given $|3t - 5| < 2$, we have

$$-2 < 3t - 5 < 2$$

Add 5 to each part and simplify.

$$-2 + 5 < 3t - 5 + 5 < 2 + 5$$
$$3 < 3t < 7$$

Divide each part by 3 and simplify.

$$\frac{3}{3} < \frac{3t}{3} < \frac{7}{3}$$
$$1 < t < \frac{7}{3}$$

The solution for $|3t - 5| < 2$ is

$$\therefore 1 < t < \frac{7}{3}$$

NOTE

We can split $-2 < 3t - 5 < 2$ into two separate inequalities and simplify, as earlier shown in Example 6.

ALTERNATIVE METHOD

Given $|3t - 5| < 2$, we have

$$
\begin{array}{lcl}
3t - 5 < 2 & \text{and} & -(3t - 5) < 2 \\
3t < 2 + 5 & & 3t - 5 > -2 \\
3t < 7 & & 3t > -2 + 5 \\
t < \dfrac{7}{3} & & 3t > 3 \\
& & t > 1
\end{array}
$$

that is

$$1 < t$$

We can combine the two inequalities

$$\therefore 1 < t < \frac{7}{3}$$

The above satisfies the original inequality $|3t - 5| < 2$

---

b) $|10 - 7t| \leq 5$

Solution

We will be quick with this, as the only difference is the sign. Given $|10 - 7t| \leq 5$, we have

$$-5 \leq 10 - 7t \leq 5$$

Subtract 10 from each part and simplify.

$$-15 \leq -7t \leq -5$$

Divide each part by -7 and simplify. Note that the inequality sign will be reversed.

$$-\frac{15}{-7} \geq -\frac{7t}{-7} \geq -\frac{5}{-7}$$

$$\frac{15}{7} \geq t \geq \frac{5}{7}$$

This can be written as

$$\frac{5}{7} \leq t \leq \frac{15}{7}$$

The solution for $|10 - 7t| \leq 5$ is

$$\therefore \frac{5}{7} \leq t \leq \frac{15}{7}$$

Let's try another method to confirm this.

ALTERNATIVE METHOD

Given $|10 - 7t| \leq 5$, we have

$$
\begin{array}{lll}
10 - 7t \leq 5 & \text{and} & -(10 - 7t) \leq 5 \\
10 - 5 \leq 7t & & -10 + 7t \leq 5 \\
5 \leq 7t & & 7t \leq 10 + 5 \\
\frac{5}{7} \leq t & & 7t \leq 15 \\
& & t \leq \frac{15}{7}
\end{array}
$$

We can combine the two inequalities

$$\therefore \frac{5}{7} \leq t \leq \frac{15}{7}$$

---

c) $|3t + 4| \geq 2$

Solution

Given $|3t + 4| \geq 2$, we have

$$
\begin{array}{lll}
3t + 4 \leq -2 & \text{and} & 3t + 4 \geq 2 \\
3t \leq -6 & & 3t \geq -2 \\
t \leq -2 & & t \geq -\frac{2}{3}
\end{array}
$$

The solution for $|3t + 4| \geq 2$ is

$$\therefore t \leq -2 \quad \textbf{AND} \quad t \geq -\frac{2}{3}$$

10.7 CHAPTER SUMMARY

1) An inequality (pl. inequalities) is more evident in our day-to-day affairs now than ever before. Like its sister, i.e., equality, an inequality also plays a crucial role in mathematics.

2) The < and > symbols are a pair and so are ≤ and ≥ symbols. However, < and ≥ are complementary just like > complements ≤. What it means is that in any given set of numbers, if an inequality sign is used to enlist some members, then its complementary must be used to enlist the remaining.

3) Fundamental rules of inequalities include:

   - **Addition and subtraction**: The inequality sign will not change by adding or subtracting a constant positive value c (i.e., $c > 0$) from both sides of the inequality.

   - **Multiplication and division**: The inequality remains unchanged if both sides are multiplied by or divided by the same positive value c, i.e., $c > 0$. However, if the constant multiplier or divisor is negative (i.e., $c < 0$), then the inequality sign (sense or direction) is reversed.

   - **Power**: When both sides of the inequality are raised to the same positive value m (i.e., $m > 0$), the inequality remains valid. However, if the power is negative (i.e., $m < 0$), then the inequality sign is reversed.

   - **Inversion**: When the expressions are laterally swapped, the inequality sign is equally swapped. In other words, if the entire LHS expression (along with their respective signs) is moved to the RHS (and vice versa), then the equality sign is reversed, i.e., 'greater than' sign becomes 'less than' sign and vice versa.

   - **Variable**: An unknown variable, such as x, y, and a, should not be used to multiply or divide both sides of an inequality. This is because the numerical value of the variable can be positive or negative, and each has a different effect on an inequality sign as shown in the previous rules.

4) A number line is like the x-axis in an $x-y$ Cartesian plane, i.e., a horizontal line demarcated with the centre zero or reference point. Positive integers are evenly placed to the right of the origin and corresponding negative integers are placed to the left of the origin.

5) Solving inequalities generally follow similar procedures as its equivalent equations. In other words, a linear inequality will be solved in a similar way as we would solve a linear equation; the same applies to a quadratic inequality. In summary, two key tips to take away:

   - Treat an inequality as its equivalent equation when using a positive constant. Do the same with a negative constant but reverse the sense of the inequality.

   - Do not multiply or divide with a variable.

6) A linear inequality can take any of the following general forms: $ax + b < c$, $ax + b > c$, $ax + b \leq c$, or $ax + b \geq c$ where a, b, and c are real numbers.

7) A quadratic inequality in one variable can take any of the following general forms: $ax^2 + bx + c < 0$, $ax^2 + bx + c > 0$, $ax^2 + bx + c \leq 0$, or $ax^2 + bx + c \geq 0$.

8) Four common methods to solve quadratic inequalities are discussed in this chapter: the graphical method, the author's method, the interval method, and the table method.

****

10.8 FURTHER PRACTICE

To access complementary contents, including additional exercises, please go to www.dszak.com.

11 Indices

Learning Outcomes

Once you have studied the content of this chapter, you should be able to:

- Explain the term indices
- State the laws of indices
- Apply the laws of indices in simplifying expressions

11.1 INTRODUCTION

Indices (sing. index) together with logarithms are central to many scientific and engineering processes and are regularly used in simplifying expressions and solving equations. They are used in our daily activities including finance, economics and natural phenomena, and will be the focus of this chapter.

11.2 WHAT ARE INDICES?

Indices can be described as a number expressed in the form x^y where x is called the base and y is the index. The index is also called power or exponent; they are essentially the same. However, you probably have noticed that one of the terms (index, exponent, or power) is more widely used than the others.

In general, power is the most frequently used as such, x^y is read as 'x **raised to the power of y**', 'x **raised to the power y**' or simply 'x **to the power y**'. What does this imply? It means that x has been multiplied by itself y times inclusive. For example, if $x = 5$ and $y = 3$, then this can be written as 5^3 and read as '**five to the power three**' or '**five cubed**'. This implies $5 \times 5 \times 5$, which equals **125**.

So, what do you think about $10 \times 10 \times 10 \times 10$? It is **10 000** in value but written in index form as 10^4. Now let us work in the reverse order and find out what 10^9 is. This can be written as multiples of tens, i.e., $10 \times 10 \times 10 \times 10 \times 10 \times 10 \times 10 \times 10 \times 10$ and it is the same as **1 000 000 000** (a billion).

From this example, it is obvious that we need indices to express large numbers not only because it saves space and time but more importantly it is sometimes easy to remember and work with. So 10^9 is a better way of writing a cumbersome **1 000 000 000**. Calculation involving indices can be carried out on a scientific calculator, commonly via a button marked x^y or its equivalent key.

The next question that might come to mind is whether any number can be written in index form. The answer is yes. Any rational number can be expressed in this form just like any integer can be written in the form $\frac{x}{y}$. For instance, **10** can be expressed in both index and fractional forms as 10^1 and $\frac{10}{1}$ respectively, although this **1** is generally omitted. This is the same for any other number.

DOI:10.1201/9781003027928-11

11.3 LAWS OF INDICES

There are certain laws that govern these operations, which will be discussed and illustrated with examples. Note that the fundamental and derived (special) laws categories are simply for reference here.

11.3.1 FUNDAMENTAL LAWS

These can be regarded as the primary laws, and they are three in number.

11.3.1.1 Multiplication Law

It states that:

$$x^n \times x^m = x^{n+m}$$

(11.1)

where n and m are any real number.

For example, $x^3 \times x^2 = x^{3+2} = x^5$. Alternatively, the same can be expressed as:

$$x^3 \times x^2 = (x \times x \times x) \times (x \times x)$$
$$= x \times x \times x \times x \times x$$
$$= x^5$$

The following should be noted about this law:

Note 1 The terms must have the same base, otherwise the law cannot be used. For example, $5^3 \times 6^2 \neq 5^{3+2} \neq 6^{3+2}$, because 5 and 6 are not like bases.

Note 2 The sign between the terms must be a multiplication and not an addition. In other words, $2^3 \times 2^2 = 2^{3+2} = 2^5$ but $2^3 + 2^2 \neq 2^{3+2}$. This can easily be proven because $2^5 = 2 \times 2 \times 2 \times 2 \times 2 = 32$ and $2^3 + 2^2 = 8 + 4 = 12$.

Note 3 The power can either be the same or different. Also, the power can be any real number, positive or negative. This is illustrated in the worked examples.

Note 4 The base is any real number excluding zero, i.e., $x \in \mathbb{R} - \{0\}$.

Note 5 The terms can be two or more.

Let's try some examples.

Example 1

Without using a calculator, evaluate the following using the multiplication law.

a) $10^2 \times 10^3$

b) $a^2 \times a^3$

c) $7^{\frac{1}{2}} \times 7^{\frac{3}{2}}$

d) $2^6 \times 2^{-3}$

e) $5^{3\frac{1}{3}} \times 5^{-\frac{1}{3}}$

f) $3c^7 \times 5c^0 \times 2c^{-6}$

What did you get? Find the solution below to double-check your answer.

Solution to Example 1

HINT

All you need is the law of indices shown below

$$x^n \times x^m = x^{n+m}$$

a) $10^2 \times 10^3$
Solution

$$10^2 \times 10^3 = 10^{2+3}$$
$$= 10^5 = 100\,000$$
$$\therefore \mathbf{10^2 \times 10^3 = 100\,000}$$

---

b) $a^2 \times a^3$
Solution

$$a^2 \times a^3 = a^{2+3} = a^5$$
$$\therefore \boldsymbol{a^2 \times a^3 = a^5}$$

---

c) $7^{\frac{1}{2}} \times 7^{\frac{3}{2}}$
Solution

$$7^{\frac{1}{2}} \times 7^{\frac{3}{2}} = 7^{\left(\frac{1}{2}+\frac{3}{2}\right)}$$
$$= 7^2 = 49$$
$$\therefore \boldsymbol{7^{\frac{1}{2}} \times 7^{\frac{3}{2}} = 49}$$

---

d) $2^6 \times 2^{-3}$
Solution

$$2^6 \times 2^{-3} = 2^{6+(-3)}$$
$$= 2^3 = 8$$
$$\therefore \boldsymbol{2^6 \times 2^{-3} = 8}$$

---

e) $5^{3\frac{1}{3}} \times 5^{-\frac{1}{3}}$
Solution

$$5^{3\frac{1}{3}} \times 5^{-\frac{1}{3}} = 5^{3\frac{1}{3}+\left(-\frac{1}{3}\right)}$$
$$= 5^3$$
$$= 125$$
$$\therefore \boldsymbol{5^{3\frac{1}{3}} \times 5^{-\frac{1}{3}} = 125}$$

f) $3c^7 \times 5c^0 \times 2c^{-6}$

Solution

$$3c^7 \times 5c^0 \times 2c^{-6} = 3 \times c^7 \times 5 \times c^0 \times 2 \times c^{-6}$$
$$= (3 \times 5 \times 2) \times \left(c^7 \times c^0 \times c^{-6}\right)$$
$$= 30\left(c^{7+0-6}\right)$$
$$= 30c$$
$$\therefore 3c^7 \times 5c^0 \times 2c^{-6} = 30c$$

Good stuff! Let's try another set of examples.

Example 2

Simplify the following:

a) $\left(\theta^2 - \phi^2\right)^{-\frac{1}{3}} \times (\theta - \phi)^{\frac{4}{3}} \times (\theta + \phi)^{\frac{1}{3}}$

b) $\left(x^9 y^{\frac{1}{3}} z^{\frac{1}{2}}\right)^{\frac{1}{3}} \times y^{\frac{8}{9}} \times \left(3^6 x^6 y^2 z^{\frac{1}{3}}\right)^{-\frac{1}{2}}$

c) $\left(\omega^2 - 1\right)^3 \times (\omega + 1)^{\frac{1}{3}} \div (\omega - 1)^{-\frac{1}{3}}$

What did you get? Find the solution below to double-check your answer.

Solution to Example 2

HINT

Again, we will be using the multiplication law. These questions are a bit advanced, but a good grasp of the law will make these easy to handle.

a) $\left(\theta^2 - \phi^2\right)^{-\frac{1}{3}} \times (\theta - \phi)^{\frac{4}{3}} \times (\theta + \phi)^{\frac{1}{3}}$

Solution

We will employ the difference of two squares (Section 3.5.2), thus:

$$\left(\theta^2 - \phi^2\right)^{-\frac{1}{3}} \times (\theta - \phi)^{\frac{4}{3}} \times (\theta + \phi)^{\frac{1}{3}} = [(\theta - \phi)(\theta + \phi)]^{-\frac{1}{3}} \times (\theta - \phi)^{\frac{4}{3}} \times (\theta + \phi)^{\frac{1}{3}}$$

Now we can carry on with the application of the multiplication law.

$$(\theta - \phi)^{-\frac{1}{3}} \times (\theta + \phi)^{-\frac{1}{3}} \times (\theta - \phi)^{\frac{4}{3}} \times (\theta + \phi)^{\frac{1}{3}} = (\theta - \phi)^{-\frac{1}{3}+\frac{4}{3}} \times (\theta + \phi)^{-\frac{1}{3}+\frac{1}{3}}$$
$$= (\theta - \phi)^1 \times (\theta + \phi)^0 = \theta - \phi$$
$$\therefore \left(\theta^2 - \phi^2\right)^{-\frac{1}{3}} \times (\theta - \phi)^{\frac{4}{3}} \times (\theta + \phi)^{\frac{1}{3}} = \theta - \phi$$

b) $\left(x^9 y^{\frac{1}{3}} z^{\frac{1}{2}}\right)^{\frac{1}{3}} \times y^{\frac{8}{9}} \times \left(3^6 x^6 y^2 z^{\frac{1}{3}}\right)^{-\frac{1}{2}}$

Solution

$$\left(x^9 y^{\frac{1}{3}} z^{\frac{1}{2}}\right)^{\frac{1}{3}} \times y^{\frac{8}{9}} \times \left(3^6 x^6 y^2 z^{\frac{1}{3}}\right)^{-\frac{1}{2}} = x^3 y^{\frac{1}{9}} z^{\frac{1}{6}} \times y^{\frac{8}{9}} \times 3^{-3} x^{-3} y^{-1} z^{-\frac{1}{6}}$$

$$= 3^{-3} \times x^{(3-3)} y^{\left(\frac{1}{9}+\frac{8}{9}-1\right)} z^{\left(\frac{1}{6}-\frac{1}{6}\right)} = 3^{-3} x^0 y^0 z^0$$

$$= 3^{-3} = \frac{1}{3^3} = \frac{1}{27}$$

$$\therefore \left(x^9 y^{\frac{1}{3}} z^{\frac{1}{2}}\right)^{\frac{1}{3}} \times y^{\frac{8}{9}} \times \left(3^6 x^6 y^2 z^{\frac{1}{3}}\right)^{-\frac{1}{2}} = \frac{1}{27}$$

---

c) $\left(\omega^2 - 1\right)^3 \times (\omega + 1)^{\frac{1}{3}} \div (\omega - 1)^{-\frac{1}{3}}$

Solution

$$\left(\omega^2 - 1\right)^3 \times (\omega + 1)^{\frac{1}{3}} \div (\omega - 1)^{-\frac{1}{3}} = [(\omega - 1)(\omega + 1)]^3 \times (\omega + 1)^{\frac{1}{3}} \times (\omega - 1)^{\frac{1}{3}}$$

$$= (\omega - 1)^3 \times (\omega + 1)^3 \times (\omega + 1)^{\frac{1}{3}} \times (\omega - 1)^{\frac{1}{3}}$$

$$= (\omega - 1)^{3+\frac{1}{3}} \times (\omega + 1)^{3+\frac{1}{3}}$$

$$= (\omega - 1)^{\frac{10}{3}} \times (\omega + 1)^{\frac{10}{3}}$$

$$= [(\omega - 1)(\omega + 1)]^{\frac{10}{3}}$$

$$= (\omega^2 - 1)^{\frac{10}{3}}$$

$$\therefore (\omega^2 - 1)^3 \times (\omega + 1)^{\frac{1}{3}} \div (\omega - 1)^{-\frac{1}{3}} = (\omega^2 - 1)^{\frac{10}{3}}$$

11.3.1.2 Quotient (or Division) Law

It states that:

$$\boxed{x^n \div x^m = x^{n-m}} \qquad \textbf{OR} \qquad \boxed{\frac{x^n}{x^m} = x^{n-m}} \tag{11.2}$$

where n and m are any real number.

For example, $x^7 \div x^4 = x^{7-4} = x^3$. Again, this can alternatively be expressed as:

$$\frac{x^7}{x^4} = \frac{x \times x \times x \times x \times x \times x \times x}{x \times x \times x \times x}$$

$$= x \times x \times x$$

$$= x^3$$

The following should be noted about this law:

Note 1 The terms must have the same base otherwise the law cannot be used. For example, $6^5 \div 5^3 \neq 6^{5-3} \neq 5^{5-3}$.

Note 2 The sign between the terms must be a division and not a subtraction. In other words, $2^5 \div 2^3 = 2^{5-3} = 2^2$ but $2^5 - 2^3 \neq 2^{5-3}$. This is because $2^5 \div 2^3 = 32 \div 8 = 4$ and $2^5 - 2^3 = 32 - 8 = 24$.

Note 3 The power can either be the same or different. Also, the power can be any real number, positive or negative. This is illustrated in the worked examples.

Note 4 The base is any real number excluding zero, i.e., $x \in \mathbb{R} - \{0\}$.

Note 5 The terms can be two or more.

Let's illustrate this law with examples.

Example 3

Without using a calculator, evaluate the following using the division law.

a) $6^5 \div 6^3$ **b)** $11^{\frac{7}{2}} \div 11^{\frac{3}{2}}$ **c)** $e^8 \div e^5$

d) $f^{-3} \div f^{-5}$ **e)** $13^{\frac{1}{5}} \div 13^{-\frac{4}{5}}$ **f)** $32h^{11} \div 8h^4$

What did you get? Find the solution below to double-check your answer.

Solution to Example 3

HINT

In this example, we will be applying the division law of indices exclusively.

$$x^m \div x^n = x^{m-n}$$

a) $6^5 \div 6^3$
Solution

$$6^5 \div 6^3 = 6^{5-3}$$
$$= 6^2 = 36$$
$$\therefore 6^5 \div 6^3 = 36$$

---

b) $11^{\frac{7}{2}} \div 11^{\frac{3}{2}}$
Solution

$$11^{\frac{7}{2}} \div 11^{\frac{3}{2}} = 11^{\left(\frac{7-3}{2}\right)}$$
$$= 11^2 = 121$$
$$\therefore 11^{\frac{7}{2}} \div 11^{\frac{3}{2}} = 121$$

c) $e^8 \div e^5$

Solution

$$e^8 \div e^5 = e^{8-5} = e^3$$
$$\therefore e^8 \div e^5 = e^3$$

ALTERNATIVE METHOD

Remember that

$$e^8 \div e^5 = \frac{e^8}{e^5}$$
$$= \frac{e \times e \times e \times e \times e \times e \times e \times e}{e \times e \times e \times e \times e}$$
$$= e \times e \times e = e^3$$
$$\therefore e^8 \div e^5 = e^3$$

d) $f^{-3} \div f^{-5}$

Solution

$$f^{-3} \div f^{-5} = f^{-3-(-5)}$$
$$= f^{-3+5} = f^2$$
$$\therefore f^{-3} \div f^{-5} = f^2$$

e) $13^{\frac{1}{5}} \div 13^{-\frac{4}{5}}$

Solution

$$13^{\frac{1}{5}} \div 13^{-\frac{4}{5}} = 13^{\frac{1}{5}-\left(-\frac{4}{5}\right)}$$
$$= 13^{\left(\frac{1}{5}+\frac{4}{5}\right)} = 13^{\left(\frac{1+4}{5}\right)}$$
$$= 13^1 = 13$$
$$\therefore 13^{\frac{1}{5}} \div 13^{-\frac{4}{5}} = 13$$

f) $32h^{11} \div 8h^4$

Solution

$$32h^{11} \div 8h^4 = \left(32 \times h^{11}\right) \div \left(8 \times h^4\right)$$
$$= \left(32 \div 8\right)\left(h^{11} \div h^4\right)$$
$$= \left(4\right)\left(h^{11-4}\right) = 4\text{h}^7$$
$$\therefore 32h^{11} \div 8h^4 = 4h^7$$

ALTERNATIVE METHOD

Remember that

$$32h^{11} \div 8h^4 = \frac{32h^{11}}{8h^4} = \frac{32}{8}\left(h^{11-4}\right) = 4\text{h}^7$$
$$\therefore 32h^{11} \div 8h^4 = 4h^7$$

11.3.1.3 Power Law

This law states that:

$$\boxed{\left(x^m\right)^n = x^{mn}}$$ (11.3)

where n and m are any real number.

For example, $\left(x^4\right)^2 = x^{4 \times 2} = x^8$. Alternatively, this can be shown using the multiplication law as:

$$\left(x^4\right)^2 = \left(x \times x \times x \times x\right)^2$$
$$= \left[x \times x \times x \times x\right] \times \left[x \times x \times x \times x\right]$$
$$= x \times x \times x \times x \times x \times x \times x \times x$$
$$= x^8$$

This law is self-explanatory and the notes that we've made about the previous two laws are also applicable to this law.

Now let's try some examples to illustrate.

Example 4

Without using a calculator, evaluate the following using the power law.

a) $\left(10^2\right)^3$ b) $\left(3^{-2}\right)^{-2}$ c) $\left(k^3\right)^7$

d) $\left(11^4\right)^{\frac{1}{2}}$ e) $\left(-m^5\right)^4$ f) $\left(27r^6s^{12}\right)^{\frac{1}{3}}$

What did you get? Find the solution below to double-check your answer.

Solution to Example 4

a) $\left(10^2\right)^3$

Solution

$$\left(10^2\right)^3 = 10^{2\times3}$$
$$= 10^6 = 1\,000\,000$$
$$\therefore \left(10^2\right)^3 = 1\,000\,000$$

b) $\left(3^{-2}\right)^{-2}$

Solution

$$\left(3^{-2}\right)^{-2} = 3^{-2\times-2}$$
$$= 3^4 = 81$$
$$\therefore \left(3^{-2}\right)^{-2} = 81$$

c) $\left(k^3\right)^7$

Solution

$$\left(k^3\right)^7 = k^{3\times7} = k^{21}$$
$$\therefore \left(k^3\right)^7 = k^{21}$$

d) $\left(11^4\right)^{\frac{1}{2}}$

Solution

$$\left(11^4\right)^{\frac{1}{2}} = 11^{\left(4\times\frac{1}{2}\right)}$$
$$= 11^2 = 121$$
$$\therefore \left(11^4\right)^{\frac{1}{2}} = 121$$

e) $\left(-m^5\right)^4$

Solution

$$\left(-m^5\right)^4 = \left(-1 \times m^5\right)^4$$
$$= (-1)^4 \times m^{5\times4}$$
$$= 1 \times m^{20} = m^{20}$$
$$\therefore \left(-m^5\right)^4 = m^{20}$$

f) $\left(27r^6s^{12}\right)^{\frac{1}{3}}$

Solution

$$\left(27r^6s^{12}\right)^{\frac{1}{3}} = \left(3^3r^6s^{12}\right)^{\frac{1}{3}}$$
$$= \left(3^{3\times\frac{1}{3}}\right)\left(r^{6\times\frac{1}{3}}\right)\left(s^{12\times\frac{1}{3}}\right)$$
$$= 3r^2s^4$$
$$\therefore \left(27r^6s^{12}\right)^{\frac{1}{3}} = 3r^2s^4$$

Let's try another set of examples.

Example 5

Simplify the following:

a) $3a^{-1} \times (2a)^2$ 　　　　　　　　　　　**b)** $\left(2p^4q^2\right)^3 \times \left(3pq^2\right)^2$

What did you get? Find the solution below to double-check your answer.

Solution to Example 5

HINT

In this example, we will be applying the laws below exclusively.

$$x^m \times x^n = x^{m+n} \quad \textbf{AND} \quad (x^m)^n = x^{mn}$$

a) $3a^{-1} \times (2a)^2$

Solution

$$3a^{-1} \times (2a)^2 = 3 \times a^{-1} \times 2^2 \times a^2$$
$$= 3 \times 2^2 \times a^{-1} \times a^2$$
$$= 3 \times 4 \times a^{-1+2} = 12a$$
$$\therefore \boldsymbol{3a^{-1} \times (2a)^2 = 12a}$$

b) $\left(2p^4q^2\right)^3 \times \left(3pq^2\right)^2$

Solution

$$\left(2p^4q^2\right)^3 \times \left(3pq^2\right)^2 = \left(2^3p^{4\times3}q^{2\times3}\right) \times \left(3^2p^2q^{2\times2}\right)$$
$$= 8p^{12}q^6 \times 9p^2q^4$$
$$= 8 \times 9 \times p^{12+2}q^{6+4}$$
$$= 72 \times p^{14}q^{10} = 72p^{14}q^{10}$$
$$\therefore \left(2p^4q^2\right)^3 \times \left(3pq^2\right)^2 = 72p^{14}q^{10}$$

11.3.2 DERIVED OR SPECIAL LAWS

11.3.2.1 Zero Power Law

This law can be written as:

$$\boxed{x^0 = 1} \;\; x \neq 0 \qquad\qquad (11.4)$$

It can be said that '**anything**', excluding zero, to the power of zero is 1. Although this rule might look odd, it is logical and well established. This can be proved as follows.

Given that

$$\boxed{x^n \div x^m = x^{n-m}}$$

If $m = n$, then, we have:

$$x^m \div x^m = x^{m-m} = x^0 \quad ------(\text{i})$$

Similarly,

$$x^m \div x^m = 1 \quad ------(\text{ii})$$

Equating the RHS of Equations (i) and (ii)

$$\boxed{\therefore x^0 = 1}$$

Let's try some examples to illustrate this law.

Example 6

Without using a calculator, evaluate the following using the zero-power law.

a) 5^0 **b)** 0.003^0 **c)** $\left(\dfrac{1}{234}\right)^0$ **d)** $\left(-\dfrac{3}{29}\right)^0$

What did you get? Find the solution below to double-check your answer.

Solution to Example 6

a) 5^0

Solution

$$5^0 = 1$$

b) 0.003^0

Solution

$$0.003^0 = 1$$

c) $\left(\frac{1}{234}\right)^0$

Solution

$$\left(\frac{1}{234}\right)^0 = 1$$

d) $\left(-\frac{3}{29}\right)^0$

Solution

$$\left(-\frac{3}{29}\right)^0 = 1$$

Let's try another example.

Example 7

Without using a calculator, evaluate $6^{d-3} \times 6^{3-d}$.

What did you get? Find the solution below to double-check your answer.

Solution to Example 7

Solution

In this example, we will be applying two laws of indices exclusively, namely: $x^m \times x^n = x^{m+n}$ and $x^0 = 1$. We therefore have:

$$6^{d-3} \times 6^{3-d} = 6^{(d-3+3-d)}$$
$$= 6^0 = 1$$
$$\therefore 6^{d-3} \times 6^{3-d} = 1$$

11.3.2.2 Negative Power Law

It states that:

$$\boxed{x^{-n} = \frac{1}{x^n}} \text{ OR } \boxed{x^n = \frac{1}{x^{-n}}}$$ (11.5)

where n is any real number.

Provided that $x \neq 0$, it implies that a number to the power of a negative index is the reciprocal of that number to the same base but with a positive index.

Sometimes, one might need to carry out some operations that require a change from a positive index to a negative index, let us account for this and re-state the rule as 'whenever the reciprocal of a number with an index is computed, the sign of the index changes'. With this more exclusive definition, we can illustrate with the following three examples.

$$\textbf{(a) } x^{-3} = \tfrac{1}{x^3} \qquad \textbf{(b) } \tfrac{1}{x^3} = x^{-3} \qquad \textbf{(c) } \tfrac{1}{x^{-3}} = x^3$$

In each of the above three cases, we have computed the reciprocal of the LHS to obtain the corresponding RHS. So, in (a) and (c), the sign of the index changes from negative to positive whilst in (b) it changes from positive three $(+3)$ to negative three (-3).

Let's try some examples.

Example 8

Without using a calculator, evaluate the following using the negative power law.

a) 5^{-3} b) $(-7)^{-1}$ c) $\left(\tfrac{1}{5}\right)^{-1}$ d) $7v^{-\frac{1}{7}}$

e) $(4z)^{-3}$ f) $\tfrac{1}{9^{-2}}$ g) $\left(\tfrac{15}{13}\right)^{-1}$ h) $\tfrac{7^{-1}}{4^{-1}}$

What did you get? Find the solution below to double-check your answer.

Solution to Example 8

a) 5^{-3}
Solution

$$5^{-3} = \frac{1}{5^3} = \frac{1}{125}$$

$$\therefore 5^{-3} = \frac{1}{125}$$

---

b) $(-7)^{-1}$

Solution

$$(-7)^{-1} = \frac{1}{(-7)^1} = -\frac{1}{7}$$

$$\therefore (-7)^{-1} = -\frac{1}{7}$$

c) $\left(\frac{1}{5}\right)^{-1}$

Solution

$$\left(\frac{1}{5}\right)^{-1} = \left(\frac{5}{1}\right)^1 = 5^1 = 5$$

$$\therefore \left(\frac{1}{5}\right)^{-1} = 5$$

d) $7v^{-\frac{1}{7}}$

Solution

$$7v^{-\frac{1}{7}} = \frac{7}{v^{\frac{1}{7}}}$$

$$\therefore 7v^{-\frac{1}{7}} = \frac{7}{v^{\frac{1}{7}}}$$

e) $(4z)^{-3}$

Solution

$$(4z)^{-3} = \left(\frac{1}{4z}\right)^3$$

$$= \frac{1}{4^3 z^3} = \frac{1}{64z^3}$$

$$\therefore (4z)^{-3} = \frac{1}{64z^3}$$

f) $\frac{1}{9^{-2}}$

Solution

$$\frac{1}{9^{-2}} = \frac{1}{\frac{1}{9^2}}$$

$$= \frac{1}{1} \times \frac{9^2}{1} = 9^2 = 81$$

$$\therefore \frac{1}{9^{-2}} = 81$$

ALTERNATIVE METHOD

We can simplify use the fact that

$$\frac{1}{x^{-m}} = x^m$$

Thus

$$\frac{1}{9^{-2}} = 9^2$$
$$= 81$$

---

g) $\left(\frac{15}{13}\right)^{-1}$

Solution

$$\left(\frac{15}{13}\right)^{-1} = \frac{13}{15}$$

---

h) $\frac{7^{-1}}{4^{-1}}$

Solution

$$\frac{7^{-1}}{4^{-1}} = \frac{4^1}{7^1} = \frac{4}{7}$$
$$\therefore \frac{7^{-1}}{4^{-1}} = \frac{4}{7}$$

In the examples to follow, in some cases we've stated the relevant laws that are applicable. It's however possible that other laws or combinations can be used. In other instances, there is no mention of the relevant laws, this is deliberate to provide flexibility of approach.

Example 9

Without using a calculator, evaluate the following giving the answer in the simplest form.

a) $7b \times 3b^{-2}$ **b)** $5c^{\frac{3}{2}} \times 2c^{-\frac{7}{2}}$ **c)** $11d^3e \times 4d^2e^{-3}$ **d)** $(-64)^{-\frac{4}{3}}$

What did you get? Find the solution below to double-check your answer.

Solution to Example 9

a) $7b \times 3b^{-2}$

Solution

$$
\begin{aligned}
7b \times 3b^{-2} &= 7 \times b \times 3 \times b^{-2} \\
&= 7 \times 3 \times b^1 \times b^{-2} \\
&= 7 \times 3 \times b^{1-2} \\
&= 21b^{-1} = \frac{21}{b} \\
\therefore 7b \times 3b^{-2} &= \frac{21}{b}
\end{aligned}
$$

b) $5c^{\frac{3}{2}} \times 2c^{-\frac{7}{2}}$

Solution

$$
\begin{aligned}
5c^{\frac{3}{2}} \times 2c^{-\frac{7}{2}} &= 5 \times 2 \times c^{\frac{3}{2} - \frac{7}{2}} \\
&= 10 \times c^{\frac{3-7}{2}} = 10 \times c^{\frac{-4}{2}} \\
&= 10c^{-2} = \frac{10}{c^2} \\
\therefore 5c^{\frac{3}{2}} \times 2c^{-\frac{7}{2}} &= \frac{10}{c^2}
\end{aligned}
$$

c) $11d^3e \times 4d^2e^{-3}$

Solution

$$
\begin{aligned}
11d^3e \times 4d^2e^{-3} &= 11 \times 4 \times d^{3+2}e^{1-3} \\
&= 44d^5e^{-2} = \frac{44d^5}{e^2} \\
\therefore 11d^3e \times 4d^2e^{-3} &= \frac{44d^5}{e^2}
\end{aligned}
$$

d) $(-64)^{-\frac{4}{3}}$

Solution

$$
\begin{aligned}
(-64)^{-\frac{4}{3}} &= \left[(-4)^3\right]^{-\frac{4}{3}} \\
&= (-4)^{-4} \\
&= \frac{1}{(-4)^4} = \frac{1}{256} \\
\therefore (-64)^{-\frac{4}{3}} &= \frac{1}{256}
\end{aligned}
$$

NOTE

The key rule to know here is that

$$(-x)^m = \begin{cases} x^m, & \text{if } m \text{ is an even number} \\ -x^m, & \text{if } m \text{ is an odd number} \end{cases}$$

Example 10

Without using a calculator, evaluate the following:

a) $\left(4q^2\right)^{-3}$ b) $\left(\frac{1}{3}r^{-2}\right)^{-3}$ c) $(10\,000)^{-\frac{3}{4}}$ d) $\left(1\frac{1}{3}\right)^{-3}$ e) $\left[\left(\frac{1}{8}\right)^{\frac{1}{6}}\right]^2$

What did you get? Find the solution below to double-check your answer.

Solution to Example 10

HINT

In this example, we will be applying the following laws of indices exclusively:

$$(x^m)^n = x^{mn} \text{ and } x^{-m} = \frac{1}{x^m}$$

a) $\left(4q^2\right)^{-3}$

Solution

$$\left(4q^2\right)^{-3} = \frac{1}{\left(4q^2\right)^3}$$

$$= \frac{1}{4^3 q^6} = \frac{1}{64q^6}$$

$$\therefore \left(4q^2\right)^{-3} = \frac{1}{64q^6}$$

b) $\left(\frac{1}{3}r^{-2}\right)^{-3}$

Solution

$$\left(\frac{1}{3}r^{-2}\right)^{-3} = \left(3^{-1} \times r^{-2}\right)^{-3}$$

$$= 3^{(-1\times -3)} \times r^{(-2\times -3)}$$

$$= 3^3 r^6 = 27r^6$$

$$\therefore \left(\frac{1}{3}r^{-2}\right)^{-3} = 27r^6$$

---

c) $(10\,000)^{-\frac{3}{4}}$

Solution

$$(10\,000)^{-\frac{3}{4}} = \left(10^4\right)^{-\frac{3}{4}} = 10^{\left(4\times -\frac{3}{4}\right)}$$

$$= 10^{-3} = \frac{1}{10^3}$$

$$= \frac{1}{1000}$$

$$\therefore (10\,000)^{-\frac{3}{4}} = \frac{1}{1000}$$

---

d) $\left(1\frac{1}{3}\right)^{-3}$

Solution

$$\left(1\frac{1}{3}\right)^{-3} = \left(\frac{4}{3}\right)^{-3}$$

$$= \left(\frac{3}{4}\right)^3 = \frac{27}{64}$$

$$\therefore \left(1\frac{1}{3}\right)^{-3} = \frac{27}{64}$$

---

e) $\left[\left(\frac{1}{8}\right)^{\frac{1}{6}}\right]^2$

Solution

$$\left[\left(\frac{1}{8}\right)^{\frac{1}{6}}\right]^2 = \left(\frac{1}{8}\right)^{\frac{1}{3}} = \left(8^{-1}\right)^{\frac{1}{3}}$$

$$= \left[\left(2^3\right)^{-1}\right]^{\frac{1}{3}} = \left(2^{-3}\right)^{\frac{1}{3}}$$

$$= 2^{-1} = \frac{1}{2}$$

$$\therefore \left[\left(\frac{1}{8}\right)^{\frac{1}{6}}\right]^2 = \frac{1}{2}$$

Another example for us to try.

Example 11

Without using a calculator, evaluate $\left(\frac{1}{3}\right)^{-1} + \left(\frac{1}{3}\right)^{-2} + \left(\frac{1}{3}\right)^{-3}$.

What did you get? Find the solution below to double-check your answer.

Solution to Example 11

HINT

This example seems to indicate a need for the law of multiplication of indices, but you will see shortly that we do not.

$$\left(\frac{1}{3}\right)^{-1} + \left(\frac{1}{3}\right)^{-2} + \left(\frac{1}{3}\right)^{-3} = 3 + 3^2 + 3^3$$

$$= 3 + 9 + 27 = 39$$

$$\therefore \left(\frac{1}{3}\right)^{-1} + \left(\frac{1}{3}\right)^{-2} + \left(\frac{1}{3}\right)^{-3} = 39$$

Another example to try.

Example 12

Without using a calculator, evaluate $\left(2\frac{1}{4}\right)^{-\frac{1}{2}}$.

What did you get? Find the solution below to double-check your answer.

Solution to Example 12

$$\left(2\frac{1}{4}\right)^{-\frac{1}{2}} = \left(\frac{9}{4}\right)^{-\frac{1}{2}} = \left(\frac{4}{9}\right)^{\frac{1}{2}}$$

$$= \left(\frac{2^2}{3^2}\right)^{\frac{1}{2}} = \frac{2^{\left(2\times\frac{1}{2}\right)}}{3^{\left(2\times\frac{1}{2}\right)}} = \frac{2}{3}$$

$$\therefore \left(2\frac{1}{4}\right)^{-\frac{1}{2}} = \frac{2}{3}$$

ALTERNATIVE METHOD

$$\left(2\frac{1}{4}\right)^{-\frac{1}{2}} = \left(\frac{9}{4}\right)^{-\frac{1}{2}} = \left(\frac{4}{9}\right)^{\frac{1}{2}}$$

$$= \sqrt{\frac{4}{9}} = \sqrt{\frac{2^2}{3^2}} = \frac{2}{3}$$

$$\therefore \left(2\frac{1}{4}\right)^{-\frac{1}{2}} = \frac{2}{3}$$

Another example to try.

Example 13

Given that $x = 3$, without using a calculator find the value of $\frac{1}{3}x^{-2}$.

What did you get? Find the solution below to double-check your answer.

Solution to Example 13

$$\frac{1}{3}x^{-2} = \frac{1}{3x^2}$$

Now, let's substitute $x = 3$, we have

$$\frac{1}{3}x^{-2} = \frac{1}{3 \times 3^2} = \frac{1}{27}$$

$$\therefore \frac{1}{3}x^{-2} = \frac{1}{27}$$

A final set of examples to try.

Example 14

If $y = -27$ without using a calculator determine the value of each of the following:

a) $\frac{1}{3}y^{\frac{2}{3}}$ b) $2y^{\frac{1}{6}}$ c) $\left(\frac{1}{9}y\right)^{\frac{3}{4}}$

What did you get? Find the solution below to double-check your answer.

Solution to Example 14

HINT

This is like the previous example except that the substituting value is negative, and this would need to be diligently handled.

a) $\frac{1}{3}y^{\frac{2}{3}}$

Solution

$$\frac{1}{3}y^{\frac{2}{3}} = 3^{-1} \times y^{\frac{2}{3}}$$

Now, let's substitute $y = -27$, we have

$$\frac{1}{3}y^{\frac{2}{3}} = 3^{-1}(-27)^{\frac{2}{3}}$$

$$= 3^{-1}(-27)^{2\times\frac{1}{3}} = 3^{-1}\left[(-27)^2\right]^{\frac{1}{3}}$$

Note that $(-27)^2 = (27)^2$, then we have

$$3^{-1}\left[(27)^2\right]^{\frac{1}{3}} = 3^{-1}\left[(3^3)^2\right]^{\frac{1}{3}}$$

$$= 3^{-1}\left[3^{(3\times2\times\frac{1}{3})}\right] = 3^{-1} \times 3^2 = 3$$

$$\therefore \frac{1}{3}y^{\frac{2}{3}} = 3$$

b) $2y^{\frac{1}{6}}$

Solution

$$2y^{\frac{1}{6}} = 2 \times y^{\frac{1}{6}}$$

Now, let's substitute $y = -27$, we have

$$2y^{\frac{1}{6}} = 2(-27)^{\frac{1}{6}}$$

$$= 2\left[(-3)^3\right]^{\frac{1}{6}} = 2(-3)^{\frac{1}{2}}$$

$$= 2\sqrt{-3}$$

$$\therefore \textbf{No solution}$$

This is because the square root of a negative number does not give a real number.

c) $\left(\frac{1}{9}y\right)^{\frac{3}{4}}$

Solution

$$\left(\frac{1}{9}y\right)^{\frac{3}{4}} = \left(\frac{1}{9} \times y\right)^{\frac{3}{4}}$$

Now, let's substitute $y = -27$, we have

$$\left(\frac{1}{9}y\right)^{\frac{3}{4}} = \left(\frac{1}{9} \times -27\right)^{\frac{3}{4}} = (-3)^{\frac{3}{4}}$$

$$= \left(\sqrt[4]{-3}\right)^3 = \sqrt[4]{-27}$$

$$\therefore \textbf{No solution}$$

In the last two examples, we stated that there are no solutions to the questions. What this means is that the answers are not real values. The two questions have answers, which are complex numbers. For example, $2\sqrt{-3} = j2\sqrt{3} = j3.4641$; similarly, $\sqrt[4]{-27} = 1.6119 + j1.6119$. Complex numbers are covered in *Advanced Mathematics for Engineers and Scientists with Worked Examples* by the same author. Now, let's discuss more laws.

11.3.2.3 Fractional Power (or Root) Law

It states that:

$$\boxed{x^{\frac{1}{n}} = \sqrt[n]{x}} \tag{11.6}$$

where n is any positive integer.

This means that 'anything' raised to the power of one-*n*th (i.e., $1/n$) is equal to the *n*th root of 'the same thing'. Note that $\sqrt{x} = \sqrt[2]{x}$ but 2 is generally omitted. It is read as '**square root of x**'. Similarly, $\sqrt[3]{x}$ is the '**cube root of x**'. While $\sqrt[4]{x}$ and $\sqrt[5]{x}$ are the '**fourth root of x**' and 'fifth root of x' respectively.

Remember that a power is the inverse of a root just like a whole number is the opposite of a fraction. For example, $9^{\frac{1}{2}} = \sqrt[2]{9}$ and $8^{\frac{1}{3}} = \sqrt[3]{8}$. We know that the answers are **3** and **2** respectively.

What about a situation when the numerator of the fractional index is not 1? Yes, there is a rule for this case, which states that:

$$\boxed{x^{\frac{m}{n}} = \left(\sqrt[n]{x}\right)^m} \textbf{ OR } \boxed{x^{\frac{m}{n}} = \sqrt[n]{x^m}} \tag{11.7}$$

This is because

$$x^{\frac{m}{n}} = x^{\left(\frac{1}{n} \times m\right)}$$

$$= \left(x^{\frac{1}{n}}\right)^m = \left(\sqrt[n]{x}\right)^m$$

Let's try some examples to illustrate this.

Example 15

Without using a calculator, evaluate the following using the fractional power (or root) law.

a) $1^{\frac{1}{5}}$

b) $9^{\frac{1}{2}}$

c) $256^{\frac{1}{4}}$

d) $8^{\frac{2}{3}}$

e) $\left(\frac{16}{81}\right)^{\frac{1}{4}}$

f) $0.04^{\frac{1}{2}}$

g) $27^{\frac{4}{3}}$

h) $\left(3\frac{3}{8}\right)^{\frac{2}{3}}$

i) $32^{0.4}$

j) $(-32)^{\frac{1}{5}}$

What did you get? Find the solution below to double-check your answer.

Solution to Example 15

HINT

In this example, we will be applying the fractional power of indices exclusively.

$$x^{\frac{1}{n}} = \sqrt[n]{x} \text{ and } x^{\frac{m}{n}} = \left(\sqrt[n]{x}\right)^{m}$$

a) $1^{\frac{1}{5}}$
Solution

$$1^{\frac{1}{5}} = \sqrt[5]{1}$$

But $1 = 1^2 = 1^3 = \cdots = 1^{n-1} = 1^n$, where n is any real number. Therefore

$$1^{\frac{1}{5}} = \sqrt[5]{1^5} = 1$$
$$\therefore 1^{\frac{1}{5}} = 1$$

NOTE
nth root of 1 is always 1 irrespective of the value of n, just like 1 to the power of anything is always 1.

$$\sqrt[5]{1} = \sqrt[5]{1^5} = 1$$

b) $9^{\frac{1}{2}}$
Solution

$$9^{\frac{1}{2}} = \sqrt{9}$$
$$= \sqrt{3^2} = 3$$
$$\therefore 9^{\frac{1}{2}} = 3$$

NOTE
In general,

$$\sqrt[n]{a^n} = a$$

Hence, for ease of evaluation, one can express the radicand (the number under the root symbol) in index form when determining roots of numbers. We will need this in the subsequent examples.

---

c) $256^{\frac{1}{4}}$

Solution

$$256^{\frac{1}{4}} = \sqrt[4]{256}$$
$$= \sqrt[4]{4^4} = 4$$
$$\therefore 256^{\frac{1}{4}} = 4$$

---

d) $8^{\frac{2}{3}}$

Solution

$$8^{\frac{2}{3}} = \left(\sqrt[3]{8}\right)^2$$
$$= \left(\sqrt[3]{2^3}\right)^2$$
$$= 2^2 = 4$$
$$\therefore 8^{\frac{2}{3}} = 4$$

---

e) $\left(\dfrac{16}{81}\right)^{\frac{1}{4}}$

Solution

$$\left(\frac{16}{81}\right)^{\frac{1}{4}} = \sqrt[4]{\frac{16}{81}}$$
$$= \sqrt[4]{\frac{2^4}{3^4}} = \sqrt[4]{\left(\frac{2}{3}\right)^4} = \frac{2}{3}$$
$$\therefore \left(\frac{16}{81}\right)^{\frac{1}{4}} = \frac{2}{3}$$

---

f) $0.04^{\frac{1}{2}}$

Solution

$$0.04^{\frac{1}{2}} = \left(\frac{4}{100}\right)^{\frac{1}{2}}$$
$$= \sqrt{\frac{4}{100}} = \frac{\sqrt{4}}{\sqrt{100}}$$
$$= \frac{\sqrt{2^2}}{\sqrt{10^2}} = \frac{2}{10} = \frac{1}{5}$$
$$\therefore 0.04^{\frac{1}{2}} = \frac{1}{5}$$

NOTE

In general,

$$\frac{\sqrt[n]{a}}{\sqrt[n]{b}} = \sqrt[n]{\frac{a}{b}}$$

and

$$\sqrt[n]{a} \times \sqrt[n]{b} = \sqrt[n]{ab}$$

---

g) $27^{\frac{4}{3}}$

Solution

$$27^{\frac{4}{3}} = \left(\sqrt[3]{27}\right)^4$$

$$= \left(\sqrt[3]{3^3}\right)^4 = 3^4$$

$$\therefore 27^{\frac{4}{3}} = 81$$

---

h) $\left(3\frac{3}{8}\right)^{\frac{2}{3}}$

Solution

$$\left(3\frac{3}{8}\right)^{\frac{2}{3}} = \left(\frac{27}{8}\right)^{\frac{2}{3}}$$

$$= \left(\sqrt[3]{\frac{27}{8}}\right)^2 = \left(\sqrt[3]{\frac{3^3}{2^3}}\right)^2 = \frac{3^2}{2^2} = \frac{9}{4}$$

$$\therefore \left(3\frac{3}{8}\right)^{\frac{2}{3}} = \frac{9}{4}$$

ALTERNATIVE METHOD

$$\left(3\frac{3}{8}\right)^{\frac{2}{3}} = \left(\frac{27}{8}\right)^{\frac{2}{3}} = \left(\frac{3^3}{2^3}\right)^{\frac{2}{3}}$$

$$= \left(\frac{3^3}{2^3}\right)^{\frac{2}{3}} = \frac{3^{\left(3\times\frac{2}{3}\right)}}{2^{\left(3\times\frac{2}{3}\right)}}$$

$$= \frac{3^2}{2^2} = \frac{9}{4}$$

$$\therefore \left(3\frac{3}{8}\right)^{\frac{2}{3}} = \frac{9}{4}$$

i) $32^{0.4}$

Solution

$$32^{0.4} = 32^{\frac{4}{10}}$$
$$= 32^{\frac{2}{5}} = \left(\sqrt[5]{32}\right)^2$$
$$= \left(\sqrt[5]{2^5}\right)^2 = 2^2$$
$$\therefore \mathbf{32^{0.4} = 4}$$

ALTERNATIVE METHOD

$$32^{0.4} = 32^{\frac{4}{10}}$$
$$= 32^{\frac{2}{5}} = (2^5)^{\frac{2}{5}}$$
$$= 2^{\left(5 \times \frac{2}{5}\right)} = 2^2$$
$$\therefore \mathbf{32^{0.4} = 4}$$

j) $(-32)^{\frac{1}{5}}$

Solution

$$(-32)^{\frac{1}{5}} = \sqrt[5]{(-32)}$$
$$= \sqrt[5]{(-2)^5} = -2$$
$$\therefore \mathbf{(-32)^{\frac{1}{5}} = -2}$$

This example shows that the nth root of a negative number is possible, or we can evaluate the nth root of a negative number to obtain an answer which is a real number provided n is odd. The answer in this case is always a negative real number.

Another example for us to try.

Example 16

Given that $x = 3$, without using a calculator determine the value of $\sqrt[3]{\dfrac{8}{(x^3)^{-1}}}$.

What did you get? Find the solution below to double-check your answer.

Solution to Example 16

$$\sqrt[3]{\frac{8}{(x^3)^{-1}}} = \sqrt[3]{\frac{8}{x^{-3}}} = \sqrt[3]{8x^3}$$

$$= \sqrt[3]{2^3 x^3} = \sqrt[3]{(2x)^3}$$

$$= 2x = 2 \times 3 = 6$$

$$\therefore \sqrt[3]{\frac{8}{(x^3)^{-1}}} = 6$$

11.3.2.4 Same Power Law

So far, we have introduced rules relating to terms that share a common base. We will now look at rules that can be used when the base of the term differs. In general, the operations are carried out as normal, though with one exception, which is that the index of the terms must be the same. Consequently, the following rules should be applied:

Rule 1

$$\boxed{x^n \times y^n = (x \times y)^n = (xy)^n}$$ (11.8)

Rule 2

$$\boxed{x^n \div y^n = (x \div y)^n} \text{ OR } \boxed{\frac{x^n}{y^n} = \left[\frac{x}{y}\right]^n}$$ (11.9)

Rule 3

$$\boxed{(x^a y^b)^n = x^{an} y^{bn}} \text{ AND } \boxed{\left(\frac{x^a}{y^b}\right)^n = \frac{x^{an}}{y^{bn}}}$$ (11.10)

where a, b, and n are any real number.

The rules above can be easily verified. It should be noted that Rules 1 and 2 are not valid for addition and subtraction. Consequently,

$$\boxed{x^n + y^n \neq (x + y)^n} \text{ AND } \boxed{x^n - y^n \neq (x - y)^n}$$ (11.11)

This last law (same power law) will be a useful reference when solving complex problems involving indices. In fact, we've used it in our previous examples.

Now that we've covered all the relevant laws of indices, we will try examples that will require a combination of these laws.

Example 17

Without using a calculator, evaluate the following and present the answers in the form $a\sqrt[n]{b}$, where a, b, and n are real numbers. State the values of a, b, and n.

a) $\sqrt[3]{32} \times \sqrt[3]{14}$ **b)** $\sqrt[5]{81} \times \sqrt[5]{6}$ **c)** $7^{\frac{7}{3}} + 7^{\frac{1}{3}} - 7^{\frac{4}{3}}$ **d)** $5^{-\frac{3}{2}} + 5^{-\frac{1}{2}} + 5^{\frac{1}{2}}$

What did you get? Find the solution below to double-check your answer.

Solution to Example 17

HINT

These questions can also be solved using Surds – covered in Chapter 8; however, employing indices may prove to be easier. We will show the application of rule 1 in the first two examples.

a) $\sqrt[3]{32} \times \sqrt[3]{14}$
Solution

$$\sqrt[3]{32} \times \sqrt[3]{14} = (32)^{\frac{1}{3}} \times (14)^{\frac{1}{3}} = (2^5)^{\frac{1}{3}} \times (2 \times 7)^{\frac{1}{3}}$$

$$= 2^{\frac{5}{3}} \times 2^{\frac{1}{3}} \times 7^{\frac{1}{3}} = 2^{\frac{5}{3}+\frac{1}{3}} \times 7^{\frac{1}{3}}$$

$$= 2^2 \times 7^{\frac{1}{3}} = 4 \times \sqrt[3]{7} = 4\sqrt[3]{7}$$

$$\therefore \sqrt[3]{32} \times \sqrt[3]{14} = 4\sqrt[3]{7}$$

Hence

$$a = 4, \, b = 7 \text{ and } n = 3$$

ALTERNATIVE METHOD

Using the same power law, in conjunction with one or more law(s), we have:

$$\sqrt[3]{32} \times \sqrt[3]{14} = (32)^{\frac{1}{3}} \times (14)^{\frac{1}{3}} = (32 \times 14)^{\frac{1}{3}} = (2^5 \times 2 \times 7)^{\frac{1}{3}}$$

$$= (2^6 \times 7)^{\frac{1}{3}} = (2^6)^{\frac{1}{3}} \times (7)^{\frac{1}{3}}$$

$$= 2^2 \times 7^{\frac{1}{3}} = 4 \times \sqrt[3]{7} = 4\sqrt[3]{7}$$

$$\therefore \sqrt[3]{32} \times \sqrt[3]{14} = 4\sqrt[3]{7}$$

b) $\sqrt[5]{81} \times \sqrt[5]{6}$

Solution

$$\sqrt[5]{81} \times \sqrt[5]{6} = (81)^{\frac{1}{5}} \times (6)^{\frac{1}{5}}$$

$$= \left(3^4\right)^{\frac{1}{5}} \times (3 \times 2)^{\frac{1}{5}} = 3^{\frac{4}{5}} \times 3^{\frac{1}{5}} \times 2^{\frac{1}{5}}$$

$$= 3^{\frac{4}{5}+\frac{1}{5}} \times 2^{\frac{1}{5}} = 3 \times 2^{\frac{1}{5}}$$

$$= 3 \times \sqrt[5]{2} = 3\sqrt[5]{2}$$

$$\therefore \sqrt[5]{81} \times \sqrt[5]{6} = 3\sqrt[5]{2}$$

Hence

$$\therefore a = 3, \ b = 2 \text{ and } n = 5$$

ALTERNATIVE METHOD

Using the same power law, in conjunction with one or more law(s), we have:

$$\sqrt[5]{81} \times \sqrt[5]{6} = (81)^{\frac{1}{5}} \times (6)^{\frac{1}{5}} = (81 \times 6)^{\frac{1}{5}}$$

$$= \left(3^4 \times 3 \times 2\right)^{\frac{1}{5}} = \left(3^5 \times 2\right)^{\frac{1}{5}}$$

$$= \left(3^5\right)^{\frac{1}{5}} \times (2)^{\frac{1}{5}} = 3 \times 2^{\frac{1}{5}}$$

$$= 3 \times \sqrt[5]{2} = 3\sqrt[5]{2}$$

$$\therefore \sqrt[5]{81} \times \sqrt[5]{6} = 3\sqrt[5]{2}$$

---

c) $7^{\frac{7}{3}} + 7^{\frac{1}{3}} - 7^{\frac{4}{3}}$

Solution

$$7^{\frac{7}{3}} + 7^{\frac{1}{3}} - 7^{\frac{4}{3}} = 7^{\left(2+\frac{1}{3}\right)} + 7^{\left(0+\frac{1}{3}\right)} - 7^{\left(1+\frac{1}{3}\right)}$$

$$= 7^{\frac{1}{3}} \times 7^2 + 7^{\frac{1}{3}} \times 7^0 - 7^{\frac{1}{3}} \times 7^1$$

$$= 7^{\frac{1}{3}} \left(7^2 + 7^0 - 7^1\right)$$

$$= 7^{\frac{1}{3}} (49 + 1 - 7)$$

$$= 7^{\frac{1}{3}} (43) = 43\sqrt[3]{7}$$

$$\therefore 7^{\frac{7}{3}} + 7^{\frac{1}{3}} - 7^{\frac{4}{3}} = 43\sqrt[3]{7}$$

Hence

$$\therefore a = 43, \ b = 7 \text{ and } n = 3$$

d) $5^{-\frac{3}{2}} + 5^{-\frac{1}{2}} + 5^{\frac{1}{2}}$

Solution

$$5^{-\frac{3}{2}} + 5^{-\frac{1}{2}} + 5^{\frac{1}{2}} = 5^{\left(-2+\frac{1}{2}\right)} + 5^{\left(-1+\frac{1}{2}\right)} + 5^{\left(0+\frac{1}{2}\right)}$$

$$= 5^{\frac{1}{2}}\left(5^{-2} + 5^{-1} + 5^{0}\right)$$

$$= 5^{\frac{1}{2}}\left(\frac{1}{25} + \frac{1}{5} + 1\right) = 5^{\frac{1}{2}}\left(\frac{1+5+25}{25}\right)$$

$$= 5^{\frac{1}{2}}\left(\frac{31}{25}\right) = \frac{31}{25}\sqrt{5}$$

$$\therefore 5^{-\frac{3}{2}} + 5^{-\frac{1}{2}} + 5^{\frac{1}{2}} = \frac{31}{25}\sqrt{5}$$

Hence

$$a = \frac{31}{25}, b = 5 \text{ and } n = 2$$

Another set of examples to try.

Example 18

Without using a calculator, evaluate the following using appropriate laws of indices.

a) $0.008^{-\frac{1}{3}}$ **b)** $\sqrt[5]{36^{2.5}}$ **c)** $33 \times 10^3 \div (11 \times 10^{-2})$

d) $\left(3d^{-1}\right)^{\frac{1}{2}} \times \left(3d^5\right)^{\frac{3}{2}}$ **e)** $\left(24x^4y^2\right)^{\frac{1}{3}} \div \left(3xy^2\right)^{\frac{1}{3}}$

What did you get? Find the solution below to double-check your answer.

Solution to Example 18

HINT

We've only shown a sample of combinations of the laws in these examples; there could be other possible combinations or methods to solve the problems.

a) $0.008^{-\frac{1}{3}}$

Solution

$$0.008^{-\frac{1}{3}} = \left(\frac{8}{1000}\right)^{-\frac{1}{3}}$$

$$= \left(\frac{1000}{8}\right)^{\frac{1}{3}} = \sqrt[3]{\frac{1000}{8}}$$

We won't simplify $\frac{1000}{8}$ in the root yet, but will change them to index form as

$$= \sqrt[3]{\frac{10^3}{2^3}} = \sqrt[3]{\left(\frac{10}{2}\right)^3} = \frac{10}{2} = 5$$

$$\therefore 0.008^{-\frac{1}{3}} = 5$$

b) $\sqrt[5]{36^{2.5}}$

Solution

$$\sqrt[5]{36^{2.5}} = \left(36^{2.5}\right)^{1/5}$$

$$= \left(36^{\frac{5}{2}}\right)^{\frac{1}{5}} = 36^{\frac{1}{2}}$$

$$= \left(6^2\right)^{\frac{1}{2}} = 6$$

$$\therefore \sqrt[5]{36^{2.5}} = 6$$

c) $33 \times 10^3 \div (11 \times 10^{-2})$

Solution

$$33 \times 10^3 \div \left(11 \times 10^{-2}\right) = \frac{33 \times 10^3}{11 \times 10^{-2}}$$

Expressing the question in the above format makes it easier to visualise the problem.

$$\frac{33}{11} \times \frac{10^3}{10^{-2}} = 3 \times 10^{3+2}$$

$$= 3 \times 10^5 = 300\,000$$

$$\therefore 33 \times 10^3 \div \left(11 \times 10^{-2}\right) = 300\,000$$

d) $\left(3d^{-1}\right)^{\frac{1}{2}} \times \left(3d^5\right)^{\frac{3}{2}}$

Solution

$$\left(3d^{-1}\right)^{\frac{1}{2}} \times \left(3d^5\right)^{\frac{3}{2}} = \left(3^{\frac{1}{2}} \times d^{-\frac{1}{2}}\right) \times \left(3^{\frac{3}{2}} \times d^{\frac{15}{2}}\right)$$

$$= 3^{\frac{1}{2}} \times 3^{\frac{3}{2}} \times d^{-\frac{1}{2}} \times d^{\frac{15}{2}}$$

$$= 3^{\left(\frac{1}{2}+\frac{3}{2}\right)} \times d^{\left(-\frac{1}{2}+\frac{15}{2}\right)}$$

$$= 3^{\left(\frac{1+3}{2}\right)} \times d^{\left(\frac{-1+15}{2}\right)}$$

$$= 3^2 \times d^7 = 9d^7$$

$$\therefore \left(3d^{-1}\right)^{\frac{1}{2}} \times \left(3d^5\right)^{\frac{3}{2}} = 9d^7$$

e) $\left(24x^4y^2\right)^{\frac{1}{3}} \div \left(3xy^2\right)^{\frac{1}{3}}$

Solution

$$\left(24x^4y^2\right)^{\frac{1}{3}} \div \left(3xy^2\right)^{\frac{1}{3}} = \left(24x^4y^2 \div 3xy^2\right)^{\frac{1}{3}}$$

$$= \left(\frac{24x^4y^2}{3xy^2}\right)^{\frac{1}{3}} = \left(8x^3\right)^{\frac{1}{3}}$$

$$= \left(2^3x^3\right)^{\frac{1}{3}} = \left(2^3\right)^{\frac{1}{3}} \times \left(x^3\right)^{\frac{1}{3}} = 2x$$

$$\therefore \left(24x^4y^2\right)^{\frac{1}{3}} \div \left(3xy^2\right)^{\frac{1}{3}} = 2x$$

Example 19

Rewrite the following expressions using index notation in the form ax^n, where a and n are rational numbers.

a) $3x^2\left(\sqrt[3]{x}\right)$ **b)** $\sqrt[5]{32x^2}$ **c)** $\dfrac{1}{\sqrt[3]{x^2}}$ **d)** $\dfrac{6\sqrt{x}}{x\left(\sqrt[3]{x}\right)}$

What did you get? Find the solution below to double-check your answer.

Solution to Example 19

a) $3x^2\left(\sqrt[3]{x}\right)$

Solution

$$3x^2\left(\sqrt[3]{x}\right) = 3x^2 \times x^{\frac{1}{3}} = 3 \times x^2 \times x^{\frac{1}{3}}$$

$$= 3 \times x^{\left(2+\frac{1}{3}\right)} = 3 \times x^{\frac{7}{3}}$$

$$= 3x^{\frac{7}{3}}$$

$$\therefore 3x^2\left(\sqrt[3]{x}\right) = 3x^{\frac{7}{3}}$$

---

b) $\sqrt[5]{32x^2}$

Solution

$$\sqrt[5]{32x^2} = \left(32x^2\right)^{\frac{1}{5}} = \left(32\right)^{\frac{1}{5}}\left(x^2\right)^{\frac{1}{5}}$$

$$= \left(2^5\right)^{\frac{1}{5}} \times x^{\frac{2}{5}} = 2^1 \times x^{\frac{2}{5}}$$

$$= 2x^{\frac{2}{5}}$$

$$\therefore \sqrt[5]{32x^2} = 2x^{\frac{2}{5}}$$

c) $\dfrac{1}{\sqrt[3]{x^2}}$

Solution

$$\frac{1}{\sqrt[3]{x^2}} = \frac{1}{(x^2)^{\frac{1}{3}}}$$

$$= \frac{1}{x^{\frac{2}{3}}} = x^{-\frac{2}{3}}$$

$$\therefore \frac{1}{\sqrt[3]{x^2}} = x^{-\frac{2}{3}}$$

d) $\dfrac{6\sqrt{x}}{x\left(\sqrt[3]{x}\right)}$

Solution

$$\frac{6\sqrt{x}}{x\left(\sqrt[3]{x}\right)} = \frac{6x^{\frac{1}{2}}}{x \times x^{\frac{1}{3}}} = \frac{6x^{\frac{1}{2}}}{x^{\left(1+\frac{1}{3}\right)}}$$

$$= \frac{6x^{\frac{1}{2}}}{x^{\frac{4}{3}}} = 6 \times x^{\frac{1}{2}-\frac{4}{3}}$$

$$= 6 \times x^{\frac{3-8}{6}} = 6x^{-\frac{5}{6}}$$

$$\therefore \frac{6\sqrt{x}}{x\left(\sqrt[3]{x}\right)} = 6x^{-\frac{5}{6}}$$

Example 20

Evaluate the following:

a) $16t^3 \times \left(2t^2\right)^{-3}$

b) $\left(3xy^3z^5\right)^2 \div \left(3xy^3z^5\right)$

c) $\dfrac{10a^5 \times 4a^3}{8a^{-2}}$

d) $\dfrac{(-c)^2 \times c^6}{-c^7}$

What did you get? Find the solution below to double-check your answer.

Solution to Example 20

a) $16t^3 \times \left(2t^2\right)^{-3}$

Solution

$$16t^3 \times \left(2t^2\right)^{-3} = 16t^3 \times 2^{-3}t^{-6} = 16 \times t^3 \times 2^{-3} \times t^{-6}$$

$$= 16 \times 2^{-3} \times t^3 \times t^{-6}$$

$$= 16 \times 2^{-3} \times t^{3-6}$$

$$= \frac{16}{2^3} \times t^{-3} = \frac{16}{8} \times \frac{1}{t^3}$$

$$= 2 \times \frac{1}{t^3} = \frac{2}{t^3}$$

$$\therefore 16t^3 \times \left(2t^2\right)^{-3} = \frac{2}{t^3}$$

ALTERNATIVE METHOD

$$16t^3 \times \left(2t^2\right)^{-3} = \frac{16t^3}{\left(2t^2\right)^3} = \frac{16t^3}{2^3 \times t^6}$$

$$= \frac{16t^3}{8t^6} = \frac{2t^3}{t^6}$$

$$= 2 \times t^{3-6}$$

$$= 2t^{-3} = \frac{2}{t^3}$$

$$\therefore 16t^3 \times \left(2t^2\right)^{-3} = \frac{2}{t^3}$$

---

b) $\left(3xy^3z^5\right)^2 \div \left(3xy^3z^5\right)$

Solution

$$\left(3xy^3z^5\right)^2 \div \left(3xy^3z^5\right) = 3^2x^2y^{3\times2}z^{5\times2} \div 3xy^3z^5$$

$$= 9x^2y^6z^{10} \div 3xy^3z^5 = (9 \div 3) \times \left(x^2y^6z^{10} \div xy^3z^5\right)$$

$$= 3 \times \left(x^{2-1}y^{6-3}z^{10-5}\right)$$

$$= 3 \times xy^3z^5 = 3xy^3z^5$$

$$\therefore \left(3xy^3z^5\right)^2 \div \left(3xy^3z^5\right) = 3xy^3z^5$$

NOTE

We can consider the expression in the brackets as a unit and evaluate it as:

$$\left(3xy^3z^5\right)^2 \div \left(3xy^3z^5\right) = \left(3xy^3z^5\right)^{2-1}$$

$$= 3xy^3z^5$$

---

c) $\frac{10a^5 \times 4a^3}{8a^{-2}}$

Solution

$$\frac{10a^5 \times 4a^3}{8a^{-2}} = \frac{10 \times 4 \times a^5 \times a^3}{8 \times a^{-2}}$$

$$= \left(\frac{10 \times 4}{8}\right) \times \left(\frac{a^5 \times a^3}{a^{-2}}\right) = \left(\frac{40}{8}\right) \times \left(a^{5+3-(-2)}\right)$$

$$= 5 \times a^{10} = 5a^{10}$$

$$\therefore \frac{10a^5 \times 4a^3}{8a^{-2}} = 5a^{10}$$

d) $\dfrac{(-c)^2 \times c^6}{-c^7}$

Solution

$$\frac{(-c)^2 \times c^6}{-c^7} = \frac{c^2 \times c^6}{-1 \times c^7} = \frac{c^8}{-1 \times c^7}$$
$$= -1 \times c^{8-7} = -c$$

$$\therefore \frac{(-c)^2 \times c^6}{-c^7} = -c$$

Example 21

Simplify the following, giving each answer in the form 3^n.

a) $\left(\dfrac{1}{3^2}\right)^{-8} \times \left(3^4\right)^{-3}$ **b)** $3^2 \div 27 \times 9^3$ **c)** $\dfrac{\left(3^{-2}\right)^3 \times 3^9}{3^{11}}$

What did you get? Find the solution below to double-check your answer.

Solution to Example 21

a) $\left(\dfrac{1}{3^2}\right)^{-8} \times \left(3^4\right)^{-3}$

Solution

$$\left(\frac{1}{3^2}\right)^{-8} \times \left(3^4\right)^{-3} = \left(3^{-2}\right)^{-8} \times \left(3^4\right)^{-3}$$
$$= 3^{-2\times-8} \times 3^{4\times-3}$$
$$= 3^{16} \times 3^{-12}$$
$$= 3^{16-12} = 3^4$$

$$\therefore \left(\frac{1}{3^2}\right)^{-8} \times \left(3^4\right)^{-3} = 3^4$$

---

b) $3^2 \div 27 \times 9^3$

Solution

$$3^2 \div 27 \times 9^3 = 3^2 \div 3^3 \times \left(3^2\right)^3$$
$$= 3^{2-3} \times 3^6$$
$$= 3^{-1} \times 3^6$$
$$= 3^{-1+6} = 3^5$$
$$\therefore 3^2 \div 27 \times 9^3 = 3^5$$

c) $\dfrac{\left(3^{-2}\right)^3 \times 3^9}{3^{11}}$

Solution

$$\frac{\left(3^{-2}\right)^3 \times 3^9}{3^{11}} = \frac{3^{-6} \times 3^9}{3^{11}}$$

$$= \frac{3^{-6+9}}{3^{11}} = \frac{3^3}{3^{11}}$$

$$= 3^{3-11} = 3^{-8}$$

$$\therefore \frac{\left(3^{-2}\right)^3 \times 3^9}{3^{11}} = 3^{-8}$$

A new set of examples for us to try.

Example 22

Simplify the following, giving each answer in the form 3^n.

a) $3^{50} + 3^{50} + 3^{50}$

b) $81^{0.5} + 81^{0.5} + 81^{0.5}$

c) $27^{11} - 9^{16} - 3^{32}$

What did you get? Find the solution below to double-check your answer.

Solution to Example 22

a) $3^{50} + 3^{50} + 3^{50}$

Solution

$$3^{50} + 3^{50} + 3^{50} = 3^{50}\left(1 + 1 + 1\right)$$

$$= 3^{50}\left(3\right) = 3^{50+1} = 3^{51}$$

$$\therefore 3^{50} + 3^{50} + 3^{50} = 3^{51}$$

b) $81^{0.5} + 81^{0.5} + 81^{0.5}$

Solution

$$81^{0.5} + 81^{0.5} + 81^{0.5} = 81^{0.5}\left(1 + 1 + 1\right)$$

$$= 81^{0.5}\left(3\right) = \left(3^4\right)^{0.5}\left(3\right)$$

$$= \left(3^2\right)\left(3\right) = 3^{2+1}$$

$$= 3^3$$

$$\therefore 81^{0.5} + 81^{0.5} + 81^{0.5} = 3^3$$

c) $27^{11} - 9^{16} - 3^{32}$

Solution

$$27^{11} - 9^{16} - 3^{32} = \left(3^3\right)^{11} - \left(3^2\right)^{16} - 3^{32}$$
$$= 3^{33} - 3^{32} - 3^{32}$$
$$= 3\left(3^{32}\right) - 3^{32} - 3^{32}$$
$$= 3^{32}\left(3 - 1 - 1\right)$$
$$= 3^{32}\left(1\right) = 3^{32}$$
$$\therefore \mathbf{27^{11} - 9^{16} - 3^{32} = 3^{32}}$$

Example 23

Without using a calculator simplify $\left(\sqrt{5}\right)^{-2} + \left(\sqrt{5}\right)^{-1} + \left(\sqrt{5}\right)^{0} + \left(\sqrt{5}\right)^{1} + \left(\sqrt{5}\right)^{2}$, giving the answer in surd form.

What did you get? Find the solution below to double-check your answer.

Solution to Example 23

Solution

$$\left(\sqrt{5}\right)^{-2} + \left(\sqrt{5}\right)^{-1} + \left(\sqrt{5}\right)^{0} + \left(\sqrt{5}\right)^{1} + \left(\sqrt{5}\right)^{2}$$
$$= \sqrt{5}\left\{\left(\sqrt{5}\right)^{-3} + \left(\sqrt{5}\right)^{-2} + \left(\sqrt{5}\right)^{-1} + \left(\sqrt{5}\right)^{0} + \left(\sqrt{5}\right)\right\}$$
$$= \sqrt{5}\left\{\frac{1}{\left(\sqrt{5}\right)^3} + \frac{1}{\left(\sqrt{5}\right)^2} + \frac{1}{\sqrt{5}} + 1 + \sqrt{5}\right\}$$
$$= \sqrt{5}\left\{\frac{1}{5\sqrt{5}} + \frac{1}{5} + \frac{1}{\sqrt{5}} + 1 + \sqrt{5}\right\}$$
$$= \sqrt{5}\left\{\frac{1 + \sqrt{5} + 5 + 5\sqrt{5} + 25}{5\sqrt{5}}\right\}$$
$$= \sqrt{5}\left\{\frac{31 + 6\sqrt{5}}{5\sqrt{5}}\right\} = \frac{31 + 6\sqrt{5}}{5} = \frac{1}{5}\left(31 + 6\sqrt{5}\right)$$
$$\therefore \left(\sqrt{5}\right)^{-2} + \left(\sqrt{5}\right)^{-1} + \left(\sqrt{5}\right)^{0} + \left(\sqrt{5}\right)^{1} + \left(\sqrt{5}\right)^{2} = \frac{1}{5}\left(31 + 6\sqrt{5}\right)$$

A final example to try.

Example 24

Determine the value of a in terms of b for which 8^{2-6a} is equal to $\frac{1}{4^b}$.

What did you get? Find the solution below to double-check your answer.

Solution to Example 24

$$8^{2-6a} = \left(2^3\right)^{2-6a}$$
$$= 2^{3(2-6a)} = 2^{6-18a}$$

Similarly,

$$\frac{1}{4^b} = \frac{1}{2^{2b}} = 2^{-2b}$$

Given that

$$8^{2-6a} = \frac{1}{4^b}$$

we have

$$2^{6-18a} = 2^{-2b}$$

From this, we have

$$6 - 18a = -2b$$
$$6 + 2b = 18a$$
$$a = \frac{6 + 2b}{18}$$
$$= \frac{2(3 + b)}{18} = \frac{3 + b}{9}$$
$$\therefore a = \frac{1}{9}(3 + b)$$

We would like to finish this section with this remark that, from our discussion, it is evident that indices can be exclusively numbers, wholly letters or a combination of both and the laws will still be valid.

11.4 CHAPTER SUMMARY

1) Indices refer to the expression of a number in the form x^y, where x is called the base and y is the index.

2) The term 'index' is also referred to as 'power' or 'exponent'; essentially, they all mean the same thing. In general, power is the most frequently used as such, x^y is read as 'x **raised to the power of y**', 'x **raised to the power y**' or simply 'x **to the power y**'.

3) Laws of indices

  - Multiplication law

$$x^n \times x^m = x^{n+m}$$

  - Quotient (or Division) law

$$x^n \div x^m = x^{n-m} \quad \text{OR} \quad \frac{x^n}{x^m} = x^{n-m}$$

  - Power law

$$(x^m)^n = x^{mn}$$

  - Zero power law

$$x^0 = 1$$

It can be regarded that '**anything**' (excluding zero) to the power of zero is 1.

  - Negative power law

$$x^{-n} = \frac{1}{x^n} \quad \text{OR} \quad x^n = \frac{1}{x^{-n}}$$

That is '**whenever the reciprocal of a number with an index is computed, the sign of the index changes**'.

  - Fractional power (or root) law

$$x^{\frac{1}{n}} = \sqrt[n]{x}$$

This means that 'anything' raised to the power of one-nth (i.e., $1/n$) is equal to the nth root of 'the same thing'. In general, we have

$$x^{\frac{m}{n}} = \left(\sqrt[n]{x}\right)^m \quad \text{OR} \quad x^{\frac{m}{n}} = \sqrt[n]{x^m}$$

This is because

$$x^{\frac{m}{n}} = x^{\left(\frac{1}{n} \times m\right)} = \left(x^{\frac{1}{n}}\right)^m = \left(\sqrt[n]{x}\right)^m$$

4) Same power law

- $\boxed{x^n \times y^n = (x \times y)^n = (xy)^n}$

- $\boxed{x^n \div y^n = (x \div y)^n}$ OR $\dfrac{x^n}{y^n} = \left[\dfrac{x}{y}\right]^n$

- $\boxed{\left(x^a y^b\right)^n = x^{an} y^{bn}}$ AND $\boxed{\left(\dfrac{x^a}{y^b}\right)^n = \dfrac{x^{an}}{y^{bn}}}$

- $\boxed{x^n + y^n \neq (x + y)^n}$ AND $\boxed{x^n - y^n \neq (x - y)^n}$

****

11.5 FURTHER PRACTICE

To access complementary contents, including additional exercises, please go to www.dszak.com.

12 Logarithms

Learning Outcomes

Once you have studied the content of this chapter, you should be able to:

- Explain the term logarithm
- State the laws of logarithm
- Apply the laws of logarithm in simplifying expressions

12.1 INTRODUCTION

Logarithm is a derived term from two Greek words, namely, **logos** (expression) and **arithmos** (number) and refers to a technique of expressing numbers. In fact, it is a system of evaluating multiplication, division, powers, and roots by appropriately converting them to addition and subtraction. In continuation of our discussion in the preceding chapter on indices, this chapter will focus on logarithms, covering their meaning, types, and laws.

12.2 WHAT ARE LOGARITHMS?

Technically, the logarithm (commonly shortened to **log** and sometimes as **lg**) of a number to a given base is the value of the power to which the base must be raised to produce the number.

Let x be the number and y be its associated base. If c equals 'the logarithm to base y of x' or 'the logarithm of x to base y', then we can write this as:

$$\log_y x = c \tag{12.1}$$

y must be a positive real number excluding **1**. This is because **1** raised to the power of anything is **1** and so $\log_1 x$ will only be valid for $x = 1$ since $1^c = 1$ for $c \in \mathbb{R}$. x must also be positive for c to be a real number. If the log is such that $0 < x < 1$, then c will be negative, i.e., $c < 0$. However, if x is negative (i.e., $x < 0$), then c will be a complex number, which is beyond the scope of this book.

From this, we can state that

$$\log_y x = \begin{cases} +ve & \text{for } x > 1 \\ -ve & \text{for } 0 < x < 1 \end{cases} \tag{12.2}$$

For example, the logarithm of 100 to base 10 is 2 because, if the base is raised to the power of 2 we will get the number (100). In other words, $\mathbf{100 = 10^2}$.

It is evident that indices and logarithms are inversely related; it is therefore possible to change from one notation to the other. This relationship can be written as:

$$\boxed{\log_y x = c} \iff \boxed{y^c = x} \tag{12.3}$$

12.3 TYPES OF LOGARITHMS

Logarithms can be classified according to the value of their base. Essentially, there are two (or three) types.

a) Common Logarithm

This is a logarithm to the base of 10, written as $\log_{10} N$. In general, when the base is 10, it is usually omitted. In other words, $\log_{10} N$ is simply written as $\log N$. Common logarithm is also called decimal logarithm or Briggsian logarithm, named after Henry Briggs.

b) Natural Logarithm

This is the logarithm to the base of an irrational number denoted as e, where e is approximately 2.718281 (correct to 6 decimal places). The constant e is also called exponential constant, and it's similar to the constant π, as the latter is also irrational. Why '**natural**'? It could be due to the behaviour of certain natural phenomena (e.g., radioactive decay, charging–discharging a capacitor, frequency response, biological functions, interest rate, etc.) being dependent on the functions of e.

Natural logarithm is also called hyperbolic or Napierian logarithm, named after John Napier. It is usually written as *ln N* (read as 'ell-en of N') instead of $\log_e N$ as one might have expected.

c) The third category (if we may say) is any other logarithm with a base other than **10** and e.

Let's try some examples.

Example 1

Change each of the following index forms into their equivalent logarithmic forms.

a) $7^3 = 343$ b) $125^{\frac{1}{3}} = 5$ c) $2^{-3} = \frac{1}{8}$ d) $5^{\frac{t}{2}} = R$ e) $e^{x+1} = 4$

f) $3^x = 0.12$ g) $\sqrt[3]{64} = 4$ h) $2.5e^{-Rt/L} = 5$ i) $6^{2t} - 4 = 2x$ j) $(0.01)^{2.5} = 0.00001$

What did you get? Find the solution below to double-check your answer.

Solution to Example 1

a) $7^3 = 343$
Solution

$$7^3 = 343$$
$$\therefore \log_7 343 = 3$$

b) $125^{\frac{1}{3}} = 5$

Solution

$$125^{\frac{1}{3}} = 5$$

$$\therefore \log_{125} 5 = \frac{1}{3}$$

---

c) $2^{-3} = \frac{1}{8}$

Solution

$$2^{-3} = \frac{1}{8}$$

$$\therefore \log_2 \left(\frac{1}{8}\right) = -3$$

---

d) $5^{\frac{t}{2}} = R$

Solution

$$5^{\frac{t}{2}} = R$$

$$\therefore \log_5 R = \frac{t}{2}$$

NOTE

This can be simplified to $2 \log_5 R = t$.

---

e) $e^{x+1} = 4$

Solution

$$e^{x+1} = 4$$

$$\therefore \log_e 4 = x + 1$$

NOTE

This is commonly written as $\ln 4 = x + 1$.

---

f) $3^x = 0.12$

Solution

$$3^x = 0.12$$

$$\therefore \log_3 0.12 = x$$

---

g) $\sqrt[3]{64} = 4$

Solution

$$\sqrt[3]{64} = 4$$

which implies that

$$64^{\frac{1}{3}} = 4$$

$$\therefore \log_{64} 4 = \frac{1}{3}$$

h) $2.5e^{-Rt/L} = 5$
Solution

$$2.5e^{-Rt/L} = 5$$

Divide both sides by 2.5 and then convert as

$$e^{-Rt/L} = 2$$
$$\therefore \log_e 2 = -Rt/L$$

NOTE
This can be written as **ln 2 = −Rt/L**.

i) $6^{2t} - 4 = 2x$
Solution

$$6^{2t} - 4 = 2x$$

Re-arrange the above equation and then convert as

$$6^{2t} = 2x + 4$$
$$\therefore \log_6 (2x + 4) = 2t$$

j) $(0.01)^{2.5} = 0.00001$
Solution

$$(0.01)^{2.5} = 0.00001$$
$$\therefore \log_{0.01} (0.00001) = 2.5$$

Let's try another set of examples.

Example 2

Express the following logarithmic forms in their corresponding index notation.

a) $\log_{13} 1 = 0$ **b)** $\log_{39} 39 = 1$ **c)** $\log_{2.6} 6.76 = 2$
d) $\log_4 \left(\frac{1}{64}\right) = -3$ **e)** $\log_e 0.3Q = -t/RC$ **f)** $\log_2 (2a + b) = 3.6$

What did you get? Find the solution below to double-check your answer.

Solution to Example 2

a) $\log_{13} 1 = 0$
Solution

$$\log_{13} 1 = 0$$
$$\therefore 13^0 = 1$$

b) $\log_{39} 39 = 1$
Solution

$$\log_{39} 39 = 1$$
$$\therefore 39^1 = 39$$

---

c) $\log_{2.6} 6.76 = 2$
Solution

$$\log_{2.6} 6.76 = 2$$
$$\therefore 2.6^2 = 6.76$$

---

d) $\log_4 \left(\frac{1}{64}\right) = -3$
Solution

$$\log_4 \left(\frac{1}{64}\right) = -3$$
$$\therefore 4^{-3} = \frac{1}{64}$$

---

e) $\log_e 0.3Q = -t/RC$
Solution

$$\log_e 0.3Q = -t/RC$$
$$\therefore e^{-t/RC} = 0.3Q$$

---

f) $\log_2 (2a + b) = 3.6$
Solution

$$\log_2 (2a + b) = 3.6$$
$$\therefore 2^{3.6} = 2a + b$$

Another example to try.

Example 3

Given that $\log_5 (y + 2) = x$, express y in terms of x.

What did you get? Find the solution below to double-check your answer.

Solution to Example 3

$$\log_5 (y + 2) = x$$

This implies that

$$y + 2 = 5^x$$
$$\therefore y = 5^x - 2$$

12.4 LAWS OF LOGARITHM

Like indices, there are certain laws governing the operation of logarithms and these will be discussed under the following headings.

12.4.1 FUNDAMENTAL LAWS

Essentially, there are three main laws of logarithm.

12.4.1.1 Addition–Product Law

This can be written as:

$$\boxed{\log_x M + \log_x N = \log_x (MN)} \tag{12.4}$$

where x, M, and N are positive real numbers.

In other words, the sum of logs of numbers to the same base is equal to the log of their products and vice versa. The following should be noted about this law:

Note 1 The logs must have the same base, otherwise this law cannot be used. For example, $\log_5 2 + \log_3 8 \neq \log_5 16 \neq \log_3 16$.

Note 2 The sign between the terms must be addition and not multiplication. In other words, $\log_2 5 + \log_2 10 = \log_2 50$ but $\log_2 5 \times \log_2 10 \neq \log_2 50$.

This is also called multiplication law.

For natural logarithm, Equation 12.4 can be written as

$$\boxed{\ln (M) + \ln (N) = \ln (MN)} \tag{12.5}$$

Let's illustrate this law with examples.

Example 4

Without using a calculator, simplify the following using the addition–product law. Present the final answer as a single log.

a) $\log_5 10 + \log_5 2$ **b)** $\log_3 6 + \log_3 1.5 + \log_3 10$ **c)** $\log_7 \sqrt{2} + \log_7 \sqrt{18}$

What did you get? Find the solution below to double-check your answer.

Solution to Example 4

HINT

In this example, we will apply the law of logarithms shown below, although the application of the laws of indices will also be necessary.

$$\log_x M + \log_x N = \log_x (MN)$$

a) $\log_5 10 + \log_5 2$
Solution

$$\log_5 10 + \log_5 2 = \log_5 (10 \times 2)$$
$$= \log_5 20$$
$$\therefore \ \log_5 10 + \log_5 2 = \log_5 20$$

b) $\log_3 6 + \log_3 1.5 + \log_3 10$
Solution

$$\log_3 6 + \log_3 1.5 + \log_3 10 = \log_3 (6 \times 1.5 \times 10)$$
$$= \log_3 90$$
$$\therefore \ \log_3 6 + \log_3 1.5 + \log_3 10 = \log_3 90$$

c) $\log_7 \sqrt{2} + \log_7 \sqrt{18}$
Solution

$$\log_7 \sqrt{2} + \log_7 \sqrt{18} = \log_7 \left(\sqrt{2} \times \sqrt{18} \right)$$
$$= \log_7 \left(\sqrt{2 \times 18} \right)$$
$$= \log_7 \left(\sqrt{36} \right) = \log_7 6$$
$$\therefore \ \log_7 \sqrt{2} + \log_7 \sqrt{18} = \log_7 6$$

12.4.1.2 Subtraction–Quotient Law

This can be written as:

$$\log_x M - \log_x N = \log_x \left(\frac{M}{N} \right) \tag{12.6}$$

where x, M, and N are positive real numbers.

In other words, the difference of logs of numbers to the same base is equal to the log of their quotient and vice versa. The log of the minuend (M in this case) in the difference will be the numerator and subtrahend (N in this case) the denominator.

The following should be noted about this law:

Note 1 The logs must have the same base, otherwise this law cannot be used. For example,
$\log_5 32 - \log_3 8 \neq \log_5 4 \neq \log_3 4$.

Note 2 The coefficient of the log must be 1.

Note 3 The sign between the terms must be subtraction and not division. In other words, $\log_2 50 - \log_2 10 = \log_2 \left(\frac{50}{10}\right) = \log_2 5$ but $\log_2 50 \div \log_2 10 \neq \log_2 5$.

This is also called division law.

For natural logarithm, Equation 12.6 can be written as

$$\boxed{\ln(M) - \ln(N) = \ln\left(\frac{M}{N}\right)} \tag{12.7}$$

Let's illustrate this law with some examples.

Example 5

Without using a calculator, simplify the following using the subtraction–quotient law. Present the final answer as a single log.

a) $\log_{10} 35 - \log_{10} 5$ b) $\log_9 5^{\frac{5}{2}} - \log_9 25$ c) $\log_6 \sqrt[3]{250} - \log_6 \sqrt[3]{2}$

What did you get? Find the solution below to double-check your answer.

Solution to Example 5

HINT

In this example, we will apply the law of logarithms shown below, although the application of the laws of indices will also be necessary.

$$\log_x M - \log_x N = \log_x \left(\frac{M}{N}\right)$$

a) $\log_{10} 35 - \log_{10} 5$
Solution

$$\log_{10} 35 - \log_{10} 5 = \log_{10}\left(\frac{35}{5}\right)$$
$$= \log_{10} 7$$
$$\therefore \log_{10} 35 - \log_{10} 5 = \log_{10} 7$$

b) $\log_9 5^{\frac{5}{2}} - \log_9 25$

Solution

$$\log_9 5^{\frac{5}{2}} - \log_9 25 = \log_9 \left(\frac{5^{\frac{5}{2}}}{25} \right)$$

$$= \log_9 \left(\frac{5^{\frac{5}{2}}}{5^2} \right) = \log_9 \left(5^{\frac{5}{2}-2} \right)$$

$$= \log_9 5^{\frac{1}{2}} = \log_9 \sqrt{5}$$

$$\therefore \log_9 5^{\frac{5}{2}} - \log_9 25 = \log_9 \left(\sqrt{5} \right)$$

---

c) $\log_6 \sqrt[3]{250} - \log_6 \sqrt[3]{2}$

Solution

$$\log_6 \sqrt[3]{250} - \log_6 \sqrt[3]{2} = \log_6 \left(\frac{\sqrt[3]{250}}{\sqrt[3]{2}} \right)$$

$$= \log_6 \left(\sqrt[3]{\frac{250}{2}} \right) = \log_6 \left(\sqrt[3]{125} \right)$$

$$= \log_6 \left(\sqrt[3]{5^3} \right) = \log_6 5$$

$$\therefore \log_6 \sqrt[3]{250} - \log_6 \sqrt[3]{2} = \log_6 5$$

12.4.1.3 Power Law

This can be written as:

$$\boxed{\log_x M^N = N \; \log_x M} \tag{12.8}$$

For natural logarithm, Equation 12.8 can be written as

$$\boxed{\ln (M^n) = n \ln (M)} \tag{12.9}$$

There are four special formulas or properties resulting from the power law, namely:

1) $\boxed{\log_M M^N = N}$ \hfill (12.10)

2) $\boxed{\log_{(M)^N} M = \dfrac{1}{N}}$ \hfill (12.11)

3) $\boxed{\log_M \dfrac{1}{N} = - \log_M N}$ \hfill (12.12)

4) $\boxed{\log_{(M)^x} N^x = \log_M N}$ \hfill (12.13)

Let's try some examples to illustrate this law.

Example 6

Without using a calculator, simplify the following using the power law. Present the final answer as a single log in the form $a\log_b c$, where a is a rational number and b and c are positive integers.

a) $\log_3 16$ **b)** $\log_2 \sqrt{125}$ **c)** $\log_5 \sqrt[3]{81}$

What did you get? Find the solution below to double-check your answer.

Solution to Example 6

HINT

In this example, we will apply the law of logarithms shown below, although the application of the laws of indices will also be necessary.

$$\log_x M^N = N \log_x M$$

a) $\log_3 16$
Solution

$$\log_3 16 = \log_3 16$$
$$= \log_3 2^4 = 4\log_3 2$$
$$\therefore \ \log_3 16 = 4 \ \log_3 2$$

b) $\log_2 \sqrt{125}$
Solution

$$\log_2 \sqrt{125} = \log_2 125^{\frac{1}{2}} = \log_2 (5^3)^{\frac{1}{2}}$$
$$= \log_2 5^{\frac{3}{2}} = \frac{3}{2}\log_2 5$$
$$\therefore \ \log_2 \sqrt{125} = \frac{3}{2} \ \log_2 5$$

c) $\log_5 \sqrt[3]{81}$
Solution

$$\log_5 \sqrt[3]{81} = \log_5 81^{\frac{1}{3}}$$
$$= \log_5 (3^4)^{\frac{1}{3}}$$
$$= \log_5 3^{\frac{4}{3}} = \frac{4}{3}\log_5 3$$
$$\therefore \ \log_5 \sqrt[3]{81} = \frac{4}{3} \ \log_5 3$$

12.4.2 Derived or Special Laws

12.4.2.1 Unity (or Log of Unity) Law

This can be written as:

$$\boxed{\log_x 1 = 0 \mid x > 0} \tag{12.14}$$

This rule states that the logarithm of unity (1) to any base is zero. This is because any number raised to the power of zero produces one, which aligns with the power of zero law in indices.

12.4.2.2 Logarithm to the Same Base Law

This can be written as:

$$\boxed{\log_x x = 1} \tag{12.15}$$

The logarithm of any number to the same base is 1. This is because $x^1 = x$. In other words, if the log and the base are the same the answer will be 1.

12.4.2.3 Change of Base Law

This is defined as:

$$\boxed{\log_N M = \frac{\log_x M}{\log_x N}} \tag{12.16}$$

The above implies that the logarithm of a number to a certain base is the same as the logarithm of the number divided by the logarithm of the base such that both are given a new but the same base. For example, $\log_5 125$ can be written in any of the following three forms:

$$\log_5 125 = \frac{\log_{10} 125}{\log_{10} 5} \quad \textbf{OR} \quad \log_5 125 = \frac{\log_2 125}{\log_2 5} \quad \textbf{OR} \quad \log_5 125 = \frac{\log_x 125}{\log_x 5}$$

There is a special application of this rule when one needs to multiply two or more logs together. This states that:

$$\boxed{\left(\log_x M\right) \times \left(\log_y N\right) = \left(\log_x N\right)\left(\log_y M\right)} \tag{12.17}$$

Notice the swap of the base or the log. The proof for this is as follows:

$$\left(\log_x M\right)\left(\log_y N\right) = \frac{\log M}{\log x} \times \frac{\log N}{\log y}$$

$$= \frac{\log N}{\log x} \times \frac{\log M}{\log y}$$

$$= \left(\log_x N\right)\left(\log_y M\right)$$

For instance

$$\left(\log_5 100\right) \times \left(\log_{10} 25\right) = \left(\log_5 25\right) \times \left(\log_{10} 100\right)$$

$$= 2 \times 2 = 4$$

By using the above law, we can solve the above problem, otherwise it would not have been easy without a calculator. This is because $\log_5 100$ cannot be expressed as 5^x where x is an integer. Using this rule, we can also show that:

$$\log_N M = \frac{1}{\log_M N} \qquad (12.18)$$

Before moving to the final section, we will try examples to apply the laws we've covered.

Example 7

Without using a calculator, simplify the following using appropriate laws. Present the final answer in log form.

a) $\log_5 125 + 3\log_5 25 - \log_5 250$ **b)** $\log_2 \sqrt{2} + \log_3 \sqrt[3]{3} + \log_4 \sqrt[4]{4}$

What did you get? Find the solution below to double-check your answer.

Solution to Example 7

a) $\log_5 125 + 3\log_5 25 - \log_5 250$
Solution

$$\log_5 125 + 3\log_5 25 - \log_5 250 = \log_5 5^3 + 3\log_5 5^2 - \log_5(125 \times 2)$$
$$= 3\log_5 5 + 2 \times 3\log_5 5 - (\log_5 125 + \log_5 2)$$
$$= 3\log_5 5 + 2 \times 3\log_5 5 - (\log_5 5^3 + \log_5 2)$$
$$= 3\log_5 5 + 6\log_5 5 - 3\log_5 5 - \log_5 2$$
$$= 6\log_5 5 - \log_5 2 = 6(1) - \log_5 2 = 6 - \log_5 2$$
$$\therefore\ \log_5 125 + 3\log_5 25 - \log_5 250 = 6 - \log_5 2$$

b) $\log_2 \sqrt{2} + \log_3 \sqrt[3]{3} + \log_4 \sqrt[4]{4}$
Solution

$$\log_2 \sqrt{2} + \log_3 \sqrt[3]{3} + \log_4 \sqrt[4]{4} = \log_2 2^{\frac{1}{2}} + \log_3 3^{\frac{1}{3}} + \log_4 4^{\frac{1}{4}}$$
$$= \frac{1}{2}\log_2 2 + \frac{1}{3}\log_3 3 + \frac{1}{4}\log_4 4$$
$$= \frac{1}{2}(1) + \frac{1}{3}(1) + \frac{1}{4}(1)$$
$$= \frac{1}{2} + \frac{1}{3} + \frac{1}{4} = \frac{6+4+3}{12} = \frac{13}{12}$$
$$\therefore\ \log_2 \sqrt{2} + \log_3 \sqrt[3]{3} + \log_4 \sqrt[4]{4} = \frac{13}{12}$$

Example 8

Given that $\log_{10} 2 = 0.3010$, $\log_{10} 3 = 0.4771$, $\log_{10} 5 = 0.6990$, and $\log_{10} 7 = 0.8451$, determine the value of the following correct to 4 significant figures.

a) $\log_{10} 14$ **b)** $\log_{10} 40$ **c)** $\log_2 25$ **d)** $\log_{10} 3.75 + \log_{10} 4$

What did you get? Find the solution below to double-check your answer.

Solution to Example 8

a) $\log_{10} 14$
Solution

$$\begin{aligned}
\log_{10} 14 &= \log_{10}(2 \times 7) \\
&= \log_{10} 2 + \log_{10} 7 \\
&= 0.3010 + 0.8451 = 1.1461
\end{aligned}$$

$$\therefore \ \log_{10} 14 = \mathbf{1.146}$$

b) $\log_{10} 40$
Solution

$$\begin{aligned}
\log_{10} 40 &= \log_{10}(8 \times 5) \\
&= \log_{10}(2^3 \times 5) \\
&= \log_{10} 2^3 + \log_{10} 5 \\
&= 3 \log_{10} 2 + \log_{10} 5 \\
&= 3(0.3010) + 0.6990 \\
&= 0.9030 + 0.6990 = 1.6020
\end{aligned}$$

$$\therefore \ \log_{10} 40 = \mathbf{1.602}$$

ALTERNATIVE METHOD

$$\begin{aligned}
\log_{10} 40 &= \log_{10}(4 \times 10) = \log_{10}(2^2 \times 10) \\
&= \log_{10} 2^2 + \log_{10} 10 = 2\log_{10} 2 + 1 \\
&= 2(0.3010) + 1 = 0.6020 + 1 = 1.6020
\end{aligned}$$

$$\therefore \ \log_{10} 40 = \mathbf{1.602}$$

c) $\log_2 25$

Solution

$$\log_2 25 = \frac{\log 25}{\log 2} = \frac{\log 5^2}{\log 2}$$

$$= \frac{2\log 5}{\log 2} = \frac{2\,(0.6990)}{0.3010}$$

$$= 4.6445$$

$$\therefore \ \mathbf{\log_2 25 = 4.645}$$

---

d) $\log_{10} 3.75 + \log_{10} 4$

Solution

$$\log_{10} 3.75 + \log_{10} 4 = \log_{10}\,(3.75 \times 4)$$

$$= \log_{10} 15 = \log_{10}\,(3 \times 5)$$

$$= \log_{10} 3 + \log_{10} 5$$

$$= 0.4771 + 0.6990 = 1.1761$$

$$\therefore \ \mathbf{\log_{10} 3.75 + \log_{10} 4 = 1.176}$$

Another set of examples for us to try.

Example 9

Show that

a) $\log_P Q \times \log_Q P = 1$
                                                             b) $\log_b Q^{x-n} - \log_b Q^{y+n} = (x - y)\log_b Q$

What did you get? Find the solution below to double-check your answer.

Solution to Example 9

a) $\log_P Q \times \log_Q P = 1$

Solution

LHS

$$\log_P Q \times \log_Q P = \frac{\log Q}{\log P} \times \frac{\log P}{\log Q}$$

$$= 1 \ \mathbf{(RHS)}$$

NOTE

This is a practical application of Equation 12.17.

ALTERNATIVE METHOD

Alternatively (though longer), let

$$\log_P Q = y$$

which implies that

$$Q = P^y \qquad ------- (i)$$

and

$$\log_Q P = x$$

which implies that

$$P = Q^x \qquad ------- (ii)$$

Substitute for Q in the left-hand side of the original equation as:

$$\log_P Q \times \log_Q P = \log_P (P^y) \times \log_{(P^y)} P$$
$$= y \log_P P \times \left(\frac{1}{y}\right) \log_P P$$
$$= y(1) \times \left(\frac{1}{y}\right)(1) = y \times \frac{1}{y}$$
$$= 1 \ (\textbf{RHS})$$

---

b) $\log_b Q^{x-n} - \log_b Q^{y+n} = (x-y)\log_b Q$
Solution
LHS

$$\log_b Q^{x-n} - \log_b Q^{y+n} = \log_b \left\{ \frac{(Q^{x-n})}{(Q^{y+n})} \right\}$$
$$= \log_b \left[Q^{(x-n)-(y+n)} \right]$$
$$= \log_b Q^{(x-y)}$$
$$= (x-y)\log_b Q$$
$$= \textbf{RHS}$$

Example 10

Rewrite the following in their simplest log form.

a) $E = \frac{x^5}{y^{0.6} z^{0.3}}$ **b)** $f_c = \frac{1}{2\pi\sqrt{LC}}$ **c)** $V = \frac{\pi h}{3}(l+h)(l-h)$ **d)** $Q = \frac{1}{\beta}(2\alpha-1)^2 \left(\sqrt[3]{\gamma}\right)$

What did you get? Find the solution below to double-check your answer.

Solution to Example 10

a) $E = \dfrac{x^5}{y^{0.6}z^{0.3}}$

Solution

$$E = \frac{x^5}{y^{0.6}z^{0.3}}$$

This implies that

$$\log E = \log\left[\frac{x^5}{y^{0.6}z^{0.3}}\right]$$
$$= \log x^5 - \log y^{0.6}z^{0.3}$$
$$= \log x^5 - \left[\log y^{0.6} + \log z^{0.3}\right]$$
$$= 5\log x - [0.6\log y + 0.3\log z]$$
$$\therefore \log E = 5\log x - 0.6\log y - 0.3\log z$$

---

b) $f_c = \dfrac{1}{2\pi\sqrt{LC}}$

Solution

$$f_c = \frac{1}{2\pi\sqrt{LC}}$$

This implies that

$$\log f_c = \log\left[\frac{1}{2\pi\sqrt{LC}}\right] = \log 1 - \log 2\pi\sqrt{L \times C}$$
$$= 0 - \left[\log 2 + \log \pi + \log\sqrt{L} + \log\sqrt{C}\right]$$
$$= -\left[\log 2 + \log \pi + \log L^{\frac{1}{2}} + \log C^{\frac{1}{2}}\right]$$
$$= -\left[\log 2 + \log \pi + \frac{1}{2}\log L + \frac{1}{2}\log C\right]$$
$$\therefore \log f_c = -\left[\log 2 + \log \pi + \frac{1}{2}\log L + \frac{1}{2}\log C\right]$$

---

c) $V = \dfrac{\pi h}{3}(l + h)(l - h)$

Solution

$$V = \frac{\pi h}{3}(l + h)(l - h)$$

This implies that

$$\log V = \log\left[\frac{\pi h}{3}(l + h)(l - h)\right]$$
$$= \log\frac{\pi h}{3} + \log(l + h) + \log(l - h)$$

$$= \log \pi h - \log 3 + \log(l+h) + \log(l-h)$$

$$\therefore \log V = \log \pi + \log h - \log 3 + \log(l+h) + \log(l-h)$$

---

d) $Q = \frac{1}{\beta}(2\alpha - 1)^2 \left(\sqrt[3]{\gamma}\right)$

Solution

$$Q = \frac{1}{\beta}(2\alpha - 1)^2 \left(\sqrt[3]{\gamma}\right)$$

This implies that

$$\ln Q = \ln\left[\frac{1}{\beta}(2\alpha - 1)^2 \left(\sqrt[3]{\gamma}\right)\right]$$

$$= \ln\left(\frac{1}{\beta}\right) + \ln(2\alpha - 1)^2 + \ln\sqrt[3]{\gamma}$$

$$= \ln 1 - \ln\beta + 2\ln(2\alpha - 1) + \ln\gamma^{\frac{1}{3}}$$

$$= 0 - \ln\beta + 2\ln(2\alpha - 1) + \frac{1}{3}\ln\gamma$$

$$\therefore \ln Q = 2\ln(2\alpha - 1) + \frac{1}{3}\ln\gamma - \ln\beta$$

NOTE

Here, we have taken the natural logarithm of both sides of the equation instead of the common logarithm. This not only follows the question's instructions but is also correct for any base when solving this type of problem.

A final set of examples to try.

Example 11

Express the following without logarithms.

a) $\ln P = \frac{1}{2}\ln(V^2 + 3) - 3\ln R + 1$

b) $\log_3 S = 2\left(\log_3 L + \log_3 2\right) - \left(\log_3 25 + 2\log_3 \pi + 2\log_3 r\right)$

What did you get? Find the solution below to double-check your answer.

Solution to Example 11

a) $\ln P = \dfrac{1}{2} \ln (V^2 + 3) - 3 \ln R + 1$

Solution

$$\ln P = \frac{1}{2}\ln(V^2 + 3) - 3\ln R + 1$$

$$= \ln(V^2 + 3)^{\frac{1}{2}} - \ln R^3 + \ln e$$

$$= \ln\left[\frac{(V^2 + 3)^{\frac{1}{2}}e}{R^3}\right]$$

So we have

$$\ln P = \ln\left[\frac{(V^2 + 3)^{\frac{1}{2}}e}{R^3}\right]$$

Both sides are expressed to the same base, therefore:

$$P = \frac{(V^2 + 3)^{\frac{1}{2}}e}{R^3}$$

b) $\log_3 S = 2\left(\log_3 L + \log_3 2\right) - \left(\log_3 25 + 2\log_3 \pi + 2\log_3 r\right)$

Solution

$$\log_3 S = 2\left(\log_3 L + \log_3 2\right) - \left(\log_3 25 + 2\log_3 \pi + 2\log_3 r\right)$$

$$= 2\left(\log_3 2L\right) - \left(\log_3 25\pi^2 r^2\right)$$

$$= 2\log_3 2L - \log_3 25\pi^2 r^2$$

$$= \log_3 2^2 L^2 - \log_3 25\pi^2 r^2$$

$$= \log_3\left(\frac{4L^2}{25\pi^2 r^2}\right)$$

So we have

$$\log_3 S = \log_3\left(\frac{4L^2}{25\pi^2 r^2}\right)$$

Both sides are expressed to the same base, therefore:

$$S = \frac{4L^2}{25\pi^2 r^2}$$

12.5 CHAPTER SUMMARY

1) Logarithm (commonly shortened to **log** and sometimes as **lg**) is derived from two Greek words, namely, **logos** (expression) and **arithmos** (number).

2) The logarithm of a number to a given base is the value of the power to which the base must be raised to produce the number.

3) Given that x is a number and y its associated base, if c equals '**the logarithm to base y of x**', then we can write this as:

$$\boxed{\log_y x = c}$$

y must be a positive real number excluding **1** and x must also be positive for c to be a real number. From this we can state that

$$\boxed{\log_y x = \begin{cases} +ve & for\ x > 1 \\ -ve & for\ 0 < x < 1 \end{cases}}$$

4) It is evident that indices and logarithms are inversely related. This relationship is given as:

$$\boxed{\log_y x = c} \quad \Longleftrightarrow \quad \boxed{y^c = x}$$

5) Common logarithm is a logarithm to the base of 10, i.e., $\log_{10} N$. In general, when the base is 10, it is usually omitted. In other words, $\log_{10} N$ is simply written as $\log N$.

6) Natural logarithm is the logarithm to the base of an irrational number denoted as e, where e is approximately 2.718281 (correct to 6 decimal places). It is usually written as $\ln N$ (read as 'ell-en of N') instead of $\log_e N$ as one might have expected.

7) Fundamental laws of logarithms

- Addition–product law

$$\boxed{\log_x M + \log_x N = \log_x (MN)}$$

- Subtraction–quotient law

$$\boxed{\log_x M - \log_x N = \log_x \left(\frac{M}{N}\right)}$$

- Power law

$$\boxed{\log_x M^N = N \log_x M}$$

- There are four special formulas or properties resulting from the power law, namely:

$$\boxed{\log_M M^N = N}$$

$$\log_{(M)^N} M = \frac{1}{N}$$

$$\log_M \frac{1}{N} = -\log_M N$$

$$\log_{(M)^x} N^x = \log_M N$$

- Unity (or log of unity) law

$$\log_x 1 = 0$$

- Logarithm to the same base law

$$\log_x x = 1$$

- Change of base law

$$\log_N M = \frac{\log_x M}{\log_x N}$$

There is a special application of the change of base law for multiplying two or more logarithms together. It states that:

$$\left(\log_x M\right) \times \left(\log_y N\right) = \left(\log_x N\right)\left(\log_y M\right)$$

Consequently, we can show that:

$$\log_N M = \frac{1}{\log_M N}$$

****

12.6 FURTHER PRACTICE

To access complementary contents, including additional exercises, please go to www.dszak.com.

13 Indicial Equations

13.1 INTRODUCTION

Indicial equations are equations involving powers, where either the base or the exponent is the unknown variable to be determined. For this case, we will need concepts covered in the previous two chapters and Chapters 6 and 7.

13.2 SOLVING INDICIAL EQUATIONS

Indicial equations can generally be solved using any known method of solving polynomials, in conjunction with the laws of indices and logarithms. The method of solving an indicial equation is determined by the nature of the equation, which can be broadly grouped into two:

Option 1 When the unknown is the base

This is often solved by simplifying the expressions and then applying a suitable method of solving polynomials.

Option 2 When the unknown is the index (or exponent)

If the base of the two sides of the equation are the same, then equate their indices otherwise the log of both sides must be taken. In the latter case, one needs to apply the one-to-one property of logarithm which states that:

$$\text{If } \log_x M = \log_x N \text{ then } M = N \tag{13.1}$$

In the two scenarios outlined in option 2, each side of the equation must contain only a single term before equating the indices.

We will now try examples covering a range of questions relating to the two options.

Example 1

Solve the following indicial equations.

a) $x^{\frac{1}{4}} = 2$ **b)** $6z^3 = 162$ **c)** $\beta^{\frac{5}{2}} = \beta^2\sqrt{3}$ **d)** $\omega^{\frac{2}{3}} = \sqrt[3]{4\omega^{\frac{1}{3}}}$

What did you get? Find the solution below to double-check your answer.

Solution to Example 1

a) $x^{\frac{1}{4}} = 2$
Solution

$$x^{\frac{1}{4}} = 2$$

Raise both sides to the power of 4

$$\left(x^{\frac{1}{4}}\right)^4 = 2^4$$
$$x = 2^4$$
$$\therefore x = 16$$

ALTERNATIVE METHOD

$$x^{\frac{1}{4}} = 2$$

$$x^{\frac{1}{4}} = \left(2^4\right)^{\frac{1}{4}}$$

Since the power is the same, we can equate the base as

$$x = 2^4$$
$$\therefore x = 16$$

b) $6z^3 = 162$
Solution

$$6z^3 = 162$$

Divide both sides by 6

$$z^3 = 27$$

Express 27 as a power of 3 so the powers on both the LHS and RHS cancel each other. We have:

$$z^3 = 3^3$$
$$\therefore z = 3$$

---

c) $\beta^{\frac{5}{2}} = \beta^2 \sqrt{3}$

Solution

$$\beta^{\frac{5}{2}} = \beta^2 \sqrt{3}$$

Square both sides

$$\left(\beta^{\frac{5}{2}}\right)^2 = \left(\beta^2 \sqrt{3}\right)^2$$
$$\beta^5 = 3\beta^4$$
$$\beta^5 - 3\beta^4 = 0$$
$$\beta^4 (\beta - 3) = 0$$

Therefore, either

$$\beta^4 = 0 \qquad\qquad \beta - 3 = 0$$
$$\beta = 0 \qquad \text{OR} \qquad \beta = 3$$
$$\therefore \beta = 0 \quad \text{OR} \quad \beta = 3$$

ALTERNATIVE METHOD

Note that the approach shown below will give only one value of β.

$$\beta^{\frac{5}{2}} = \beta^2 \sqrt{3}$$

This implies that

$$\beta^{\frac{5}{2}} = \beta^2 \left(3^{\frac{1}{2}}\right)$$

Divide both sides by β^2

$$\frac{\beta^{\frac{5}{2}}}{\beta^2} = 3^{\frac{1}{2}}$$
$$\beta^{\frac{5}{2}-2} = 3^{\frac{1}{2}}$$
$$\beta^{\frac{1}{2}} = 3^{\frac{1}{2}}$$

Now square both sides

$$\left(\beta^{\frac{1}{2}}\right)^2 = \left(3^{\frac{1}{2}}\right)^2$$
$$\therefore \beta = 3$$

d) $\omega^{\frac{2}{3}} = \sqrt[3]{4\omega^{\frac{1}{3}}}$

Solution

$$\omega^{\frac{2}{3}} = \sqrt[3]{4\omega^{\frac{1}{3}}}$$

$$\left(\omega^{\frac{2}{3}}\right)^3 = \left(\sqrt[3]{4\omega^{\frac{1}{3}}}\right)^3$$

$$\omega^2 = 4\omega$$

$$\omega^2 - 4\omega = 0$$

$$\omega(\omega - 4) = 0$$

Therefore, either

$$\omega = 0 \qquad \textbf{OR} \qquad \begin{array}{c} \omega - 4 = 0 \\ \omega = 4 \end{array}$$

$$\therefore \omega = 0 \quad \textbf{OR} \quad \omega = 4$$

Example 2

Solve $2 + x = \sqrt{8x^{\frac{1}{2}}}$, given that $x > 0$.

What did you get? Find the solution below to double-check your answer.

Solution to Example 2

$$2 + x = \sqrt{8x^{\frac{1}{2}}}$$

Square both sides

$$(2 + x)^2 = \left(8^{\frac{1}{2}} x^{\frac{1}{2}}\right)^2$$

Open the brackets

$$4 + 4x + x^2 = 8x$$

Collect the like terms

$$x^2 - 4x + 4 = 0$$

Factorise the above equation

$$(x - 2)(x - 2) = 0$$

This implies that

$$x - 2 = 0$$
$$x = 2$$
$$\therefore x = 2 \ \textbf{(twice)}$$

ALTERNATIVE METHOD

$$2 + x = \sqrt{8}x^{\frac{1}{2}}$$

Let $y = x^{\frac{1}{2}}$, which implies $y^2 = x$. We can now substitute as

$$2 + y^2 = \sqrt{8}y$$
$$y^2 - \sqrt{8}y + 2 = 0$$

Using quadratic formula

$$\frac{-b \pm \sqrt{b^2 - 4ac}}{2a}$$

Comparing with the standard form $ax^2 + bx + c$, we note that $a = 1$, $b = -\sqrt{8}$, and $c = 2$. Hence:

$$y = \frac{\sqrt{8} \pm \sqrt{\left(-\sqrt{8}\right)^2 - 4\,(1)\,(2)}}{2 \times 1}$$

$$= \frac{\sqrt{8} \pm \sqrt{8 - 8}}{2}$$

$$= \frac{\sqrt{8}}{2} = \frac{2\sqrt{2}}{2}$$

$$= \sqrt{2}$$

Since $x = y^2$, we have

$$x = \left(\sqrt{2}\right)^2$$

$$\therefore x = 2$$

Let's try option 2 now, where the index represents the unknown variable.

Example 3

Solve the following indicial equations.

a) $25^x = 125$ b) $6^{-x} = 216$ c) $49^y = \dfrac{1}{343}$ d) $2^{y+2} = 0.125$

e) $3^{4m} \div 3^{3-m} = \dfrac{1}{243}$ f) $2^{2n+3} - 2^{2n+1} = \dfrac{3}{8}$ g) $6^{1-2n} \times 36^{2n-1} = 216$ h) $\dfrac{27^{3x}}{3^{x+4}} = \dfrac{243}{9^{x-2}}$

What did you get? Find the solution below to double-check your answer.

Solution to Example 3

a) $25^x = 125$
Solution

$$25^x = 125$$
$$5^{2x} = 5^3$$

which implies that

$$2x = 3$$
$$\therefore x = \frac{3}{2}$$

b) $6^{-x} = 216$
Solution

$$6^{-x} = 216$$
$$6^{-x} = 6^3$$

Equate the powers as:

$$-x = 3$$
$$\therefore x = -3$$

c) $49^y = \dfrac{1}{343}$
Solution

$$49^y = \frac{1}{343}$$
$$\left(7^2\right)^y = \frac{1}{7^3}$$
$$\left(7^2\right)^y = 7^{-3}$$
$$7^{2y} = 7^{-3}$$

Equate the powers as:

$$2y = -3$$
$$\therefore y = -\frac{3}{2}$$

---

d) $2^{y+2} = 0.125$

Solution

$$2^{y+2} = 0.125$$
$$2^{y+2} = \frac{125}{1000}$$
$$= \frac{1}{8} = \frac{1}{2^3}$$
$$= 2^{-3}$$

That is

$$2^{y+2} = 2^{-3}$$

Now equate the powers as:

$$y + 2 = -3$$
$$\therefore y = -5$$

---

e) $3^{4m} \div 3^{3-m} = \frac{1}{243}$

Solution

$$3^{4m} \div 3^{3-m} = \frac{1}{243}$$
$$3^{4m-(3-m)} = \frac{1}{3^5}$$
$$3^{4m-3+m} = 3^{-5}$$
$$3^{5m-3} = 3^{-5}$$

Equate the powers as:

$$5m - 3 = -5$$
$$5m = -2$$
$$\therefore m = -\frac{2}{5}$$

---

f) $2^{2n+3} - 2^{2n+1} = \frac{3}{8}$

Solution

$$2^{2n+3} - 2^{2n+1} = \frac{3}{8}$$
$$2^{2n}\left(2^3\right) - 2^{2n}\left(2\right) = \frac{3}{8}$$

Factorise

$$2^{2n}\left(2^3 - 2\right) = \frac{3}{8}$$

$$2^{2n}\left(8 - 2\right) = \frac{3}{8}$$

$$2^{2n}\left(6\right) = \frac{3}{8}$$

Divide both sides by 6, we have:

$$2^{2n} = \frac{3}{8} \div 6$$

$$2^{2n} = \frac{3}{8} \times \frac{1}{6}$$

$$2^{2n} = \frac{1}{16}$$

$$2^{2n} = \frac{1}{2^4}$$

$$2^{2n} = 2^{-4}$$

Equate the powers as:

$$2n = -4$$

$$\therefore n = -2$$

ALTERNATIVE METHOD

We will be using substitution

$$2^{2n+3} - 2^{2n+1} = \frac{3}{8}$$

using law of indices thus we have:

$$\left(2^{2n} \times 2^3\right) - \left(2^{2n} \times 2^1\right) = \frac{3}{8}$$

Let $x = 2^{2n}$, we have:

$$x\left(2^3\right) - x\left(2\right) = \frac{3}{8}$$

$$8x - 2x = \frac{3}{8}$$

$$6x = \frac{3}{8}$$

Divide both sides by 6

$$x = \frac{1}{16}$$
$$= \frac{1}{2^4}$$
$$= 2^{-4}$$

Thus

$$x = 2^{-4}$$

Substituting back, we have

$$2^{2n} = 2^{-4}$$

Therefore

$$2n = -4$$

$$\therefore n = -2$$

---

g) $6^{1-2n} \times 36^{2n-1} = 216$
Solution

$$6^{1-2n} \times 36^{2n-1} = 216$$
$$6^{1-2n} \times \left(6^2\right)^{2n-1} = 6^3$$
$$6^{1-2n} \times 6^{4n-2} = 6^3$$
$$6^{1-2n+4n-2} = 6^3$$
$$6^{2n-1} = 6^3$$

Equate the power as:

$$2n - 1 = 3$$
$$2n = 4$$
$$\therefore n = 2$$

---

h) $\frac{27^{3x}}{3^{x+4}} = \frac{243}{9^{x-2}}$
Solution

$$\frac{27^{3x}}{3^{x+4}} = \frac{243}{9^{x-2}}$$
$$\frac{\left(3^3\right)^{3x}}{3^{x+4}} = \frac{3^5}{\left(3^2\right)^{x-2}}$$

$$\frac{3^{9x}}{3^{x+4}} = \frac{3^5}{3^{2x-4}}$$
$$3^{9x-x-4} = 3^{5-2x+4}$$
$$3^{8x-4} = 3^{9-2x}$$

Equate the power as:

$$8x - 4 = 9 - 2x$$
$$10x = 13$$
$$\therefore x = \frac{13}{10}$$

Example 4

By letting $b = a^{\frac{1}{3}}$ or otherwise, determine the values of a for which $a^{\frac{1}{3}} - 2a^{-\frac{1}{3}} = 1$.

What did you get? Find the solution below to double-check your answer.

Solution to Example 4

$$a^{\frac{1}{3}} - 2a^{-\frac{1}{3}} = 1$$
$$a^{\frac{1}{3}} - \frac{2}{a^{\frac{1}{3}}} = 1$$

Substitute $b = a^{\frac{1}{3}}$ in the above equation, which gives:

$$b - \frac{2}{b} = 1$$

Multiply through by b and collect the like terms

$$b^2 - 2 = b$$
$$b^2 - b - 2 = 0$$
$$(b-2)(b+1) = 0$$

Therefore

$$b - 2 = 0 \quad \textbf{OR} \quad b + 1 = 0$$
$$b = 2 \qquad\qquad b = -1$$

Now substitute for b

When $b = 2$ **When $b = -1$**
$$a^{\frac{1}{3}} = 2 \qquad a^{\frac{1}{3}} = -1$$
$$a = 2^3 \qquad a = (-1)^3$$
$$a = 8 \qquad a = -1$$

We can therefore say that

$$a = 8 \quad \text{OR} \quad a = -1$$

ALTERNATIVE METHOD

$$a^{\frac{1}{3}} - 2a^{-\frac{1}{3}} = 1$$

$$a^{\frac{1}{3}} - \frac{2}{a^{\frac{1}{3}}} = 1$$

$$a^{\frac{1}{3}} \times a^{\frac{1}{3}} - 2 = a^{\frac{1}{3}}$$

$$\left(a^{\frac{1}{3}}\right)^2 - a^{\frac{1}{3}} - 2 = 0$$

$$\left(a^{\frac{1}{3}} - 2\right)\left(a^{\frac{1}{3}} + 1\right) = 0$$

Therefore

$$a^{\frac{1}{3}} - 2 = 0 \quad \text{OR} \quad a^{\frac{1}{3}} - 1 = 0$$
$$a^{\frac{1}{3}} = 2 \qquad\qquad a^{\frac{1}{3}} = -1$$
$$a = 8 \qquad\qquad\quad a = (-1)^3$$
$$\qquad\qquad\qquad\qquad a = -1$$

$$\therefore a = 8 \quad \text{OR} \quad a = -1$$

Example 5

Solve $v^{\frac{1}{2}} + 3v^{-\frac{1}{2}} = 4$.

What did you get? Find the solution below to double-check your answer.

Solution to Example 5

$$v^{\frac{1}{2}} + 3v^{-\frac{1}{2}} = 4$$

Multiply through by $v^{\frac{1}{2}}$

$$v^{\frac{1}{2}}\left(v^{\frac{1}{2}} + 3v^{-\frac{1}{2}}\right) = 4v^{\frac{1}{2}}$$

Open the brackets

$$v^{\frac{1}{2}} \times v^{\frac{1}{2}} + 3v^{-\frac{1}{2}} \times v^{\frac{1}{2}} = 4v^{\frac{1}{2}}$$

Apply relevant laws of indices and simplify

$$v^{\frac{1}{2}+\frac{1}{2}} + 3v^{-\frac{1}{2}+\frac{1}{2}} = 4v^{\frac{1}{2}}$$

$$v + 3v^0 = 4v^{\frac{1}{2}}$$

$$v + 3 = 4v^{\frac{1}{2}}$$

Now square both sides

$$(v+3)^2 = \left(4v^{\frac{1}{2}}\right)^2$$

Open the brackets

$$v^2 + 6v + 9 = 16v$$

Collect the like terms

$$v^2 - 10v + 9 = 0$$

Factorise the above equation

$$(v-1)(v-9) = 0$$

Therefore, either

$$\begin{array}{ccc} v - 1 = 0 & & v - 9 = 0 \\ v = 1 & \textbf{OR} & v = 9 \end{array}$$

$$\therefore v = 1 \quad \textbf{OR} \quad v = 9$$

Here's another set of examples to try Option 2, where the index is the unknown.

Example 6

For $x \in \mathbb{R}$, solve the following equations:

a) $4^x - 2^x = 2$

b) $2^{2x} - 10(2^x) + 16 = 0$

c) $3^{2x+1} - 4(3^{x+1}) + 9 = 0$

d) $6^{2x} - 9(6^x) + 18 = 0$

What did you get? Find the solution below to double-check your answer.

Solution to Example 6

---

a) $4^x - 2^x = 2$
Solution

$$4^x - 2^x = 2$$
$$4^x - 2^x - 2 = 0$$
$$\left(2^2\right)^x - 2^x - 2 = 0$$
$$\left(2^x\right)^2 - 2^x - 2 = 0$$

Let $y = 2^x$, thus

$$y^2 - y - 2 = 0$$
$$(y - 2)(y + 1) = 0$$

Therefore

$$y - 2 = 0 \qquad \text{OR} \qquad y + 1 = 0$$
$$y = 2 \qquad\qquad\qquad y = -1$$

Now substitute for y

When $y = 2$ **When $y = -1$**
$2^x = 2$ $2^x = -1$
$2^x = 2^1$ No real solution
$x = 1$

We can therefore say that

$$x = 1$$

---

b) $2^{2x} - 10\left(2^x\right) + 16 = 0$
Solution

$$2^{2x} - 10\left(2^x\right) + 16 = 0$$
$$\left(2^x\right)^2 - 10\left(2^x\right) + 16 = 0$$

Let $y = 2^x$, thus

$$y^2 - 10y + 16 = 0$$
$$(y - 8)(y - 2) = 0$$

Therefore

$$y - 8 = 0 \qquad \text{OR} \qquad y - 2 = 0$$
$$y = 8 \qquad\qquad\qquad y = 2$$

Now substitute for y

When y = 8   **When y = 2**

$2^x = 8$ $2^x = 2$

$2^x = 2^3$ $2^x = 2^1$

$x = 3$ $x = 1$

We can therefore say that

$$x = 1 \quad \textbf{OR} \quad x = 3$$

---

c) $3^{2x+1} - 4\left(3^{x+1}\right) + 9 = 0$

Solution

$$3^{2x+1} - 4\left(3^{x+1}\right) + 9 = 0$$
$$3\left(3^{2x}\right) - 4\left(3^x \times 3\right) + 9 = 0$$
$$3\left(3^x\right)^2 - 12\left(3^x\right) + 9 = 0$$

Let $v = 3^x$, thus

$$3v^2 - 12v + 9 = 0$$
$$(3v - 3)(v - 3) = 0$$

Therefore

$3v - 3 = 0$ $v - 3 = 0$

$3v = 3$ **OR** $v = 3$

$v = 1$

Now substitute for v

When v = 1   **When v = 3**

$3^x = 3^0$ $3^x = 3$

$x = 0$ $3^x = 3^1$

                                 $x = 1$

We can therefore say that

$$x = 0 \quad \textbf{OR} \quad x = 1$$

---

d) $6^{2x} - 9\left(6^x\right) + 18 = 0$

Solution

$$6^{2x} - 9\left(6^x\right) + 18 = 0$$
$$\left(6^x\right)^2 - 9\left(6^x\right) + 18 = 0$$

Let $m = 6^x$ thus

$$m^2 - 9m + 18 = 0$$
$$(m - 6)(m - 3) = 0$$

Therefore

$$m - 6 = 0 \qquad \textbf{OR} \qquad m - 3 = 0$$
$$\boldsymbol{m = 6} \qquad\qquad\qquad \boldsymbol{m = 3}$$

Now substitute for m

When $m = 6$	When $m = 3$
$6^x = 6$	$6^x = 3$
$6^x = 6^1$	$x = \log_6 3$
$x = 1$	$x = 0.6131$

We can therefore say that

$$x = \textbf{1} \quad \textbf{OR} \quad x = \textbf{0.6131}$$

Here's another set of examples to try Option 2, which involves the application of logarithm-index relationship.

Example 7

Solve the following equations and present the answers correct to 4 significant figures.

a) $3^x = 7$
c) $6\left(8^{3x+2}\right) = 5^{2x-7}$

b) $2^{2x} = 5^{x-3}$
d) $3\left(13.6^{x-3}\right) \times 2\left(4^{3x}\right) = 3.8^{x-1}$

What did you get? Find the solution below to double-check your answer.

Solution to Example 7

a) $3^x = 7$
Solution

$$3^x = 7$$

We will need to take the logarithm to base 10 of both sides.

$$\log 3^x = \log 7$$
$$x \log 3 = \log 7$$

Divide both sides by $\log 3$, we have

$$x = \frac{\log 7}{\log 3} = \frac{0.84510}{0.47712}$$
$$\therefore x = \textbf{1.771}$$

ALTERNATIVE METHOD

We could have solved this problem by converting the index form to a logarithmic form as:

$$3^x = 7$$

which implies that

$$x = \log_3 7$$
$$\therefore x = \mathbf{1.771}$$

NOTE
It is also worth noting that since 7 cannot be expressed in the form 3^n it will be helpful to solve by using a logarithmic relationship.

---

b) $2^{2x} = 5^{x-3}$
Solution

$$2^{2x} = 5^{x-3}$$

Take the log of both sides and simplify

$$\log 2^{2x} = \log 5^{x-3}$$
$$2x \log 2 = (x-3)\log 5$$
$$2x \log 2 = x \log 5 - 3 \log 5$$
$$2x \log 2 - x \log 5 = -3 \log 5$$
$$x(2 \log 2 - \log 5) = -3 \log 5$$
$$x = \frac{-3 \log 5}{2 \log 2 - \log 5}$$
$$= \frac{-3(0.69897)}{2(0.30103) - (0.69897)}$$
$$\therefore x = \mathbf{21.64}$$

---

c) $6\left(8^{3x+2}\right) = 5^{2x-7}$
Solution

$$6\left(8^{3x+2}\right) = 5^{2x-7}$$

Take the log of both sides. For this example, we will choose base e and therefore take the natural log of both sides. This implies that

$$\ln 6\left(8^{3x+2}\right) = \ln 5^{2x-7}$$
$$\ln 6 + \ln 8^{3x+2} = (2x-7)\ln 5$$
$$\ln 6 + (3x+2)\ln 8 = 2x \ln 5 - 7 \ln 5$$
$$\ln 6 + 3x \ln 8 + 2 \ln 8 = 2x \ln 5 - 7 \ln 5$$

Collect the like terms

$$\ln 6 + 2\ln 8 + 7\ln 5 = 2x\ln 5 - 3x\ln 8$$

$$\ln 6 + \ln 8^2 + \ln 5^7 = x\left(\ln 5^2 - \ln 8^3\right)$$

$$\ln\left(6 \times 8^2 \times 5^7\right) = x\left[\ln\left(\frac{5^2}{8^3}\right)\right]$$

$$x = \frac{\ln\left(6 \times 8^2 \times 5^7\right)}{\ln\left(\frac{5^2}{8^3}\right)} = \frac{17.217}{-3.0194}$$

$$\therefore x = -5.702$$

---

d) $3\left(13.6^{x-3}\right) \times 2\left(4^{3x}\right) = 3.8^{x-1}$

Solution

$$3\left(13.6^{x-3}\right) \times 2\left(4^{3x}\right) = 3.8^{x-1}$$

Take the log of both sides

$$\log 3\left(13.6^{x-3}\right) \times 2\left(4^{3x}\right) = \log 3.8^{x-1}$$

$$\log 3\left(13.6^{x-3}\right) + \log 2\left(4^{3x}\right) = (x-1)\log 3.8$$

$$\log 3 + (x-3)\log 13.6 + \log 2 + 3x\log 4 = x\log 3.8 - \log 3.8$$

Open the brackets

$$\log 3 + x\log 13.6 - 3\log 13.6 + \log 2 + 3x\log 4 = x\log 3.8 - \log 3.8$$

Collect the like terms

$$\log 3 + \log 2 + \log 3.8 - \log 13.6^3 = x\log 3.8 - x\log 13.6 - 3x\log 4$$

Simplify and make x the subject

$$\log 3 + \log 2 + \log 3.8 - \log 13.6^3 = x\left(\log 3.8 - \log 13.6 - \log 4^3\right)$$

$$\log\left(\frac{3 \times 2 \times 3.8}{13.6^3}\right) = x\left[\log\left(\frac{3.8}{13.6 \times 4^3}\right)\right]$$

$$x = \frac{\log\left(\frac{22.8}{13.6^3}\right)}{\log\left(\frac{3.8}{13.6 \times 4^3}\right)} = \frac{-2.04268}{-2.35994}$$

$$\therefore x = 0.8656$$

Example 8

Solve the following logarithmic equations and present the answer correct to 3 significant figures.

a) $\log_2 (3x - 1) = 2$ **b)** $\ln (x^2 - 2) = 1$ **c)** $\log_{11} (x^2 - 6) = \frac{1}{3}$

What did you get? Find the solution below to double-check your answer.

Solution to Example 8

a) $\log_2 (3x - 1) = 2$
Solution

$$\log_2 (3x - 1) = 2$$

Write this in index form and simplify

$$(3x - 1) = 2^2$$
$$3x - 1 = 4$$
$$3x = 5$$
$$x = \frac{5}{3}$$
$$\therefore x = 1.67$$

b) $\ln (x^2 - 2) = 1$
Solution

$$\ln (x^2 - 2) = 1$$

Write this in index form and simplify

$$x^2 - 2 = e^1$$
$$x^2 = 2 + e$$
$$x = \pm\sqrt{2 + e}$$
$$\therefore x = \pm 2.17$$

c) $\log_{11} (x^2 - 6) = \frac{1}{3}$
Solution

$$\log_{11} (x^2 - 6) = \frac{1}{3}$$

Write this in index form and simplify

$$(x^2 - 6) = 11^{\frac{1}{3}}$$
$$x^2 = 6 + \sqrt[3]{11}$$
$$x^2 = 8.2240$$
$$x = \pm\sqrt{8.2240}$$
$$\therefore x = \pm 2.87$$

Here is a final set of examples for us to try.

Example 9

Solve the following logarithmic equations.

a) $\log_3 (x + 2) + \log_3 (x + 4) = 1$

b) $\log_{10} (x^2 - 4x + 7) = \log_{10} 100$

c) $\log_8 (2x^2 - 13x + 84) = 2$

d) $\log_2 (x^2 - 4x + 12) = 3$

e) $\log_9 (x^2 + 7x + 15) = \frac{1}{2}$

f) $\log_6 x + \log_6 (x - 5) = 2$

What did you get? Find the solution below to double-check your answer.

Solution to Example 9

a) $\log_3 (x + 2) + \log_3 (x + 4) = 1$
Solution

$$\log_3 (x + 2) + \log_3 (x + 4) = 1$$

We need to express both sides in log to base 3. We know that $\log_x x = 1$. We will therefore change the RHS to $\log_3 3$.

$$\log_3 (x + 2) + \log_3 (x + 4) = \log_3 3$$

Apply the addition-product law of logarithms to the left-hand side (LHS); thus, we have:

$$\log_3 (x + 2)(x + 4) = \log_3 3$$

Therefore

$$(x + 2)(x + 4) = 3$$

Open the brackets on the LHS, simplify and solve the resulting equation

$$x^2 + 6x + 8 = 3$$
$$x^2 + 6x + 5 = 0$$
$$(x + 5)(x + 1) = 0$$

Hence

$$x + 5 = 0 \quad \textbf{OR} \quad x + 1 = 0$$
$$x = -5 \qquad\qquad x = -1$$

$$\therefore x = -5 \quad \textbf{OR} \quad x = -1$$

NOTE

$x = -5$ does not satisfy the above logarithmic equation, because log is not valid for negative numbers.

---

b) $\log_{10} (x^2 - 4x + 7) = \log_{10} 100$
Solution

$$\log_{10} (x^2 - 4x + 7) = \log_{10} 100$$

This implies that

$$x^2 - 4x + 7 = 100$$
$$x^2 - 4x - 93 = 0$$
$$x = \frac{-b \pm \sqrt{b^2 - 4ac}}{2a}$$
$$= \frac{4 \pm \sqrt{388}}{2} = \frac{4 \pm 2\sqrt{97}}{2} = 2 \pm \sqrt{97}$$

Hence

$$x = 2 + \sqrt{97} = \textbf{11.85} \quad \textbf{OR} \quad x = 2 - \sqrt{97} = \textbf{-7.85}$$

Thus

$$x = \textbf{11.85} \quad \textbf{OR} \quad x = \textbf{-7.85}$$

---

c) $\log_8 (2x^2 - 13x + 84) = 2$
Solution

$$\log_8 (2x^2 - 13x + 84) = 2$$

This implies that

$$\log_8 (2x^2 - 13x + 84) = \log_8 64$$

Therefore

$$2x^2 - 13x + 84 = 64$$
$$2x^2 - 13x + 20 = 0$$
$$(2x - 5)(x - 4) = 0$$

Hence

$$2x - 5 = 0$$
$$2x = 5 \qquad \mathbf{OR} \qquad x - 4 = 0$$
$$x = \frac{5}{2} = \mathbf{2.5} \qquad\qquad x = \mathbf{4}$$

Thus

$$x = \mathbf{2.5} \quad \mathbf{OR} \quad x = \mathbf{4}$$

---

d) $\log_2 \left(x^2 - 4x + 12 \right) = 3$
Solution

$$\log_2 \left(x^2 - 4x + 12 \right) = 3$$

Apply relevant laws of logarithm

$$\log_2 \left(x^2 - 4x + 12 \right) = \log_2 8$$

Therefore

$$x^2 - 4x + 12 = 8$$
$$x^2 - 4x + 4 = 0$$
$$(x - 2)(x - 2) = 0$$

Hence

$$x - 2 = 0 \qquad \mathbf{OR} \qquad x - 2 = 0$$
$$x = \mathbf{2} \qquad\qquad\qquad x = \mathbf{2}$$

Thus

$$x = \mathbf{2} \ (\mathbf{twice})$$

---

e) $\log_9 \left(x^2 + 7x + 15 \right) = \frac{1}{2}$
Solution

$$\log_9 \left(x^2 + 7x + 15 \right) = \frac{1}{2}$$

Apply relevant laws of logarithm

$$\log_9 \left(x^2 + 7x + 15 \right) = \log_9 3$$

Therefore

$$x^2 + 7x + 15 = 3$$
$$x^2 + 7x + 12 = 0$$
$$(x + 4)(x + 3) = 0$$

Hence

$$x + 4 = 0 \quad \textbf{OR} \quad x + 3 = 0$$
$$\textbf{x = -4} \qquad\qquad \textbf{x = -3}$$

Thus

$$x = \textbf{-3} \quad \textbf{OR} \quad x = \textbf{-4}$$

---

f) $\log_6 x + \log_6 (x - 5) = 2$

Solution

$$\log_6 x + \log_6 (x - 5) = 2$$

Apply relevant laws of logarithm

$$\log_6 x(x - 5) = \log_6 36$$

Therefore

$$x(x - 5) = 36$$
$$x^2 - 5x - 36 = 0$$
$$(x + 4)(x - 9) = 0$$

Hence

$$x + 4 = 0 \quad \textbf{OR} \quad x - 9 = 0$$
$$\textbf{x = -4} \qquad\qquad \textbf{x = 9}$$

Thus

$$x = \textbf{-4} \quad \textbf{OR} \quad x = \textbf{9}$$

NOTE

Although the values of x are -4 and 9, the only one that satisfies the original logarithmic equation is $x = 9$, because log is not valid for negative numbers. Therefore, $x = -4$ is not a solution to this equation.

13.3 CHAPTER SUMMARY

1) Indicial equations are equations involving powers, where either the base or the exponent is the unknown variable to be determined.

2) Indicial equations can generally be solved using any known method of solving polynomials in conjunction with the laws of indices and logarithms.

3) The method of solving an indicial equation is determined by the nature of the equation, which can be broadly grouped into two:

- **When the unknown is the base:** This is often solved by simplifying the expressions and then applying a suitable method of solving polynomials.
- **When the unknown is the index (or exponent):** A similar base should be derived for both sides of the equation, and then their indices equated; if not possible, the logarithm of both sides must be taken. In the latter case, one needs to apply the one-to-one property of logarithm which states that:

$$\text{If } \log_x M = \log_x N \text{ then } M = N$$

****

13.4 FURTHER PRACTICE

To access complementary contents, including additional exercises, please go to www.dszak.com.

14 Triangle, Sine Rule, and Cosine Rule

Learning Outcomes

Once you have studied the content of this chapter, you should be able to:

- Use 'degree' and 'radian' as units of angle
- Convert between 'degree' and 'radian'
- Discuss different types of triangles
- Use Pythagoras' theorem to solve right-angled triangle problems
- Use the sine rule and cosine rule to solve triangle problems
- Calculate the area of a triangle

14.1 INTRODUCTION

In this chapter, we will look at the common units of angle measurement and discuss triangles, covering their types, naming conventions, and how to determine their area. We will also cover Pythagoras' theorem, sine rule, and cosine rule.

14.2 MEASURING ANGLES

An angle is created whenever two lines or planes meet, and this represents the amount of turning a line from a reference point, such as point A in Figure 14.1.

In Figure 14.1(i), line AB is turned counter-clockwise as indicated by the arrow to overlap line AC such that the amount of turning is equivalent to θ (read as 'theta', a Greek letter that is commonly used for angles). In Figure 14.1(ii), line AC is turned clockwise to overlap line AB. In Figure 14.1(iii) however, the arrow points both ways (or sometimes the arrows can be entirely ignored) to show that it is irrelevant which way the turning is made. In other words, we are only interested in the amount of turning between line AB and line AC.

We can therefore say that the angle in Figure 14.1(i) and 14.1(ii) can be treated as vector quantities, like displacement, and the angle in Figure 14.1(iii) a scalar quantity, like distance.

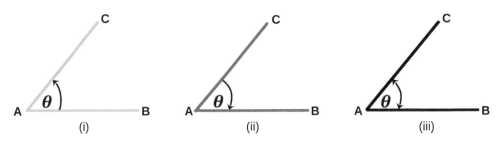

FIGURE 14.1   Angle measurement from a fixed point illustrated.

Conventionally, we use symbols ∠ or ^ to denote angles. Let's illustrate how these symbols are used:

a) ∠*BAC* is read as 'angle *BAC*', and ∠*CAB* is another way to say 'angle *CAB*', such that ∠*BAC* = ∠*CAB* = θ. We note that:

    i) the naming is a 3-point system, and

    ii) the angle of interest is at the middle point or the middle letter (i.e., letter *A*). We can also write ∠*A* = θ.

b) We also use $\widehat{A}$ = θ to denote angle. The letter, with a 'hat' sign, represents the point of reference.

Finally, you may have heard the term **'included'** in relation to an angle. This is to imply that the angle is formed by the given two sides. For example, angle θ in Figure 14.1 is said to be included **'by the sides AB and AC'**. The expression within the quote ' ' is often not written or stated.

Let's try some examples.

Example 1

Using the triangle ABC in Figure 14.2, write the following angles using the naming conventions described above.

a) Angle *ABC*   **b)** Angle *BCA*   **c)** Angle *CAB*   **d)** Angle *CBA*

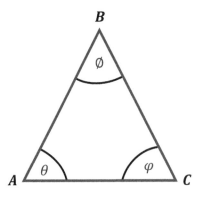

FIGURE 14.2   Example 1.

What did you get? Find the solution below to double-check your answer.

Solution to Example 1

a) Angle *ABC*
Solution

$$\angle ABC = \hat{B} = \varnothing$$

b) Angle *BCA*
Solution

$$\angle BCA = \hat{C} = \varphi$$

c) Angle *CAB*
Solution

$$\angle CAB = \hat{A} = \theta$$

d) Angle *CBA*
Solution
It should be noted that if there is no arrow indicating a particular direction, we can look at the angles as scalar quantities. Therefore, angle *CBA* is the same as angle *ABC*.

$$\angle CBA = \angle ABC = \hat{B} = \varnothing$$

Another example to try.

Example 2

Using the hexagon *ABCDEF* in Figure 14.3, write the following angles using the naming conventions described above.

a) Angle *AFE*　　　**b)** Angle *DEF*　　　**c)** Angle *EDC*　　　**d)** Angle *ABC*

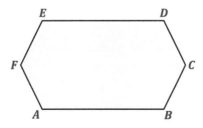

FIGURE 14.3 Example 2.

What did you get? Find the solution below to double-check your answer.

Solution to Example 2

a) Angle *AFE*
Solution

$$\angle AFE = \widehat{F}$$

b) Angle *DEF*
Solution

$$\angle DEF = \widehat{E}$$

c) Angle *EDC*
Solution

$$\angle EDC = \widehat{D}$$

d) Angle *ABC*
Solution

$$\angle ABC = \widehat{B}$$

14.2.1 DEGREES

Degree (denoted by °) is the most commonly used unit of angle measurement, and you would often find it as a default setting in most calculators. It is arbitrarily taken that the angle in one revolution (in a circle) is equal to 360 degrees. This is the angle you will get by moving from a point back to the original point clockwise or counter-clockwise. As a result, we can say that one degree is equivalent to one-360th of the angle in one full revolution.

At this point, it is pertinent to mention some special angles:

a) **Acute angle**: This is the angle that is less than 90° but more than 0, such that $0 < \theta < 90°$ (Figure 14.4). Examples include 30°, 45°, and 60°, which are special (acute) angles to be discussed later in this chapter.

b) **Right angle**: This is the angle that is equal to 90° or the angle between a horizonal and a vertical plane, where $\theta = 90°$. More generally, it is the angle between a pair of axes, i.e., $x - y$, $y - z$, and $x - z$.

Angle 90° is usually indicated by the symbol ∟, the presence of which implies a 90-degree angle, irrespective of whether it is stated or not. It is also independent of the orientation of the shape containing the angle, as shown in Figure 14.5.

A triangle in which one of its three angles is a right angle is called a **right-angled triangle** (Figure 14.5). Right angles are central to the use of Pythagoras' theorem, which will be discussed later in this chapter (Section 14.5).

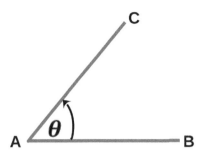

FIGURE 14.4   Acute angle illustrated.

FIGURE 14.5   Right angle illustrated.

c) **Obtuse angle**: This is the angle that is more than 90° but less than 180°, such that $90° < \theta < 180°$ (Figure 14.6).

d) **Straight-line angle**: This is the angle that is equal to 180°, such that $\theta = 180°$. It is the amount of turning in a semi-circle or in half a revolution (Figure 14.7).

e) **Reflex angle**: This is the angle that is more than 180 but less than 360 degrees, such that $180° < \theta < 360°$. Figure 14.8 shows two different reflex angles. In Figure 14.8(a), the reflex is more than 180 but less than 270 degrees, while in Figure 14.8(b), it is more than 270 but less than 360 degrees. It is worth noting that if we were to rotate the reference line AB clock-wise, it would become an obtuse angle in Figure 14.8(a) or an acute angle in Figure 14.8(b). In general, the pairs reflex-obtuse and reflex-acute always add up to 360 degrees.

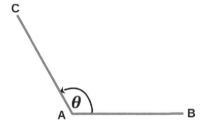

FIGURE 14.6   Obtuse angle illustrated.

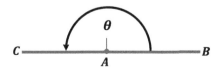

FIGURE 14.7   Straight line angle illustrated.

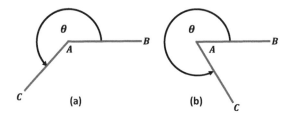

FIGURE 14.8 Reflex angles illustrated.

A null angle (also known as a zero angle) implies no rotation at all. It could also mean that there is a full rotation (or 360 degrees) or integer multiples of a complete revolution, which in effect brings the line being rotated back to its original position.

14.2.2 Radians

Radian (shortened to **rad**) is another unit of angle measurement, commonly used in mechanics, calculus, and electricity. It is the default setting in most mathematical and computational software including Excel and MATLAB. It is also second to degrees as a default unit on calculators.

In the absence of a unit attached to an angle, radian becomes the default, but in other cases the symbol ^{c} is used to imply radian. To illustrate, 2 radians can be written as **2 rad**, **2^c** or simply **2** and 'sine of 3.5 radians' can be given as **sin 3.5**, **sin 3.5^c**, or **sin 3.5 rad**. Also when an angle is a multiple of π, then it is radian. In this case, **cos 100π** can be considered to be in radian.

The definition of a radian is based on the angle bounded (or formed) at the centre of a circle by its two radii, which is opposite to a portion of the circumference, called the arc, as illustrated in Figure 14.9.

More precisely, 1 radian is defined as: '**the angle subtended at the centre of a circle by an arc length of 1 radius**'. In other words, 1 radian is the angle formed at the centre of a circle by an arc whose length is equal to the radius of the circle. 1 radian is approximately $57.3°$ rounded to 1 decimal place.

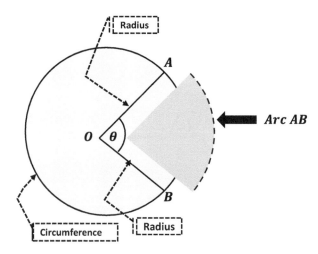

FIGURE 14.9 One radian illustrated.

In Figure 14.9, $|OA| = |OB| = r$, where r is the radius of the circle. AB is a portion of the perimeter of the circle, called the arc. It is technically said to subtend an angle θ at the centre of the circle. It is obvious that the larger the length of the arc AB, the bigger the angle θ it subtends at the centre.

Since the length of the arc AB is directly proportional to the angle subtended at the centre and r unit of arc length is equal to 1 radian, we can say that:

$$2r \rightarrow 2 \text{ radians}$$

$$3r \rightarrow 3 \text{ radians}$$

$$\pi r \rightarrow \pi \text{ radians}$$

$$2\pi r \rightarrow 2\pi \text{ radians}$$

As $2\pi r$ is the perimeter of the circle and corresponds to one full revolution, 2π radian is equivalent to angle $360°$. In summary, we arrive at the following fundamental conversion between the two units:

$$\boxed{2\pi = 360°} \text{ OR } \boxed{\pi = 180°} \tag{14.1}$$

14.2.3 Conversion Between Degrees and Radians

There are cases where we have two or more units representing the same quantity, it therefore becomes important to know the conversion between them. In this section, we will study how to convert between degrees and radians. Equation 14.1 provides us with a starting point. So, if π radians $= 180°$, then we can say that:

$$\boxed{1 \text{ radian} = \left(\frac{180}{\pi}\right)°} \text{ AND } \boxed{1° = \left(\frac{\pi}{180}\right) \text{rad}} \tag{14.2}$$

We can also use the fundamental formula above to provide common angles in degrees and radians as shown in Table 14.1.

TABLE 14.1
Conversion from Degrees to Radians Illustrated

Degree	Radian	Degree	Radian	Degree	Radian
0°	$0° \times \frac{\pi}{180°} = 0$	90°	$90° \times \frac{\pi}{180°} = \frac{\pi}{2}$	270°	$270° \times \frac{\pi}{180°} = \frac{3\pi}{2}$
15°	$15° \times \frac{\pi}{180°} = \frac{\pi}{12}$	120°	$120° \times \frac{\pi}{180°} = \frac{2\pi}{3}$	300°	$300° \times \frac{\pi}{180°} = \frac{5\pi}{3}$
30°	$30° \times \frac{\pi}{180°} = \frac{\pi}{6}$	150°	$150° \times \frac{\pi}{180°} = \frac{5\pi}{6}$	315°	$315° \times \frac{\pi}{180°} = \frac{7\pi}{4}$
45°	$45° \times \frac{\pi}{180°} = \frac{\pi}{4}$	180°	$180° \times \frac{\pi}{180°} = \pi$	330°	$330° \times \frac{\pi}{180°} = \frac{11\pi}{6}$
60°	$60° \times \frac{\pi}{180°} = \frac{\pi}{3}$	240°	$240° \times \frac{\pi}{180°} = \frac{4\pi}{3}$	360°	$360° \times \frac{\pi}{180°} = 2\pi$

Before we leave this section, let's quickly make the conversion process clear with the following notes.

Note 1 If an angle θ in degrees is to be converted to its radian equivalent, it should be multiplied by the conversion factor, which is:

$$\boxed{\theta \times \left(\frac{\pi}{180}\right) \textbf{rad}}$$ (14.3)

Thus, to convert from degrees to radians, follow these two steps:

Step 1: multiply by π, and
Step 2: divide by 180.

The order does not matter.

Note 2 If an angle φ radians is to be converted to its degree equivalent, multiply it by the conversion factor:

$$\boxed{\varphi \times \left(\frac{180}{\pi}\right)^\circ}$$ (14.4)

Thus, to convert from radian to degree, follow these steps:

Step 1: multiply by 180, and
Step 2: divide by π.

Again, the order does not matter.

Now it is time to try some examples.

Example 3

Convert the following angles from radians to degrees. Present your answer correct to 1 decimal place.

a) 1.13π b) $\frac{5}{7}\pi$ c) 4.6 rad   d) 12.7 rad   e) 1.8 rad   f) 0.45 rad

What did you get? Find the solution below to double-check your answer.

Solution to Example 3

a) 1.13π
Solution

$$1.13\pi = 1.13\pi \times \frac{180^\circ}{\pi}$$
$$= 1.13 \times 180^\circ$$
$$\therefore 3.12\pi = 203.4^\circ$$

b) $\frac{5}{7}\pi$

Solution

$$\frac{5}{7}\pi = \frac{5}{7}\pi \times \frac{180°}{\pi}$$

$$= \frac{5}{7} \times 180°$$

$$\therefore \frac{5}{7}\pi = 128.6°$$

---

c) 4.6 rad

Solution

$$4.6 \text{ rad} = 4.6 \times \frac{180°}{\pi}$$

$$\therefore 4.6 \text{ rad} = 263.6°$$

---

d) 12.7 rad

Solution

$$12.7 \text{ rad} = 12.7 \times \frac{180°}{\pi}$$

$$\therefore 12.7 \text{ rad} = 727.7°$$

---

e) 1.8 rad

Solution

$$1.8 \text{ rad} = 1.8 \times \frac{180°}{\pi}$$

$$\therefore 1.8 \text{ rad} = 103.1°$$

---

f) 0.45 rad

Solution

$$0.45 \text{ rad} = 0.45 \times \frac{180°}{\pi}$$

$$\therefore 0.45 \text{ rad} = 25.8°$$

Another set of examples to try.

Example 4

Convert the following angles from degrees to radians. Present your answer correct to 3 s.f.

a) 35°   **b)** 120°   **c)** 268°   **d)** 420°   **e)** 543°   **f)** 720°

What did you get? Find the solution below to double-check your answer.

Solution to Example 4

a) 35°
Solution

$$35° = 35° \times \frac{\pi}{180°}$$

$$= \frac{35°}{180°} \times \pi = \frac{7}{36}\pi$$

$$\therefore \mathbf{35° = 0.611\ rad}$$

b) 120°
Solution

$$120° = 120° \times \frac{\pi}{180°}$$

$$= \frac{120°}{180°} \times \pi = \frac{2}{3}\pi$$

$$\therefore \mathbf{120° = 2.09\ rad}$$

c) 268°
Solution

$$268° = 268° \times \frac{\pi}{180°}$$

$$= \frac{268°}{180°} \times \pi = \frac{67}{45}\pi$$

$$\therefore \mathbf{268° = 4.68\ rad}$$

d) 420°
Solution

$$420° = 420° \times \frac{\pi}{180°}$$

$$= \frac{420°}{180°} \times \pi = \frac{7}{3}\pi$$

$$\therefore \mathbf{420° = 7.33\ rad}$$

e) 543°
Solution

$$543° = 543° \times \frac{\pi}{180°}$$

$$= \frac{543°}{180°} \times \pi = \frac{181}{60}\pi$$

$$\therefore \mathbf{543° = 9.48\ rad}$$

f) 720°

Solution

$$720° = 720° \times \frac{\pi}{180°}$$

$$= \frac{720°}{180°} \times \pi = 4\pi$$

$$\therefore \mathbf{720° = 12.6 \ rad}$$

14.3 TRIANGLE

A triangle is a closed geometrical shape with three sides. By extension, it has three distinct angles and three vertices, as shown in Figure 14.10.

Before we proceed, it is essential to explain some terminologies that we use in relation to a triangle.

a) Vertex

A vertex (plural vertices) is the point where two lines of a triangle join or meet. Consequently, X, Y, and Z are the vertices of the triangle XYZ shown in Figure 14.10.

b) Naming convention

In general, we use the vertices to name a triangle. The triangle in Figure 14.10 will be identified as triangle **XYZ**, **YZX**, or **ZXY**. More conventionally, we use the symbol Δ to represent a triangle. The short form of naming the above triangle will then be Δ**XYZ**.

c) Length

It is the distance between two points or vertices. The three lengths (or sides) are **XY**, **YZ**, and **ZX**. Note that length is a scalar quantity, that is to say that we are only interested in the absolute value with no need for the direction. As such, length **XY** is the same as **YX**; the same applies to the other two sides.

Typically, we use double vertical lines || to imply length. Hence, instead of '**length XY**', we simply write $|XY|$. We are also allowed to use letters or symbols to reference sides instead of using two vertices. Using this nomenclature, the three sides are a, b, and c, which are respectively equivalent to **XY**, **YZ**, and **ZX**. Customarily, we use lowercase letters to represent sides, especially if we need to use the same letters to name the angles or the vertices of the same triangle.

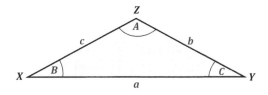

FIGURE 14.10 Angle and side naming format illustrated.

d) Angle

When we mention angles, we mean the angle subtended by a side and bounded by the two other sides of the triangle. This is usually the angle that is directly opposite to a side. For example, the angle subtended by side *a* is *A* in Figure 14.10. Similarly, angles *B* and *C* are subtended by side *b* and *c* respectively. You may have noticed that the letters for angles are in uppercase. It is conventional to use uppercase letters for angles and lowercase letters for their corresponding sides.

The above are some of the terms that we use consistently in the study of trigonometry and other related topics.

14.4 TYPES OF TRIANGLES

Our classification here will be based on two different features.

14.4.1 SIDE-BASED CLASSIFICATION

This is a classification that is based on the sides of the triangle, and we have three types, namely:

a) Equilateral triangle

This is a triangle in which all the three sides are equal. Consequently, the three angles must also be equal. Since the total angle in a triangle is 180°, we can say that each angle in an equilateral triangle is 60° (i.e., $\theta = \frac{180°}{3}$) as shown in Figure 14.11(a). An equilateral triangle also has three lines of symmetry, each of which divides the triangle into two identical right-angled triangles.

Notice that the same triangle is redrawn in Figure 14.11(b) without assigned angles; instead, a short line is drawn across each of the three sides. This is a shorthand of saying that the three sides and angles are equal. It is therefore an equilateral triangle.

b) Isosceles triangle

An isosceles triangle has only two out of the three sides equal. The two equal sides must have their corresponding angles equal. An isosceles triangle has one line of symmetry, which divides the triangle into two identical right-angled triangles.

If a triangle is isosceles, then there must be two equal angles (Figure 14.12(a)) and if a triangle has two equal sides, it is an isosceles (Figure 14.12(b)). In Figure 14.12, $|AC| = |BC|$ and also $\angle ABC = \angle BAC = \alpha$.

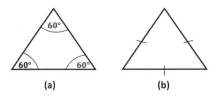

(a)   (b)

FIGURE 14.11 Equilateral triangle: (a) shows equal angles of 60 degrees, and (b) uses a line on the three sides to indicate that they are equal, with corresponding equal angles.

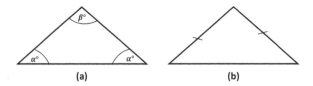

FIGURE 14.12   Isosceles triangle: (a) shows two equal angles of α, and (b) uses a line to indicate that two sides and their corresponding angles are equal.

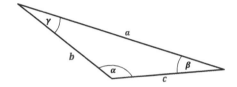

FIGURE 14.13   Scalene triangle illustrated such that $\alpha \neq \beta \neq \gamma$ and $a \neq b \neq c$.

Unlike an equilateral triangle, if an angle α, corresponding to the two equal sides of an isosceles, is given, then the third angle β can be found using $\beta = 180° - 2\alpha$ and if β is given then $\alpha = \frac{1}{2}(180° - \beta)$. This is because

$$\alpha + \alpha + \beta = 180° \quad \textbf{OR} \quad 2\alpha + \beta = 180°$$

c) Scalene triangle

It is a triangle with no two equal sides. In other words, each of the sides is distinct in length and also the angles are different (Figure 14.13). In this case, we do not expect a short line across any of its three sides as with the other two types above.

14.4.2 ANGULAR-BASED CLASSIFICATION

This classification is determined by the angles of the triangle, and we also have three types, namely:

a) Right-angled triangle

It is a triangle which has 90° as one of its angles. There can be only one of these in a triangle, as the sum of the angles must equal 180°, necessitating that the remaining two angles be acute. Figure 14.14(a) is a right-angled triangle where two sides are equal because of the presence of a short line across them, so we can say it is a right-angled isosceles triangle. The angles in this triangle will then be 90°, 45°, and 45°. On the other hand, Figure 14.14(b) is a right-angled triangle with three different sides, hence it is a right-angled scalene triangle. Note that we do not have a right-angled equilateral triangle.

Notice an inverted L-shape at the corner of the triangles in Figure 14.14. Yes, it is meant to be there to show that the angle is indeed a right-angled or 90°. We shall return to this type of triangle shortly as it is central to an important theorem known as the **Pythagoras' theorem**.

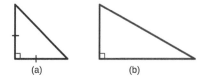

FIGURE 14.14 Right-angled triangle: (a) isosceles and (b) scalene.

b) Acute-angled triangle

It is a triangle in which all the three angles are acute. This type of triangle can belong to any of the three categories in the side-based classification. In other words, we have an acute-angled equilateral triangle, an acute-angled isosceles triangle, and an acute-angled scalene triangle.

c) Obtuse-angled triangle

It is a triangle in which one of the three angles is obtuse. This type of triangle can only be isosceles or scalene, but not equilateral for the same reason noted previously.

Before we move on, it's important to mention that the angular classification is less common, and that the last two types of angular classification are less known or used.

14.5 PYTHAGORAS' THEOREM

Let's begin this section by stating that every triangle has or must fulfil the following conditions:

Note 1 Three sides and three angles.
Note 2 Total angle must be 180 degrees. If the three angles are α, β, and γ, it follows that $\alpha + \beta + \gamma = 180°$ or $\alpha + \beta + \gamma = \pi$.
Note 3 The longest side is opposite to the biggest angle and the shortest side is opposite the smallest angle.
Note 4 The ratio of the sides is not equal to the angle ratio. In other words, doubling an angle does not necessarily imply multiplying the corresponding side by 2.
Note 5 Pythagoras' theorem is applicable to and valid only for right-angled triangles.

Given a triangle ABC (Figure 14.15), Pythagoras' theorem states that:

$$\boxed{a^2 = b^2 + c^2} \text{ OR } \boxed{\text{Hypotenuse}^2 = \text{Adjacent}^2 + \text{Opposite}^2} \qquad (14.5)$$

Note that side a or $|BC|$ is the side opposite angle 90°, it is called the **hypotenuse**. This theorem, by itself, does not look great or special but grasp its rule as you will find it again and again in your study of **STEM** (science, technology, engineering, and mathematics) subjects.

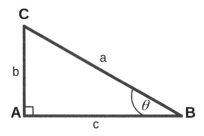

FIGURE 14.15 Pythagoras' theorem illustrated using a right-angled triangle.

It is a good time to go through some examples.

Example 5

In a triangle ABC, the three angles are A, B, and C. Calculate the missing angles in each of the following. Also, state:

   i) the longest side,

   ii) the shortest side, and

   iii) the type of the triangle.

a) $\widehat{A} = 20°$, $\widehat{B} = 50°$, $\widehat{C} = ?$ **b)** $\widehat{A} = \widehat{B} = 62°$, $\widehat{C} = ?$ **c)** $\widehat{A} = 40°$, $\widehat{B} = \widehat{C} = ?$

d) $\widehat{A} = ?$, $\widehat{B} = \frac{2}{5}\pi$, $\widehat{C} = \frac{1}{3}\pi$ **e)** $\widehat{A} = \frac{1}{2}\pi$, $\widehat{B} = 2\widehat{C} = ?$ **f)** $\widehat{A} = \frac{1}{4}\pi$, $\widehat{B} = 57°$, $\widehat{C} = ?$

What did you get? Find the solution below to double-check your answer.

Solution to Example 5

HINT

In this case, it will be helpful to sketch the triangle for visualisation.

a) $\widehat{A} = 20°$, $\widehat{B} = 50°$, $\widehat{C} = ?$
Solution

$$\widehat{A} + \widehat{B} + \widehat{C} = 180°$$
$$20° + 50° + \widehat{C} = 180°$$
$$\widehat{C} = 180° - 20° - 50°$$
$$\therefore \widehat{C} = \mathbf{110°}$$

   i) The longest side is the one opposite the biggest angle $\widehat{C}$, which is $|AB|$.

   ii) The shortest side is the one opposite the smallest angle $\widehat{A}$, which is $|BC|$.

   iii) It is an **obtuse-angled triangle** or a **scalene triangle**.

b) $\widehat{A} = \widehat{B} = 62°, \ \widehat{C} = ?$

Solution

$$\widehat{A} + \widehat{B} + \widehat{C} = 180°$$
$$62° + 62° + \widehat{C} = 180°$$
$$\widehat{C} = 180° - 62° - 62°$$
$$\therefore \widehat{C} = \mathbf{56°}$$

i) The two longer sides are the ones opposite $\widehat{A}$ and $\widehat{B}$, which are $|BC|$ and $|AC|$ respectively.

ii) The shortest side is the one opposite the smallest angle $\widehat{C}$, which is $|AB|$.

iii) It is an **acute-angled triangle** or an **isosceles triangle**.

---

c) $\widehat{A} = 40°, \ \widehat{B} = \widehat{C} = ?$

Solution

Let $\widehat{B} = \widehat{C} = x$

$$\widehat{A} + \widehat{B} + \widehat{C} = 180°$$
$$40° + x + x = 180°$$
$$2x = 180° - 40°$$
$$2x = 140°$$
$$x = 70°$$
$$\therefore \widehat{B} = \widehat{C} = \mathbf{70°}$$

i) The two longer sides are the ones opposite $\widehat{B}$ and $\widehat{C}$, which are $|AC|$ and $|AB|$ respectively.

ii) The shortest side is the one opposite the smallest angle $\widehat{A}$, which is $|BC|$.

iii) It is an **acute-angled triangle** or an **isosceles triangle**.

---

d) $\widehat{A} = ?, \ \widehat{B} = \frac{2}{5}\pi, \ \widehat{C} = \frac{1}{3}\pi$

Solution

$$\widehat{A} + \widehat{B} + \widehat{C} = \pi$$
$$\widehat{A} + \frac{2}{5}\pi + \frac{1}{3}\pi = \pi$$
$$\widehat{A} = \pi - \frac{2}{5}\pi - \frac{1}{3}\pi = (1 - \frac{2}{5} - \frac{1}{3})\pi$$
$$\therefore \widehat{A} = \frac{4}{15}\pi$$

i) The longest side is the one opposite the biggest angle $\widehat{B}$, which is $|AC|$.

ii) The shortest side is the one opposite the smallest angle $\widehat{A}$, which is $|BC|$.

iii) It is an **acute-angled triangle** or a **scalene triangle**.

---

e) $\widehat{A} = \frac{1}{2}\pi$, $\widehat{B} = 2\widehat{C} = ?$

Solution

Let $\widehat{C} = x$, thus $\widehat{B} = 2x$. Now

$$\widehat{A} + \widehat{B} + \widehat{C} = \pi$$

$$\frac{1}{2}\pi + 2x + x = \pi$$

$$3x = \pi - \frac{1}{2}\pi = \frac{1}{2}\pi$$

$$x = \frac{1}{6}\pi$$

This implies that $\widehat{C} = \frac{1}{6}\pi$ and $\widehat{B} = 2\widehat{C} = 2 \times \frac{1}{6}\pi = \frac{1}{3}\pi$.

$$\therefore \widehat{B} = \frac{1}{3}\pi, \ \widehat{C} = \frac{1}{6}\pi$$

i) The longest side is the one opposite the biggest angle $\widehat{A}$, which is $|BC|$.

ii) The shortest side is the one opposite the smallest angle $\widehat{C}$, which is $|AB|$.

iii) It is a **right-angled triangle** or a **scalene triangle**.

---

f) $\widehat{A} = \frac{1}{4}\pi$, $\widehat{B} = 57°$, $\widehat{C} = ?$

Solution

Before we proceed, we need to determine the missing angle. In this case, working in degrees would be more straightforward, as we can easily and precisely convert the given angle from radians to degrees. Thus, $\widehat{A} = \frac{1}{4}\pi = 45°$. It therefore follows that

$$\widehat{A} + \widehat{B} + \widehat{C} = 180°$$

$$45° + 57° + C = 180°$$

$$\widehat{C} = 180° - 45° - 57°$$

$$\therefore \widehat{C} = 78°$$

i) The longest side is the one opposite the biggest angle $\widehat{C}$, which is $|AB|$.

ii) The shortest side is the one opposite the smallest angle $\widehat{A}$, which is $|BC|$.

iii) It is an **acute-angled triangle** or a **scalene triangle**.

Let's try more examples.

Example 6

$\triangle ABC$ shown in Figure 14.16 is a right-angled scalene triangle such that $\widehat{A} = 90°$, $|AB| = 30$ cm and $|AC| = 40$ cm. Determine the length of the hypotenuse.

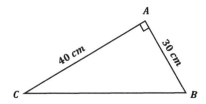

FIGURE 14.16 Example 6.

What did you get? Find the solution below to double-check your answer.

Solution to Example 6

$|AB| = c = 30$ cm, $|AC| = b = 40$ cm and $|BC| = a$
This is a right-angled triangle; therefore, we can apply Pythagoras' theorem as follows:

$$a^2 = b^2 + c^2$$
$$= 30^2 + 40^2 = 2500$$
$$a = \sqrt{2500}$$
$$\therefore a = 50 \text{ cm}$$

Example 7

The hypotenuse of a right-angled triangle is 13 cm. If the second side is 5 cm, determine the third side.

What did you get? Find the solution below to double-check your answer.

Solution to Example 7

This is a right-angled triangle so we can use Pythagoras, where

$$a^2 = b^2 + c^2$$

In this case, the hypotenuse is $a = 13$ cm. We can assume the second side to be $b = 5$ cm. Hence, the third side c can be found by re-arranging the Pythagoras' theorem as:

$$c^2 = a^2 - b^2$$
$$c = \sqrt{a^2 - b^2}$$
$$= \sqrt{13^2 - 5^2}$$
$$= \sqrt{144} = 12$$
$$\therefore c = 12 \text{ cm}$$

Example 8

$\triangle XYZ$ is such that $\hat{Y} = 90°$. Given that $|XY|$ is one-third of $|XZ|$, determine the value of $|XZ|$ in terms of $|YZ|$.

What did you get? Find the solution below to double-check your answer.

Solution to Example 8

Solution
For this example, a sketch may help visualise and solve the problem; this is provided in Figure 14.17. From Figure 14.17, we have $|XY| = z$, $|YZ| = x$ and $|XZ| = y$. If $|XY|$ is a third of $|XZ|$, then

$$y = 3z \ \textbf{OR} \ z = \frac{1}{3}y$$

We can write the Pythagoras' theorem for $\triangle XYZ$ as:

$$y^2 = x^2 + z^2$$
$$y^2 = x^2 + \left(\frac{1}{3}y\right)^2$$
$$y^2 = x^2 + \frac{1}{9}y^2$$
$$y^2 - \frac{1}{9}y^2 = x^2$$
$$y^2\left(1 - \frac{1}{9}\right) = x^2$$
$$\frac{8}{9}y^2 = x^2$$
$$y^2 = \frac{9}{8}x^2$$
$$y = \sqrt{\frac{9}{8}x^2} = \sqrt{\frac{9}{8}} \times \sqrt{x^2}$$
$$= \frac{\sqrt{9}}{\sqrt{8}} \times x = \frac{3}{2\sqrt{2}}x$$
$$\therefore |XZ| = \frac{3}{2\sqrt{2}}|YZ| \text{ or } \frac{3\sqrt{2}}{4}|YZ|$$

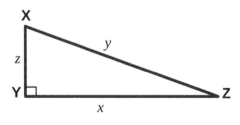

FIGURE 14.17 Solution to Example 8 Part I.

ALTERNATIVE METHOD

Given that $|XY| : |XZ|$ is $1:3$, we can draw the triangle as shown in Figure 14.18. Using Pythagoras' theorem, we can write $|YZ|$ as:

$$|YZ|^2 = 3^2 - 1^2$$

$$|YZ|^2 = 8$$

Therefore

$$|YZ| = \sqrt{8} = 2\sqrt{2} \quad ------(i)$$

Also, $|XY| : |XZ|$ is $1:3$ implies that

$$|XY| = \frac{1}{3}|XZ|$$

$$|XZ| = 3|XY| \quad ------(ii)$$

But

$$|XY| = 1$$

This can be written as

$$|XY| = \frac{2\sqrt{2}}{2\sqrt{2}} \quad ------(iii)$$

Combining (ii) and (iii), we have

$$|XZ| = 3|XY|$$

$$= 3\frac{2\sqrt{2}}{2\sqrt{2}}$$

Substituting equation (i), we have:

$$= 3\frac{|YZ|}{2\sqrt{2}} = 3\frac{1}{2\sqrt{2}}|YZ|$$

$$= \frac{3\sqrt{2}}{4}|YZ|$$

$$\therefore |XZ| = \frac{3}{2\sqrt{2}}|YZ| \quad \textbf{OR} \quad \frac{3\sqrt{2}}{4}|YZ|$$

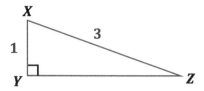

FIGURE 14.18 Solution to Example 8 Part II.

14.6 SINE RULE

One limitation of Pythagoras' theorem is that it is applicable only to right-angled triangles, even though many triangles are not right-angled. One of the techniques used to solve such non-right-angled triangles is the **sine rule**.

14.6.1 SINE RULE FORMULA

Consider $\triangle ABC$ in Figure 14.19, where the sides are a, b, and c and the angles are A, B, and C.

The sine rule states that

Variant I	**Variant II**
$\dfrac{a}{\sin A} = \dfrac{b}{\sin B} = \dfrac{c}{\sin C}$	$\dfrac{\sin A}{a} = \dfrac{\sin B}{b} = \dfrac{\sin C}{c}$

$$(14.6)$$

Let's quickly make the following notes about the sine rule.

Note 1 It can be applied to any triangle including a right-angled triangle.

Note 2 It can be used to completely solve problems involving triangles. To 'completely solve' implies determining the values of the three sides and three angles of the triangle.

Note 3 We use only a pair of the rule above, i.e., $\dfrac{a}{\sin A} = \dfrac{b}{\sin B}$, $\dfrac{a}{\sin A} = \dfrac{c}{\sin C}$ or $\dfrac{b}{\sin B} = \dfrac{c}{\sin C}$. The same can be said about the second variant.

Note 4 Use the first variant of the rule when looking for a side and the second variant when looking for an angle.

Note 5 The sine rule above can be proven or derived.

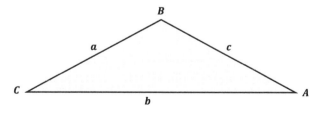

FIGURE 14.19 Sine rule illustrated using a non-right-angled triangle ABC.

14.6.2 When to Use the Sine Rule Formula

We use the sine rule in the following two instances.

Case 1 Given two angles and one side

In reference to $\triangle ABC$ in Figure 14.19, we may have the following situations:

   i) $\widehat{A}, \widehat{B}$ with a or b or c

  ii) $\widehat{A}, \widehat{C}$ with a or b or c

 iii) $\widehat{B}, \widehat{C}$ with a or b or c

The above situations are generally put under two sub-cases as:

  a) Two angles and a non-included side, which can be abbreviated as **AAS** or **SAA**. **A** stands for Angle and **S** stands for side. The two **A's** are placed next to each other to show that **S** is not included.

  b) Two angles and an included side, which can be abbreviated as **ASA**. **S** is sandwiched between the two **A's** to show that the former is included.

The sine rule can easily be applied for AAS or SAA. However, for ASA (e.g., $\widehat{B}, \widehat{C}$ with a), the sine rule cannot be used directly, as we need either side b or c to use $\frac{b}{\sin B} = \frac{c}{\sin C}$. We can however use either $\frac{a}{\sin A} = \frac{b}{\sin B}$ or $\frac{a}{\sin A} = \frac{c}{\sin C}$ to find b or c respectively, since $\widehat{A}$ can be found from the relation $\widehat{A} + \widehat{B} + \widehat{C} = 180°$.

Case 2 Given two sides and a non-included angle

This can be abbreviated as **SSA** or **ASS**. Again, using the $\triangle ABC$ in Figure 14.18, we have the following three possibilities:

    i) a, b with $\widehat{A}$, or $\widehat{B}$,   ii) a, c with $\widehat{A}$, or $\widehat{C}$,   iii) b, c with $\widehat{B}$, or $\widehat{C}$

For this case, there are two possible scenarios:

  a) When the given angle corresponds to the longer of the two sides, then the sine rule will lead to only one possible solution of the triangle. For example, if we are given $b, c,$ and $\widehat{C}$ such that $b < c$, then it is pre-determined that $\widehat{B} < \widehat{C}$.

  b) When the given angle corresponds to the shorter of the two sides, then the sine rule will lead to one of the two possible solutions of the triangle. For example, if we are given $a, b,$ and $\widehat{A}$, such that $a < b$, then it is pre-determined that $\widehat{A} < \widehat{B}$. In this case, it will be valid for $\widehat{B}$ to be an acute or obtuse angle, since the sine function has two positive values between 0 and 180 degrees. This is called the **ambiguous case**.

The above classification and cases have been summarised in Figure 14.20.

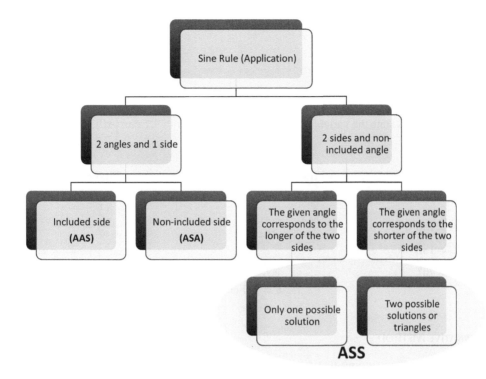

FIGURE 14.20 Cases (AAS, ASA, and ASS) when the sine rule can be used illustrated.

We will illustrate each case with an example.

Example 9

Given that $|BC| = 9.2$ cm, solve the triangle ABC in Figure 14.21. Present the answer correct to 2 s.f. (Figure 14.21).

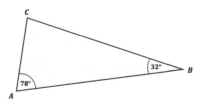

FIGURE 14.21 Example 9.

What did you get? Find the solution below to double-check your answer.

Solution to Example 9

Solution

This is case 1(a). Let's state the variables: $|AB| = c = ?$; $|BC| = a = 9.2$ cm; $|AC| = b = ?$; $\widehat{A} = 78°$; $\widehat{B} = 32°$; $\widehat{C} = ?$

- **Find $\widehat{C}$**

$$\widehat{A} + \widehat{B} + \widehat{C} = 180°$$
$$\widehat{C} = 180° - 78° - 32°$$
$$\therefore \widehat{C} = 70°$$

- **Find b**

$$\frac{b}{\sin B} = \frac{a}{\sin A}$$
$$\frac{b}{\sin 32°} = \frac{9.2}{\sin 78°}$$
$$b = \frac{9.2}{\sin 78°} \times \sin 32° = 4.984 \text{ cm}$$
$$\therefore b = 5.0 \text{ cm}$$

- **Find c**

$$\frac{c}{\sin C} = \frac{a}{\sin A}$$
$$\frac{c}{\sin 70°} = \frac{9.2}{\sin 78°}$$
$$c = \frac{9.2}{\sin 78°} \times \sin 70°$$
$$= 8.838 \text{ cm}$$
$$\therefore c = 8.8 \text{ cm}$$

ALTERNATIVE METHOD

Let's try to find $\widehat{C}$, assuming that c is already known, using the sine rule. We will use the following data: $c = 8.8$ cm; $b = 5.0$; $\widehat{B} = 32°$; $\widehat{C} = ?$

$$\frac{\sin C}{c} = \frac{\sin B}{b}$$
$$\frac{\sin C}{8.8} = \frac{\sin 32}{5.0}$$
$$\sin C = \frac{\sin 32}{5.0} \times 8.8 = 0.93266$$
$$C = \sin^{-1}(0.93266)$$
$$\therefore C = 69°$$

To obtain a more accurate value, let's use $c = 8.838$ cm; $b = 4.984$ as follows:

$$\sin C = \frac{\sin 32}{4.984} \times 8.838$$
$$C = \sin^{-1}\left(\frac{\sin 32}{4.984} \times 8.838\right)$$
$$\therefore C = 69.9999° = 70°$$

The aforementioned example underscores the importance of presenting values with precision.

Let's try case 1(b).

Example 10

Given that $|BC| = 6$ cm, solve the triangle ABC in Figure 14.22. Present the answer correct to 2 s.f.

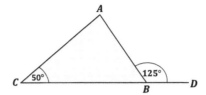

FIGURE 14.22 Example 10.

What did you get? Find the solution below to double-check your answer.

Solution to Example 10

Solution

This pertains to case 1(b), even though it may not initially appear so. The total angle in a straight line is 180 degrees. Therefore

$$\angle ABC + \angle ABD = 180°$$

$$\angle ABC = 180° - 125° = 55° = \widehat{B}$$

Now, let's state the variables as: $|AB| = c = ?$; $|BC| = a = 6$ cm; $|AC| = b = ?$; $\widehat{A} = ?$; $\widehat{B} = 55°$; $\widehat{C} = 50°$.

Since we don't have b and c, we cannot use the sine rule yet. To use it, we should find $\widehat{A}$ first using the fact that the total angle in a triangle is 180 degrees, thus:

$$\widehat{A} = 180° - \widehat{B} - \widehat{C}$$

$$= 180° - 55° - 50°$$

$$= 75°$$

- **Find b**

$$\frac{b}{\sin B} = \frac{a}{\sin A}$$

$$\frac{b}{\sin 55°} = \frac{6}{\sin 75°}$$

$$b = \frac{6}{\sin 75°} \times \sin 55°$$

$$= 5.0883 \text{ cm}$$

$$\therefore b = 5.1 \text{ cm}$$

- **Find c**

$$\frac{c}{\sin C} = \frac{a}{\sin A}$$

$$\frac{c}{\sin 50°} = \frac{6}{\sin 75°}$$

$$c = \frac{6}{\sin 75°} \times \sin 50°$$

$$= 4.7584 \text{ cm}$$

$$\therefore c = 4.8 \text{ cm}$$

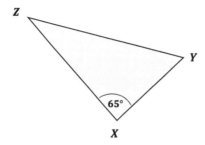

FIGURE 14.23 Example 11.

Let's try another example.

Example 11

In $\triangle XYZ$ shown in Figure 14.23, $|XY| = 4.3$ cm and $|YZ| = 7.5$ cm. Solve the triangle completely. Present the answer correct to 2 s.f.

What did you get? Find the solution below to double-check your answer.

Solution to Example 11

Here we have two sides and non-included angle (i.e., ASS or SSA) but the given angle corresponds to the longer of the two sides, so this is case 2(a). Let:

$$|XY| = a = 4.3 \text{ cm}; \ |YZ| = b = 7.5 \text{ cm}; \ |XZ| = c = ? \ \widehat{A} = ?; \ \widehat{B} = 65°; \ \widehat{C} = ?$$

- Find $\widehat{A}$

$$\frac{\sin A}{a} = \frac{\sin B}{b}$$

$$\frac{\sin A}{4.3} = \frac{\sin 65°}{7.5}$$

$$\sin A = \frac{\sin 65°}{7.5} \times 4.3 = 0.51962$$

$$A = \sin^{-1}(0.51962) = 31.3°$$

$$\therefore A = \mathbf{31°}$$

- Find $\widehat{C}$

$$\widehat{A} + \widehat{B} + \widehat{C} = 180°$$

$$\widehat{C} = 180° - 31.3° - 65° = 83.7°$$

$$\therefore \widehat{C} = \mathbf{84°}$$

- Find c

$$\frac{c}{\sin C} = \frac{b}{\sin B}$$

$$\frac{c}{\sin 83.7°} = \frac{7.5}{\sin 65°}$$

$$c = \frac{7.5}{\sin 65°} \times \sin 83.7°$$

$$= 8.2254 \text{ cm}$$

$$\therefore c = \textbf{8.2 cm}$$

Let's have one last example for this section.

Example 12

In $\triangle LMN$, $|LM| = 70$ mm and $|MN| = 50$ mm. Given that $\angle NLM = 30°$, solve the triangle completely.

What did you get? Find the solution below to double-check your answer.

Solution to Example 12

This is case 2(b), where two sides and a non-included are given, but the given angle corresponds to the shorter of the two sides.

Let: $|LM| = a = 70$ mm; $|MN| = b = 50$ mm; $|LN| = c = ?$ $\angle MNL = \widehat{A} = ?$; $\angle NLM = \widehat{B} = 30°$; $\angle LMN = \widehat{C} = ?$

- Find $\widehat{A}$

$$\frac{\sin A}{a} = \frac{\sin B}{b}$$

$$\frac{\sin A}{70} = \frac{\sin 30°}{50}$$

$$\sin A = \frac{\sin 30°}{50} \times 70 = 0.7000$$

$$A = \sin^{-1}(0.700)$$

$$\therefore A = \textbf{44.4°}$$

But this is the value of $\widehat{A}$ in the first quadrant. Its equivalent value in the second quadrant is $(\textbf{180°} - \textbf{44.4°}) = \textbf{135.6°}$. Both values are possible because $\widehat{A} + \widehat{B} = 44.4° + 30° = 74.47° < 180°$ and $\widehat{A} + \widehat{B} = 135.6° + 30° = 165.6° < 180°$. Hence, we need to use $A = \textbf{44.4°}$ and $A = \textbf{135.6°}$ to solve the triangle completely as follows.

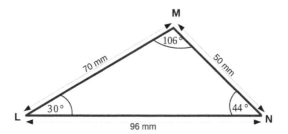

FIGURE 14.24 Solution to Example 12 – Part I.

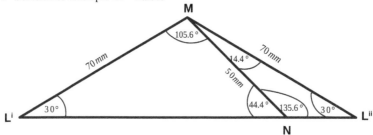

FIGURE 14.25 Solution to Example 12 – Part II.

Option 1 $A = 44.4°$

- Find $\widehat{C}$

$$\widehat{A} + \widehat{B} + \widehat{C} = 180°$$
$$\widehat{C} = 180° - 44.4° - 30°$$
$$\therefore \widehat{C} = 105.6°$$

- Find c

$$\frac{c}{\sin C} = \frac{b}{\sin B}$$
$$\frac{c}{\sin 105.6°} = \frac{50}{\sin 30°}$$
$$c = \frac{50}{\sin 30°} \times \sin 105.6° = 96.3163 \text{ mm}$$
$$\therefore c = 96 \text{ mm}$$

Hence, the complete solution for this option is:
$|LM| = a = 70$ mm; $|MN| = b = 50$ mm; $|NL| = c = 96$ mm; $\angle MNL = \widehat{A} = 44.4°$; $\angle NLM = \widehat{B} = 30°$; $\angle LMN = \widehat{C} = 105.6°$ (Figure 14.24).

Option 2 $A = 135.6°$

Hence, the complete solution for this option is:

- Find $\widehat{C}$

$$\widehat{A} + \widehat{B} + \widehat{C} = 180°$$
$$\widehat{C} = 180° - 135.6° - 30°$$
$$\therefore \widehat{C} = 14.4°$$

- Find c

$$\frac{c}{\sin C} = \frac{b}{\sin B}$$
$$\frac{c}{\sin 14.4°} = \frac{50}{\sin 30°}$$
$$c = \frac{50}{\sin 30°} \times \sin 14.4° = 24.860 \text{ mm}$$
$$\therefore c = 25 \text{ mm}$$

$|LM| = a = 70$ cm; $|MN| = b = 50$ mm; $|NL| = c = 25$ mm; $\angle MNL = \widehat{A} = 135.6$; $\angle NLM = \widehat{B} = 30°$; $\angle LMN = \widehat{C} = 14.4°$ (Figure 14.25).

14.7 COSINE RULE

This is another important rule for solving triangles including right-angled triangles. In fact, we can obtain Pythagoras' theorem from the cosine rule, which means that Pythagoras is a special case of the cosine rule. The proof of this will be demonstrated shortly in the worked examples.

14.7.1 COSINE RULE FORMULA

Consider $\triangle ABC$ in Figure 14.26 where the sides are a, b, and c and the angles are A, B, and C.

The cosine rule states that

$$\boxed{a^2 = b^2 + c^2 - 2bc \ \cos A} \ \boxed{b^2 = a^2 + c^2 - 2ac \ \cos B} \ \boxed{c^2 = a^2 + b^2 - 2ab \ \cos C} \qquad (14.7)$$

where a, b, and c are the three sides and $\widehat{A}$, $\widehat{B}$, and $\widehat{C}$ are their respective angles.

We have provided a formula for each side of the $\triangle ABC$ but this is not really required as the choice of the letters is arbitrary. We can also write the angle version of each formula as shown below, though it is obtained through transposing the above formulas.

$$\boxed{\cos A = \frac{b^2 + c^2 - a^2}{2bc}} \quad \boxed{\cos B = \frac{a^2 + c^2 - b^2}{2ac}} \quad \boxed{\cos C = \frac{a^2 + b^2 - c^2}{2ab}} \qquad (14.8)$$

The same (or similar) notes made in respect to the sine rule also apply to the cosine rule.

14.7.2 WHEN TO USE THE COSINE RULE FORMULA

We use this rule for the following cases:

Case 1 Given three sides

In reference to the $\triangle ABC$ above, given a, b, and c, we should employ two of the three angle formulas to determine two angles. Subsequently, we can apply the relation $\widehat{A} + \widehat{B} + \widehat{C} = 180°$ to find the third angle. However, we also have the option to use all three formulas to completely solve the triangle.

Case 2 Given two sides and included angle

Unlike with the sine rule, the given angle is the included angle, i.e., the angle between the two given sides. This is a straightforward case – we just need to apply the side formula of the cosine rule.

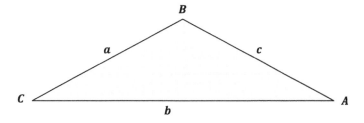

FIGURE 14.26 Cosine rule illustrated using a non-right-angled triangle ABC.

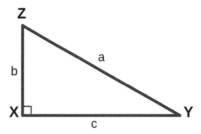

FIGURE 14.27 Example 13.

Very brief! Let's try some examples.

Example 13

Using cosine rule, show that $a^2 = b^2 + c^2$ in triangle XYZ, as depicted in Figure 14.27.

What did you get? Find the solution below to double-check your answer.

Solution to Example 13

Let

$$\angle ZXY = \widehat{A} = 90°$$

Since we are interested in the hypotenuse, i.e. the cosine rule to use is

$$a^2 = b^2 + c^2 - 2bc \cos A$$

Substituting $\widehat{A} = 90°$, we have:

$$a^2 = b^2 + c^2 - 2bc \cos 90°$$

because $\cos 90° = 0$, we therefore have:

$$a^2 = b^2 + c^2 - 2bc\,(0)$$

$$\therefore a^2 = b^2 + c^2$$

NOTE
This example shows that Pythagoras' theorem is a specific case of the cosine rule.

Another example to try.

Example 14

In a triangle ABC, $|AB| = 53$ mm, $|BC| = 68$ mm, and $|AC| = 79$ mm. Use this information to completely solve the triangle. Present the answers correct to 1 d.p.

What did you get? Find the solution below to double-check your answer.

Solution to Example 14

Let: $|AB| = c = 53$ mm; $|BC| = a = 68$ mm; $|AC| = b = 79$ mm; $\widehat{A} = ?$; $\widehat{B} = ?$ $\widehat{C} = ?$
Since we are interested in the angles, we need to use the angle formula of the cosine rule as follows:

• Find $\widehat{A}$	• Find $\widehat{B}$	• Find $\widehat{C}$
$\cos A = \dfrac{b^2+c^2-a^2}{2bc}$	$\cos B = \dfrac{a^2+c^2-b^2}{2ac}$	$\cos C = \dfrac{a^2+b^2-c^2}{2ab}$
$A = \cos^{-1}\left(\dfrac{b^2+c^2-a^2}{2bc}\right)$	$B = \cos^{-1}\left(\dfrac{a^2+c^2-b^2}{2ac}\right)$	$C = \cos^{-1}\left(\dfrac{a^2+b^2-c^2}{2ab}\right)$
$= \cos^{-1}\left(\dfrac{79^2+53^2-68^2}{2\times79\times53}\right)$	$= \cos^{-1}\left(\dfrac{68^2+53^2-79^2}{2\times68\times53}\right)$	$= \cos^{-1}\left(\dfrac{68^2+79^2-53^2}{2\times68\times79}\right)$
$= \cos^{-1}\left(\dfrac{2213}{4187}\right) = 58.0931°$	$= \cos^{-1}\left(\dfrac{149}{901}\right) = 80.4812°$	$= \cos^{-1}\left(\dfrac{1007}{1343}\right) = 41.4257°$
$\therefore A = \mathbf{58.1°}$	$\therefore B = \mathbf{80.5°}$	$\therefore C = \mathbf{41.4°}$

Alternatively, to find $\widehat{C}$, we have:

$$\widehat{A} + \widehat{B} + \widehat{C} = 180°$$
$$\widehat{C} = 180° - 58.1° - 80.5°$$
$$\therefore \widehat{C} = \mathbf{41.4°}$$

The complete solution is shown in Figure 14.28.

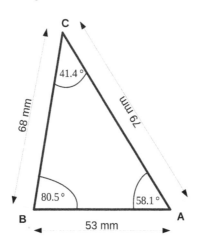

FIGURE 14.28 Solution to Example 14.

One final example to try.

Example 15

In a triangle XYZ, $|XY| = 4.5$ cm, $|YZ| = 3.7$ cm and $\angle XYZ = 105°$. Use this information to completely solve the triangle.

What did you get? Find the solution below to double-check your answer.

Solution to Example 15

Let:
$|XY| = a = 4.5$ cm; $|YZ| = b = 3.7$ cm; $|XZ| = c = ?$; $\widehat{A} = ?$; $\widehat{B} = ?$; $\widehat{C} = 105°$

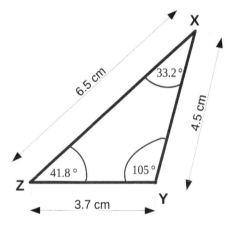

FIGURE 14.29 Solution to Example 15 – Part I.

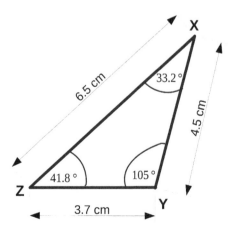

FIGURE 14.30 Solution to Example 15 – Part II.

Since we are interested in a side and two angles, we need to use both formulas of the cosine rule as:

• Find c

$c^2 = a^2 + b^2 - 2ab \cos C$

$= 4.5^2 + 3.7^2 - 2(4.5)(3.7)\cos 105°$

$= 20.25 + 13.69 - 33.3 \cos 105°$

$= 33.94 - 33.3 \cos 105°$

$c = \sqrt{33.94 - 33.3 \cos 105°} = 6.5237$ cm

$\therefore c = \mathbf{6.52}$ **cm**

• Find $\widehat{B}$

$\widehat{A} + \widehat{B} + \widehat{C} = 180°$

$\widehat{B} = 180° - 41.8° - 105°$

$\therefore \widehat{B} = \mathbf{33.2°}$

• Find $\widehat{A}$

$\cos A = \frac{b^2 + c^2 - a^2}{2bc}$

$A = \cos^{-1}\left(\frac{b^2 + c^2 - a^2}{2bc}\right)$

$= \cos^{-1}\left(\frac{3.7^2 + 6.52^2 - 4.5^2}{2 \times 3.7 \times 6.52}\right)$

$= \cos^{-1}(0.74512) = 41.8309°$

$\therefore A = \mathbf{41.8°}$

ALTERNATIVE METHOD

Alternatively, to find $\widehat{B}$, we have:

$$\cos B = \frac{a^2 + c^2 - b^2}{2ac}$$

$$B = \cos^{-1}\left(\frac{a^2 + c^2 - b^2}{2ac}\right)$$

$$= \cos^{-1}\left(\frac{4.5^2 + 6.52^2 - 3.7^2}{2 \times 4.5 \times 6.52}\right)$$

$$= \cos^{-1}(0.83624) = 33.2548°$$

$$\therefore \widehat{B} = \mathbf{33.3°}$$

The complete solution is shown in Figure 14.30.

14.8 AREA OF A TRIANGLE

While the sine and cosine rules are useful for determining sides and angles, there may also be instances where we need to calculate the area of a triangle (Figure 14.31). Fundamentally, the area A of a triangle is given as:

$$\boxed{A = \frac{1}{2}bh} \tag{14.9}$$

where:

• b is the base of the triangle, which is $|AC|$ in this case.

- **h** is the height of the triangle. This is the length of a perpendicular line from the vertex to the base. A perpendicular line is a line that meets another line at 90 degrees. Recall that 90° is represented by an inverted L.

It is important to mention that any of the three sides of a triangle can be taken as the base and as such |AB|, |AC|, and |BC| are each a base in Figure 14.31. Each has its corresponding height drawn from the directly opposite vertex, which meets the base at right angle.

Sometimes, we are not given the height and we are unable to obtain it. To determine the area of the given triangle, we need to replace the height with one side and an angle. We can carry this out using the trigonometric ratios (not covered here but can be found in *Advanced Mathematics for Engineers and Scientists with Worked Examples* by the same author); it suffices now to say that the height in the $\triangle ABC$ is equal to $|BC| \times \sin C$. Hence, we can write this new area formula as:

$$A = \frac{1}{2}bh = \frac{1}{2}b\,(a\sin C) = \frac{1}{2}ab\sin C \tag{14.10}$$

In fact, there are two more such formulas. All the three formulas are listed below for ease of reference.

$$\boxed{A = \frac{1}{2}ab\sin C} \quad \boxed{A = \frac{1}{2}ac\sin B} \quad \boxed{A = \frac{1}{2}bc\sin A} \tag{14.11}$$

The formula is easy to remember as:

$$\textbf{Half} \times \textbf{Product of any two sides} \times \textbf{Included angle} \tag{14.12}$$

There is another formula that uses only the sides, which is:

$$A = \sqrt{s\,(s-a)\,(s-b)\,(s-c)} \tag{14.13}$$

where

$$s = \frac{1}{2}\,(a+b+c)$$

It is as simple as that. Let's try some examples.

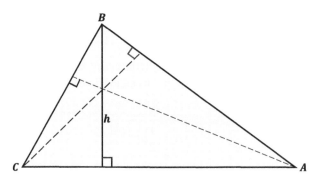

FIGURE 14.31 Determining the area of a triangle illustrated.

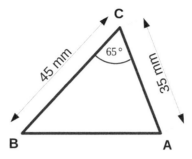

FIGURE 14.32 Solution to Example 16.

Example 16

Determine the area of a triangle ABC given that $|AC| = 35$ mm, $|BC| = 45$ mm, and $\angle ACB = 65°$. Present the answer correct to 3 s.f.

What did you get? Find the solution below to double-check your answer.

Solution to Example 16

A quick sketch may be helpful, as is shown in Figure 14.32.

Now, let $|BC| = a = 45$ mm; $|AC| = b = 35$ mm; $\angle ACB = \widehat{C} = 65°$. With the above information, we have:

$$A = \frac{1}{2}ab \sin C$$
$$= 0.5 \times 45 \times 35 \times \sin 65°$$
$$= 713.717$$
$$\therefore A = \textbf{714 mm}^2$$

Another example to try.

Example 17

Using $\triangle ABC$, show that the area of the triangle ABC in Figure 14.33 is such that

$$\frac{1}{2}ab \sin C = \frac{1}{2}ac \sin B = \frac{1}{2}bc \sin A.$$

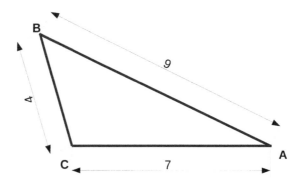

FIGURE 14.33 Example 17.

What did you get? Find the solution below to double-check your answer.

Solution to Example 17

From the diagram, we have $|BC| = a = 4$ unit; $|AC| = b = 7$ unit; $|AB| = c = 9$ unit. Since we need to find $\widehat{A}$, $\widehat{B}$, and $\widehat{C}$, we will use the cosine rule.

- Find $\widehat{A}$

$$\cos A = \frac{b^2 + c^2 - a^2}{2bc}$$

$$A = \cos^{-1}\left(\frac{b^2 + c^2 - a^2}{2bc}\right)$$

$$= \cos^{-1}\left(\frac{7^2 + 9^2 - 4^2}{2 \times 7 \times 9}\right)$$

$$= \cos^{-1}\left(\frac{19}{21}\right) = 25.2088°$$

$$\therefore A = \mathbf{25.21°}$$

- Find $\widehat{B}$

Now we will find $\widehat{B}$ and use the cosine rule again

$$\cos B = \frac{a^2 + c^2 - b^2}{2ac}$$

$$B = \cos^{-1}\left(\frac{a^2 + c^2 - b^2}{2ac}\right)$$

$$= \cos^{-1}\left(\frac{4^2 + 9^2 - 7^2}{2 \times 4 \times 9}\right)$$

$$= \cos^{-1}\left(\frac{48}{72}\right) = 48.19°$$

$$\therefore B = \mathbf{48.19°}$$

- Find $\widehat{C}$

We can still use the cosine or sine rule for the third angle but let's find $\widehat{C}$ using

$$\widehat{A} + \widehat{B} + \widehat{C} = 180°$$
$$\widehat{C} = 180° - 25.21° - 48.19°$$
$$\therefore \widehat{C} = \mathbf{106.60°}$$

Now let's find the area of the triangle ABC using the three formulas as follows:

Option 1 $\frac{1}{2}ab \sin C$ **Option 2** $\frac{1}{2}ac \sin B$ **Option 3** $\frac{1}{2}bc \sin A$

$A = \frac{1}{2}ab \sin C$ $A = \frac{1}{2}ac \sin B$ $A = \frac{1}{2}bc \sin A$

$= 0.5 \times 4 \times 7 \times \sin 106.6°$ $= 0.5 \times 4 \times 9 \times \sin 48.19$ $= 0.5 \times 7 \times 9 \times \sin 25.21$

$= 13.4165$ $= 13.4165$ $= 13.4170$

$\therefore A = \mathbf{13.4 \ unit^2}$ $\therefore A = \mathbf{13.4 \ unit^2}$ $\therefore A = \mathbf{13.4 \ unit^2}$

This shows that the formulas are equal, i.e., $\frac{1}{2}ab \sin C = \frac{1}{2}ac \sin B = \frac{1}{2}bc \sin A$, and can be used to find the area of a triangle.

Another example for us to try.

Example 18

Determine the area of a triangle ABC shown in Figure 14.34. Present the answer correct to 1 d.p.

What did you get? Find the solution below to double-check your answer.

Solution to Example 18

From the diagram, we have $|BC| = a = 15$ mm; $|AC| = b = 12$ mm; $\widehat{A} = 70°$.

We can use $\frac{1}{2}ab \sin C$ to find the area of $\triangle ABC$, but we need to find $\widehat{C}$. We will use the sine rule to find $\widehat{B}$ and then find $\widehat{C}$.

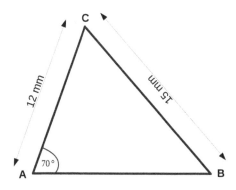

FIGURE 14.34 Example 18.

- **Find $\hat{B}$**

$$\frac{\sin B}{b} = \frac{\sin A}{b}$$

$$\frac{\sin B}{12} = \frac{\sin 70°}{15}$$

$$\sin B = \frac{\sin 70°}{15} \times 12 = 0.75175$$

$$\hat{B} = \sin^{-1}(0.75175) = 48.74°$$

$$\therefore \hat{A} = \mathbf{48.7°}$$

- **Find $\hat{C}$**

Now let's find $\hat{C}$ using

$$\hat{A} + \hat{B} + \hat{C} = 180°$$

$$\hat{C} = 180° - 48.74°$$

$$-70° = 61.26°$$

$$\therefore \ \hat{C} = \mathbf{61.3°}$$

- **Determine area**

Now it is time to find the area

$$A = \frac{1}{2}ab \sin C$$

$$= 0.5 \times 15 \times 12 \times \sin 61.26°$$

$$= 78.9129$$

$$\therefore \ \hat{B} = \mathbf{78.9 \ mm^2}$$

Another example to try.

Example 19

In $\triangle XYZ$, $|XY| = 3x$ cm, $|YZ| = 2x$ cm, $|XZ| = \sqrt{14}$ cm, and $\hat{Y} = 60°$.

  a) Determine the value of x.

  b) Calculate the area of the triangle XYZ.

What did you get? Find the solution below to double-check your answer.

Solution to Example 19

a) Value of x
Solution
From the question, we have two sides and their included angle. Hence, we will use the cosine rule as

$$a^2 = b^2 + c^2 - 2bc \cos A$$

$$|XZ|^2 = |XY|^2 + |YZ|^2 - 2|XY||YZ| \cos \hat{Y}$$

$$\left(\sqrt{14}\right)^2 = (3x)^2 + (2x)^2 - 2(3x)(2x)\cos 60°$$

$$14 = 9x^2 + 4x^2 - 6x^2$$

$$14 = 7x^2$$

$$2 = x^2$$

$$x = \sqrt{2}$$

$$\therefore x = \pm\sqrt{2}$$

Length cannot be negative, so the only valid value of x is $x = \sqrt{2}$.

b) Area
Solution

$$A = \frac{1}{2}ab \sin C = \frac{1}{2}|XY|\,|YZ|\sin \hat{Y}$$

$$= \frac{1}{2}(3x)(2x)\sin 60°$$

$$= \frac{1}{2}(6x^2)\sin 60°$$

$$= 3x^2 \times \frac{\sqrt{3}}{2} = 3 \times \left(\sqrt{2}\right)^2 \times \frac{\sqrt{3}}{2}$$

$$\therefore A = 3\sqrt{3} \text{ cm}^2$$

One last example to try.

Example 20

In a right-angled triangle ABC, $|AB| = 40$ mm, $|BC| = 50$ mm, $|AC| = 30$ mm, and $\hat{A} = 90°$. Determine the area of $\triangle ABC$ using the formula $\sqrt{s(s-a)(s-b)(s-c)}$. Confirm this using another suitable formula for finding the area.

What did you get? Find the solution below to double-check your answer.

Solution to Example 20

Solution
Let's quickly sketch the triangle as shown in Figure 14.35.
We have $a = 30$ mm, $b = 50$ mm, and $c = 40$ mm. Therefore

$$s = \frac{1}{2}(a+b+c) = \frac{1}{2}(30+50+40) = 60 \text{ mm}$$

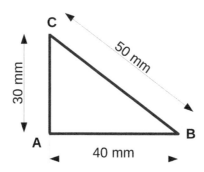

FIGURE 14.35 Solution to Example 20.

The area is

$$A = \sqrt{s(s-a)(s-b)(s-c)}$$
$$= \sqrt{60(60-30)(60-50)(60-40)}$$
$$= \sqrt{60(30)(10)(20)}$$
$$\therefore A = 600 \text{ mm}^2$$

ALTERNATIVE METHOD

As this is a right-angled triangle, it will be much easier to use the $A = \frac{1}{2}bh$. For this case, base and height are 40 mm and 30 mm respectively. Therefore

$$A = \frac{1}{2}bh$$
$$= \frac{1}{2} \times 40 \times 30$$
$$\therefore A = 600 \text{ mm}^2$$

14.9 CHAPTER SUMMARY

1) Angle measurement involves the amount of turning a line from a reference point.

2) The symbol θ (read as 'theta') is a Greek letter, which is commonly used to denote angles.

3) Conventionally, we use symbol $\angle$ or $\hat{}$ to denote angles. Thus, $\angle BAC$ is read as 'angle BAC' and $\hat{A}$ is 'angle A'.

4) Degree (symbolised by °) is a unit of angle and you would often find it as a default setting in most calculators. It is arbitrarily taken that the angle in one revolution (in a circle) is equal to 360 degrees.

5) Some special angles include:

   - Acute angle
   - Right angle
   - Obtuse angle
   - Straight line angle
   - Reflex angle

6) Radian (shortened to **rad**) is another unit used to measure angles. In the absence of a unit attached to an angle, radian becomes the default, but in other cases the symbol ^{c} is used to imply radian.

7) One radian is defined as: '**the angle subtended at the centre of a circle by an arc length of 1 radius**'.

8) The relationship between degrees and radians is such that:

$$\boxed{2\pi \text{ radians} = 360°}$$

- One radian in degrees is given by:

$$1 \text{ radian} = \left(\frac{180}{\pi}\right)^{\circ}$$

- One degree in radians is given by

$$1^{\circ} = \left(\frac{\pi}{180}\right) \text{ rad}$$

9) A triangle is a shape that has three and only three sides. By extension, it has three distinct angles and three vertices.

10) The longest side in a right-angled triangle is called the **hypotenuse**.

11) Pythagoras' theorem states that

$$\text{Hypotenuse}^2 = \text{Adjacent}^2 + \text{Opposite}^2$$

12) Based on the sides, we have three types of triangle, namely:

- **Equilateral triangle:** This is a triangle in which all the three sides are equal. Consequently, the three angles must also be equal. Since the total angle in a triangle is 180°, it follows that each angle in an equilateral triangle is 60°.
- **Isosceles triangle:** An isosceles triangle has only two out of the three sides equal. The two equal sides must have their corresponding angles equal.
- **Scalene triangle:** It is a triangle with no two equal sides. In other words, each of the side is distinct in length and also the angles are different.

13) The sine rule states that

$$\frac{a}{\sin A} = \frac{b}{\sin B} = \frac{c}{\sin C}$$

14) The cosine rule states that

$$a^2 = b^2 + c^2 - 2bc \cos A$$

where a, b, and c are the three sides and $\widehat{A}, \widehat{B}$, and $\widehat{C}$ are their respective angles. Alternatively,

$$\cos A = \frac{b^2 + c^2 - a^2}{2bc}$$

15) The area A of a triangle is given as:

$$A = \frac{1}{2}bh$$

Alternatively,

$$A = \frac{1}{2}ab \sin C$$

****

14.10 FURTHER PRACTICE

To access complementary contents, including additional exercises, please go to www.dszak.com.

15 Circle

Learning Outcomes

Once you have studied the content of this chapter, you should be able to:

- Calculate the length of an arc and a chord
- Determine the circumference (or perimeter) of a circle
- Determine the area of a circle
- Calculate the area of a sector and a segment of a circle

15.1 INTRODUCTION

As part of our exploration of plane geometry (a branch of mathematics concerned with 2-dimensional shapes), this final chapter will delve into the study of circles. We will cover various aspects including the perimeter of a circle, the length of an arc and a chord, as well as the area of a circle, a sector, and a segment.

15.2 CIRCLE

The circle, an important shape in geometry, is defined as a locus of points that are equidistant from a specific reference point. This reference point is known as the centre of the circle, and the distance from it to any point on the circumference is referred to as the **radius**. This is illustrated in Figure 15.1. We can therefore say that a circle is completely defined by its centre and radius. Hence, two circles are equal if the length of their radius is the same.

15.3 CIRCUMFERENCE

The circumference of a circle is the length of the line (or boundary) that forms the circle, and it is also called the perimeter of the circle.

The length of the circumference of a circle C is given by

$$\boxed{C = 2\pi r} \text{ OR } \boxed{C = \pi d} \tag{15.1}$$

where:

- r is the radius of the circle.
- d is the diameter of the circle, i.e. the length of a straight line which passes the centre of the circle and connects two points on the circumference. Therefore, $d = 2r$.

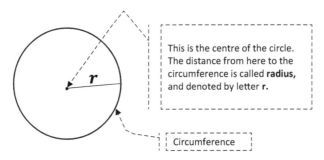

FIGURE 15.1 Key parts of a circle illustrated.

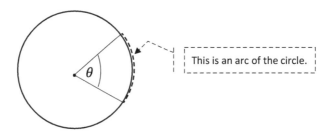

FIGURE 15.2 Arc length illustrated.

- π or pi is a Greek letter (pronounced as pie, and sounds like 'pai'), often used in science and engineering. We've encountered it earlier in this book (Section 14.2), where we mentioned that it is equal to 180 degrees. Generally, π is an irrational number, approximated to four significant figures as 3.142. It is sometimes represented as the fraction 22/7.

15.4 ARC LENGTH

An arc is a curved portion of the perimeter of a circle and there are several (or better still, unlimited numbers of) arcs in a circle. An arc is usually said to subtend an angle at the centre of the circle bounded by the arc and two radii, as shown in Figure 15.2.

15.4.1 FORMULA

It is clear that as the length of the arc increases, the angle formed at the centre also increases. If we denote the length of an arc by L, this relationship can be expressed as $L \propto \theta$. Since a circumference has 360° or 2π radians, we can write the length of an arc in two forms:

in degrees as   **OR**   in radians as

$$\frac{L}{C} = \frac{\theta}{360°}$$ $$\frac{L}{C} = \frac{\theta}{2\pi}$$

$$\therefore L = \frac{\theta}{360°} \times 2\pi r$$ $$\therefore L = \frac{\theta}{2\pi} \times 2\pi r = r\theta$$

(15.2)

15.4.2 TYPES

We have two types of arcs, namely:

a) **Minor arc**: If the angle θ subtended by an arc is such that $\theta < 180°$ or $\theta < \pi$, then the arc so formed is called a minor arc.

b) **Major arc**: If the angle θ subtended by an arc is such that $180° < \theta < 360°$ or $\pi < \theta < 2\pi$, then the arc so formed is called a major arc.

The sum of the lengths of the minor and major arcs of a given circle equals the length of the circumference of the same circle. Consequently:

$$\boxed{\textbf{Circumference of a circle = Major arc length + Minor arc lengeth}} \qquad (15.3)$$

15.5 PERIMETER OF A SECTOR

A sector is a portion of a circle and there are several (or better still, unlimited numbers of) sectors in a circle. The perimeter of a sector is the length around a sector which consists of an arc and two radii, as shown in Figure 15.3.

If we denote the perimeter of a sector by s, then s is given in degrees as:

$$\boxed{\therefore s = 2r + \frac{\theta}{360°} \times 2\pi r = 2r\left(1 + \frac{\pi\theta}{360°}\right)} \qquad (15.4)$$

or in radians as:

$$\boxed{\therefore s = 2r + r\theta = r(2 + \theta)} \qquad (15.5)$$

15.6 LENGTH OF A CHORD

A chord is the length of a straight line connecting two opposite points on the circumference (Figure 15.4). Note that the diameter is a special chord which passes through the centre.

If we denote the length of a chord by L, then we have:

$$\boxed{L = 2r\sin\left(\frac{\theta}{2}\right)} \qquad (15.6)$$

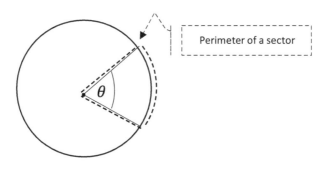

FIGURE 15.3 Perimeter of a sector illustrated.

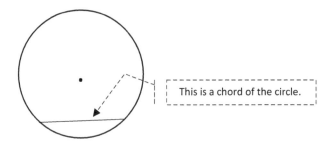

This is a chord of the circle.

FIGURE 15.4 Length of a chord illustrated.

It is a good time to try some examples.

Example 1

A circle has a radius of 5 cm. Calculate the length of an arc, correct to 2 decimal places, if the angle subtended by the arc at the centre of the circle is:

a) 20°   **b)** 115°   **c)** 305°   **d)** $\frac{1}{5}\pi$ **e)** 2.5 rad   **f)** 5.6 rad

State whether it is a minor or major arc.

What did you get? Find the solution below to double-check your answer.

Solution to Example 1

a) 20°
Solution

$$L = \frac{\theta}{360°} \times 2\pi r$$
$$= \frac{20°}{360°} \times 2\pi (5)$$
$$= \frac{1}{18} \times 10\pi = \frac{5}{9}\pi$$
$$\therefore L = 1.75 \text{ cm}$$

Because $\theta < 180°$

It's a minor arc.

b) 115°
Solution

$$L = \frac{\theta}{360°} \times 2\pi r$$

$$= \frac{115°}{360°} \times 2\pi(5)$$

$$= \frac{23}{72} \times 10\pi = \frac{115}{36}\pi$$

$$\therefore L = \textbf{10.04 cm}$$

Because $\theta < 180°$

It's a minor arc.

---

c) 305°
Solution

$$L = \frac{\theta}{360°} \times 2\pi r$$

$$= \frac{305°}{360°} \times 2\pi(5)$$

$$= \frac{61}{72} \times 10\pi = \frac{305}{36}\pi$$

$$\therefore L = \textbf{26.62 cm}$$

Because $\theta > 180°$

It's a major arc.

---

d) $\frac{1}{5}\pi$
Solution

$$L = r\theta = 5 \times \frac{1}{5}\pi = \pi$$

$$\therefore L = \textbf{3.14 cm}$$

Because $\theta < \pi$

It's a minor arc.

---

e) 2.5 rad
Solution

$$L = r\theta = 5 \times 2.5$$

$$= 12.5$$

$$\therefore L = \textbf{12.50 cm}$$

Because $\theta < \pi$

It's a minor arc.

f) 5.6 rad
Solution

$$L = r\theta = 5 \times 5.6$$
$$= 28$$
$$\therefore L = \mathbf{28.00 \text{ cm}}$$

Because $\theta > \pi$

It's a major arc.

Let's try another example.

Example 2

A circle has a radius of 8 cm. Calculate the angle subtended by an arc of length 26 cm. Give the answer in:

a) degrees   **b)** radians

What did you get? Find the solution below to double-check your answer.

Solution to Example 2

a) In degrees
Solution
We have $r = 8$ cm, $L = 26$ cm. Given that

$$L = \frac{\theta}{360°} \times 2\pi r$$

Let's make the angle θ as the subject of the formula, thus

$$\theta = \frac{L}{2\pi r} \times 360°$$

Now substitute

$$L = \frac{26}{2\pi \times 8} \times 360° = \frac{585}{\pi}$$
$$\therefore \theta = \mathbf{186°}$$

---

b) In radians
Solution

$$L = r\theta$$

which implies

$$\theta = \frac{L}{r}$$

$$= \frac{26}{8} = \frac{13}{4}$$

$$\therefore \theta = 3.25 \text{ rad}$$

ALTERNATIVE METHOD

Alternatively, convert the degrees in (a) to radians using

$$186° = 186° \times \frac{\pi}{180°}$$

$$= \frac{186°}{180°} \times \pi = \frac{31}{30}\pi$$

$$\therefore 186° = 3.25 \text{ rad}$$

15.7 AREA OF A CIRCLE

The area of a circle A is given by

$$\boxed{A = \pi r^2} \tag{15.7}$$

Since $r = \frac{d}{2}$, we can re-write the above formula as:

$$\boxed{A = \frac{1}{4}\pi d^2} \tag{15.8}$$

Because

$$A = \pi r^2 = \pi \left(\frac{d}{2}\right)^2 = \pi \frac{d^2}{4} = \frac{1}{4}\pi d^2$$

Let's try an example.

Example 3

A hollow circle, which consists of two concentric circles, is shown in Figure 15.5. The radius of the inner circle, measured from the centre O, is 2.5 mm and the distance between the two circumferences is 2 mm. Determine the area of the solid (or shaded) part. Present the answer correct to 1 d.p.

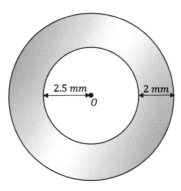

FIGURE 15.5 Example 3.

What did you get? Find the solution below to double-check your answer.

Solution to Example 3

Let's A_1 and A_2 be the area of the outer and inner circles respectively. If the area of the shaded part is A, then

$$A = A_1 - A_2$$
$$= \pi r_1^2 - \pi r_2^2 = \pi \left(r_1^2 - r_2^2 \right)$$
$$= \pi \left(r_1 + r_2 \right) \left(r_1 - r_2 \right)$$

where $r_1 = 2 + 2.5 = 4.5$ mm and $r_2 = 2.5$ mm. Thus

$$A = \pi \left(4.5 + 2.5 \right) \left(4.5 - 2.5 \right)$$
$$= \pi \left(7.0 \right) \left(2.0 \right) = 14\,\pi$$
$$= 43.98$$
$$\therefore A = \mathbf{44.0\ mm^2}\ (\textbf{1 d.p.})$$

15.8 AREA OF A SECTOR

A sector subtends an angle at the centre of the circle bounded by an arc and two radii as shown in Figure 15.6.

15.8.1 FORMULA

It is obvious that the area of the sector is directly related to the angle it subtends at the centre, i.e., $A \propto \theta$. As we did for the arc above, we can write the area of a sector in two forms as:

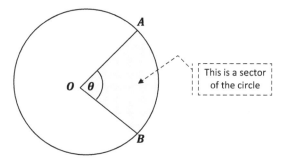

FIGURE 15.6 Area of a sector illustrated.

in degrees as   OR   in radians as

$$\frac{A_s}{A} = \frac{\theta}{360°} \qquad\qquad \frac{A_s}{A} = \frac{\theta}{2\pi} \qquad (15.9)$$

$$\boxed{\therefore A_s = \frac{\theta}{360°} \times \pi r^2} \qquad \boxed{\therefore A_s = \frac{\theta}{2\pi} \times \pi r^2 = \frac{1}{2}r^2\theta}$$

15.8.2 Types

We also have two types of sectors, namely:

a) **Minor sector**: If the angle θ subtended by a sector is such that $\theta < 180°$ or $\theta < \pi$, then the sector that is formed is called a minor sector.

b) **Major sector**: If the angle θ subtended by a sector is such that $180° < \theta < 360°$ or $\pi < \theta < 2\pi$, then the sector that is formed is called a major sector.

The sum of the areas of the minor and major sectors of a given circle equals the area of the same circle. Consequently:

$$\boxed{\textbf{Area of a circle = Area of the major sector + Area of the minor sector}} \qquad (15.10)$$

Let's try some examples.

Example 4

A circle has a radius of 5 cm. Calculate the area of a sector which subtends the following angles at the centre of the circle:

a) 30°   **b)** 135°   **c)** 280°   **d)** $\frac{3}{5}\pi$ **e)** 3.1 rad   **f)** 4.7 rad

State whether it is a minor or major sector. Present the answers correct to 2 d.p.

What did you get? Find the solution below to double-check your answer.

Solution to Example 4

a) 30°
Solution

$$A_{sector} = \frac{\theta}{360°} \times \pi r^2$$

$$= \frac{30°}{360°} \times \pi (5)^2$$

$$= \frac{1}{12} \times 25\pi = \frac{25}{12}\pi$$

$$\therefore A_{sector} = 6.54 \text{ cm}^2$$

Because $\theta < 180°$

It's a minor sector.

---

b) 135°
Solution

$$A_{sector} = \frac{\theta}{360°} \times \pi r^2$$

$$= \frac{135°}{360°} \times \pi (5)^2$$

$$= \frac{3}{8} \times 25\pi = \frac{75}{8}\pi$$

$$\therefore A_{sector} = 29.45 \text{ cm}^2$$

Because $\theta < 180°$

It's a minor sector.

---

c) 280°
Solution

$$A_{sector} = \frac{\theta}{360°} \times \pi r^2$$

$$= \frac{280°}{360°} \times \pi (5)^2$$

$$= \frac{7}{9} \times 25\pi = \frac{175}{9}\pi$$

$$\therefore A_{sector} = 61.09 \text{ cm}^2$$

Because $\theta > 180°$

It's a major sector.

---

d) $\frac{3}{5}\pi$

Solution

$$A_{\text{sector}} = \frac{1}{2}r^2\theta = \frac{1}{2}(5)^2 \times \frac{3}{5}\pi$$

$$= \frac{25}{2} \times \frac{3}{5}\pi = \frac{15}{2}\pi$$

$$\therefore A_{\text{sector}} = \mathbf{23.56 \ cm^2}$$

Because $\theta < \pi$

It's a minor sector.

---

e) 3.1 rad

Solution

$$A_{\text{sector}} = \frac{1}{2}r^2\theta$$

$$= \frac{1}{2}(5)^2 \times 3.1$$

$$= \frac{25}{2} \times 3.1$$

$$\therefore A_{\text{sector}} = \mathbf{38.75 \ cm^2}$$

Because $\theta < \pi$

It's a minor sector.

---

f) 4.7 rad

Solution

$$A_{\text{sector}} = \frac{1}{2}r^2\theta$$

$$= \frac{1}{2}(5)^2 \times 4.7$$

$$= \frac{25}{2} \times 4.7$$

$$\therefore A_{\text{sector}} = \mathbf{58.75 \ cm^2}$$

Because $\theta > \pi$

It's a major sector.

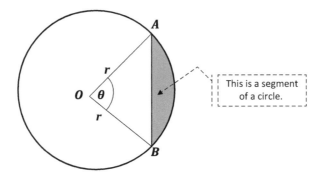

FIGURE 15.7   Area of a segment illustrated.

15.9 AREA OF A SEGMENT

The segment of a circle is an area bounded by a chord AB and arc AB as shown in Figure 15.7.

From the figure above, it is obvious that the segment can be found using the following equation.

Area of Segment AB = Area of Sector OAB − Area of Triangle OAB

The formula to calculate this is given in degrees as:

$$A_{\text{segment}} = \left(\frac{\theta}{360°} \times \pi r^2 \right) - \left(\frac{1}{2} r^2 \sin \theta \right) \tag{15.11}$$

or in radians as:

$$A_{\text{segment}} = \left(\frac{1}{2} r^2 \theta \right) - \left(\frac{1}{2} r^2 \sin \theta \right) = \frac{1}{2} r^2 \left(\theta - \sin \theta \right) \tag{15.12}$$

15.10 CHAPTER SUMMARY

1) A circle is a locus of points equidistant from a given or reference point. This reference point is the centre of the circle and the length from it to any other point on its circumference is called the radius.

2) A circle is completely defined by its centre and radius, which implies that if we know these two items then we can distinctly know a circle.

3) The circumference of a circle is the length of the line that forms the circle, and it is also called the perimeter of the circle.

4) The length of the circumference of a circle C is given by

$$\boxed{C = 2\pi r} \ \text{OR} \ \boxed{C = \pi d}$$

where r is the radius of the circle.

5) An arc is a curved portion of the perimeter of a circle. Its length is given by:

- In degrees:

$$L = \frac{\theta}{360°} \times 2\pi r$$

- In radians:

$$L = r\theta$$

6) There are two types of arcs, namely: minor and major arcs. The relationship between them and the circumference is such that:

Circumference of a circle = Major arc length + Minor arc length

7) The perimeter of a sector is given by:

- In degrees:

$$s = 2r + \frac{\theta}{360°} \times 2\pi r = 2r\left(1 + \frac{\pi\theta}{360°}\right)$$

- In radians:

$$s = 2r + r\theta = r(2 + \theta)$$

8) A chord is the length of a straight line connecting two opposite points on the circumference. Its length is given by:

$$L = 2r \sin\left(\frac{\theta}{2}\right)$$

9) The diameter is a special chord which passes through the centre.

10) The area of a circle A is given by:

- In degrees:

$$A = \pi r^2$$

- In radians:

$$A = \frac{1}{4}\pi d^2$$

11) A sector is a portion of a circle, which subtends an angle at the centre of the circle bounded by an arc and two radii. Its area is given by:

- In degrees:

$$A_s = \frac{\theta}{360°} \times \pi r^2$$

- In radians:

$$A_s = \frac{1}{2}r^2\theta$$

12) There are two types of sectors, namely: minor and major sectors. The relationship between them and the circle is given by:

Area of a circle = Area of the major sector + Area of the minor sector

13) The area of a segment is given by:

- In degrees:

$$A_{\text{segment}} = \left(\frac{\theta}{360°} \times \pi r^2\right) - \left(\frac{1}{2}r^2 \sin \theta\right)$$

- In radians:

$$A_{\text{segment}} = \left(\frac{1}{2}r^2\theta\right) - \left(\frac{1}{2}r^2 \sin \theta\right) = \frac{1}{2}r^2\left(\theta - \sin \theta\right)$$

****

15.11 FURTHER PRACTICE

To access complementary contents, including additional exercises, please go to www.dszak.com.

Glossary

S/N	Term	Explanation
1.	Absolute value	The magnitude (or size) of a quantity without the direction. In other words, its value without a positive or negative sign attached. It is indicated using two vertical straight lines. In this case, the absolute value of -5 and $+5$ is 5. Absolute value is also referred to as modulus.
2.	Acute angle	An angle that is less than $90°$, but more than 0 such that $0 < \theta < 90°$.
3.	Adjacent (adj)	Linguistically it means 'next to' or 'close to'. In a triangle, it refers to the side that is close to the angle being referred to.
4.	Asymptotes	A situation where a line (or curve) approaches another line (or curve) but never touches it. In this context, the line (or curve) is said to be asymptotic to the line (or curve) it cannot touch.
5.	Average	Strictly speaking, it is one of the parameters used to measure the location of a dataset. When used in language, it is synonymous with arithmetic mean or simply mean.
6.	Base	This is the number to which a power (or index) is added. If there is no apparent index on a number, one is assumed as the power. Base is also used to refer to the number system, which shows the maximum number of digits used in counting within that system. For example, base 10 (or decimal system) has ten different digits from 0 to 9 included and base 2 (or binary system) has only two digits, namely 0 and 1.
7.	Bisector	A line that divides another line or angle into two equal parts.
8.	Cartesian coordinate system	This is a system that is used to describe the position of a point on a plane or in a space, and is measured from a reference point or axis. The reference point is given as $(0,0)$ for a plane or $(0,0,0)$ for a space.

9.	Cartesian coordinates	The coordinates (x,y) and (x,y,z) describe the position of a point in terms of its distance from the origin $(0,0)$ and $(0,0,0)$ respectively. The order of stating the coordinates is to write x-coordinate, then y-coordinate, and finally the z-coordinate (if it's a point in space).
10.	Common logarithm	A logarithm to the base of 10, i.e., $\log_{10}N$. In general, when the base is 10, it is usually omitted. In other words, $\log_{10}N$ is simply written as $\log N$.
11.	Decimal point	A dot that is used to separate a whole number from a fraction.
12.	Element	It is used in set theory to refer to the members of a set or in a matrix to denote the members of the matrix.
13.	Empty or null set	A set (or collection of something) without a single member.
14.	Exponent	See 'power'.
15.	Function	An expression that maps one variable (the independent variable) and another variable (the dependent variable). It takes inputs and transforms them to produce the corresponding outputs. For example, $y = f(x)$ is read as 'y is a function of x' and implies that the value of y varies as the value of x changes.
16.	Gradient	See 'slope'.
17.	Hypotenuse (hyp)	The longest side in a right-angled triangle, and it is directly opposite to the right-angle corner or vertex.
18.	Improper fraction	A fraction where the numerator is greater than the denominator.
19.	Included angle	The angle formed by two sides.
20.	Index	See power.
21.	Inverse	It is sometimes used to mean 'reciprocal' and at other times to imply 'opposite'.
22.	Line of symmetry	It is also called the 'mirror line'. It is a line that divides (or can be used to divide) a plane into two equal and identical parts. For example, a square has four lines of symmetry, and a parallelogram has none.
23.	Modulus	See 'absolute'.
24.	Multiples	Multiples of a number N are numbers obtained when the original number is multiplied by natural numbers $1, 2, 3$, etc.

25.	Natural logarithm	Natural logarithm is the logarithm to the base of an irrational number denoted as e. Usually, natural logarithm is written as $\ln N$.
26.	Obtuse angle	This is the angle that is more than $90°$, but less than $180°$, such that $90° < \theta < 180°$.
27.	Opposite (opp)	Linguistically, the term refers to being 'directly across from each other' or 'facing each other'. In the context of a given triangle, it refers to the side that is facing the angle being referred to.
28.	Origin	The point on Cartesian plane where the x-axis crosses the y-axis and with coordinates $(0, 0)$.
29.	Parallel	Two or more lines (or planes) are parallel if they are always at the same separation distance. It is also correct that there is 0-degree angle between them.
30.	Perfect number	A number that is exactly half of the sum of its factors.
31.	Perfect square	A number whose square root is a whole number.
32.	Perimeter	It is the length (or distance) around the edge of a shape.
33.	Perpendicular	A line or plane that meets another line or plane at exactly $90°$.
34.	Positive angle	The angle that is measured in an anti-clockwise direction.
35.	Power	Power (or index or exponent) is the superscript of a number or term called the base. Power is a shorthand way of saying that a number is being multiplied by itself a number of times dictated by the value of the power.
36.	Product	The result of the multiplication of two or more numbers or variables. Thus, the product of 2 and 3 is 6, and this can be written as $2 \times 3 = 6$.
37.	Proper fraction	A fraction where the numerator is less than the denominator.
38.	Quadrilateral	'Quad' implies 'four', so a quadrilateral is a four-sided polygon. Examples include rectangle, square, parallelogram, rhombus, kite, and trapezium.
39.	Radicand	A number or expression whose square root is to be determined. For example, 5 is the radicand of $\sqrt{5}$. The square root itself is called the radical sign.

40.	Rationalisation	A method of simplifying a fraction having a surd either as its denominator or as both the denominator and numerator such that it can be rewritten without a surd in its denominator. This is achieved by multiplying the numerator and denominator with the conjugate of the denominator.
41.	Reciprocal	The reciprocal of a number is the number obtained when **1** is divided by the number or when the number in a fraction is flipped such that the denominator becomes the numerator and vice versa. It can be indicated by raising the number to the power of $-\mathbf{1}$. For example, the reciprocal of **3** is $\mathbf{3^{-1}}$ and that of $\frac{5}{7}$ is $\left(\frac{5}{7}\right)^{-1}$.
42.	Recurring decimal	This is a fraction with an infinite decimal, such that the digit after the decimal point repeats endlessly.
43.	Reflex angle	An angle that is more than $\mathbf{180°}$ but less than $\mathbf{360°}$, such that $\mathbf{180° < \theta < 360°}$.
44.	Scientific form	A number that is expressed in the form $A \times \mathbf{10}^{x}$, where A is a number such that $\mathbf{1} \leq A < \mathbf{10}$, and x is any integer, positive or negative. It is also called scientific notation.
45.	Set	A collection of items or members.
46.	Slope	It is also called gradient, denoted by m, and is a measure of the steepness of a line. The more the value of m, the greater the steepness. Slope can be positive or negative. A positive slope ramps up from left to right and a negative slope ramps down from left to right.
47.	Standard notation	See 'Scientific form'.
48.	Straight line angle	This is the angle that is equal to $\mathbf{180°}$ or $\theta = \mathbf{180°}$. It is the amount of turning in a semi-circle.
49.	Sum	The result of adding two or more numbers or variables. Thus, the sum of **5** and **7** is **12**, and this can be written as $\mathbf{5 + 7 = 12}$. In this case, **5** is called augend and **7** is called addend.

50.	Supplementary angles	These are two angles that add up to make **180°**, and each of the angles is said to be the supplement of the other.
51.	Surd	An irrational number or expression of the form $\pm k\sqrt[n]{x}$, which cannot be expressed exactly as a fraction of two integers.
52.	Terminating decimal	A fraction which has a finite number of digits after the decimal point.
53.	Trapezium	Also called trapezoid, it is a quadrilateral with a pair of parallel sides and one or two sloping sides. If the sloping sides are the same in length, the resulting trapezium is called an isosceles trapezium.
54.	Vertex	It is the highest point where two lines, sides, or edges of a shape meet.
55.	x-coordinate	The distance of the point from the origin along the x-axis, which is the first number given in a coordinate system notation.
56.	x-intercept	This is where a curve crosses the x-axis.
57.	y-coordinate	The distance of the point from the origin along the y-axis, which is the second number given in a coordinate system notation.
58.	y-intercept	This is where a curve crosses the y-axis.
59.	z-coordinate	The distance of the point from the origin along the z-axis, which is the third number given in a coordinate system notation.

Index